Environmental Footprints and Eco-design of Products and Processes

Series Editor

Subramanian Senthilkannan Muthu, Green Story Inc.
Toronto, ON, Canada

This series aims to broadly cover all the aspects related to environmental assessment of products, development of environmental and ecological indicators and eco-design of various products and processes. Below are the areas fall under the aims and scope of this series, but not limited to: Environmental Life Cycle Assessment; Social Life Cycle Assessment; Organizational and Product Carbon Footprints; Ecological, Energy and Water Footprints; Life cycle costing; Environmental and sustainable indicators; Environmental impact assessment methods and tools; Eco-design (sustainable design) aspects and tools; Biodegradation studies; Recycling; Solid waste management; Environmental and social audits; Green Purchasing and tools; Product environmental footprints; Environmental management standards and regulations; Eco-labels; Green Claims and green washing; Assessment of sustainability aspects.

Indexed by Scopus

Amilton José Vieira Arruda
Felipe Luis Palombini
Editors

Biomimetics, Biodesign and Bionics

Technological Advances Toward Sustainable Development

Springer

Editors
Amilton José Vieira Arruda
Department of Design
Federal University of Pernambuco
Recife, Pernambuco, Brazil

Felipe Luis Palombini
Department of Industrial Design (DDI),
Graduate Program of Architecture,
Urbanism, Landscaping (PPGAUP)
Federal University of Santa Maria (UFSM)
Santa Maria, Rio Grande do Sul, Brazil

ISSN 2345-7651 ISSN 2345-766X (electronic)
Environmental Footprints and Eco-design of Products and Processes
ISBN 978-3-031-51310-7 ISBN 978-3-031-51311-4 (eBook)
https://doi.org/10.1007/978-3-031-51311-4

This Springer imprint is published by the registered company Springer Nature Switzerland AG
The registered company address is: Gewerbestrasse 11, 6330 Cham, Switzerland

Paper in this product is recyclable.

Contents

Didactic/Methodological Proposal: Use of the Natural Classification as an Element of Bionic Studies

Amilton José Vieira Arruda

1 Introduction

This research method was developed internally in the master's degree in design and bionics at the Center for *Research and Natural Structures at the Istituto Europeo di Design in Milan* in 1992 and translated into Portuguese in April 2019, and English motivated to be part of a book chapter.

A preamble is necessary: all the images found here were made using a B&P photocopy process and pasted into the original. Therefore, it is impossible 30 years after identifying their origin. I agree that it is always necessary when it is possible to cite the source and give due credit. But I believe that this does not take away the merit of the work, since it was designed and completed for academic and scientific purposes and never for commercial purposes. I apologize in advance and thank Prof. Carmelo, since this set of images, even though they did not belong to the CRSN at the time, was found and traced in the various handouts, master's theses, and applied research available in these libraries.

At the beginning of the IED CRSN master's course, in addition to the practical aspects proposed as a research subject, we sought to identify new didactic modalities to communicate each step to follow in the same argumentation typology (Fig. 1). As an initial research project, starting from a universal vision of the problem, fundamental points for this discipline were highlighted – bionics – which until now has been concerned with identifying aspects of nature to apply to the world of design. We prefer to analyze these arguments from a more academic and less marketing point of view.

A. J. V. Arruda (✉)
BioDesign Lab, Federal University of Pernambuco, Rio de Janeiro, Brazil
e-mail: amilton.arruda@ufpe.br

A. J. V. Arruda, F. L. Palombini (eds.), *Biomimetics, Biodesign and Bionics*, Environmental Footprints and Eco-design of Products and Processes,
https://doi.org/10.1007/978-3-031-51311-4_1

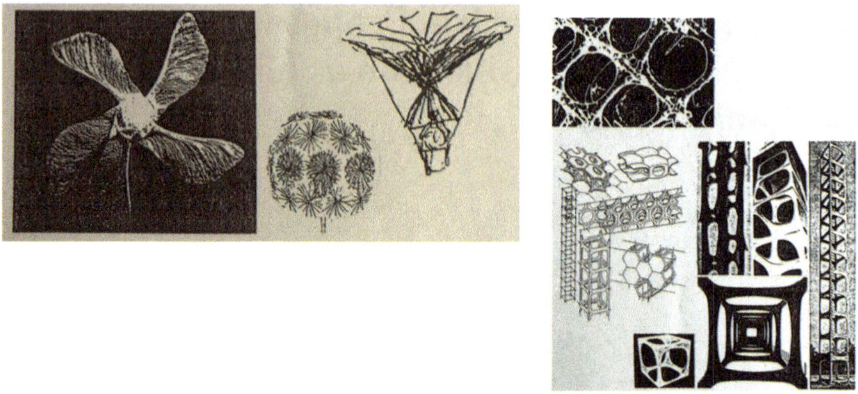

Fig. 1 Images provided from the CRSN photographic collection coordinated by Prof. Carmelo Di Bartolo, who allowed us to use them

As a main element, with regard to theoretical research, we seek to demonstrate to those who research the natural and biological world, useful ideas for design, a process of applied research method based mainly on the *taxonomy of living beings*.

As a second objective, this research had the purpose of comparing natural themes with the synthesis capacity that we industrial designers have in an experimental experience of a bionic type.

Additionally, we seek to:

1. Motivate all people interested in a possible intervention within the *scope of bionic design*
2. Demonstrate to designers the possibility of obtaining useful *elements from nature for design purposes*
3. Organize a logical sequence of work, which begins with the *observation of natural sciences as an element of investigation*
4. Develop applied research of academic interest with *particular references to design courses*
5. Build, in the medium and long term, a *database structure on subjects related to bionics*, organized according to the taxonomic model used by natural sciences

2 The Bionics

Regarding the birth of this science, it is quite recent, and it is possible to specify when it began; however, we cannot forget that in the last 3 million years that man has inhabited this planet and been part of this system, there has always been a very close connection with the environment and nature. It is possible to verify from the writings in the caves and to clearly state that the various instruments developed by that civilization were made not only with materials of a rustic nature but above all with shapes and characteristics very similar to those of the natural elements of that time.

It was, however, at a congress in the city of Dayton in Ohio (1959), where scientists from different sectors including engineers, mathematicians, physicists, psychologists, and biologists, among others, were promoted by the United States Air Force, which at the same time at that moment a research program had been created at the Wright Patterson Center called "bionics." It was when this new science was officially presented for the first time, and there, through *Major Jack E. Stelle*, the first definition of this new sector of scientific knowledge was drawn up. "Bionics is the science of systems where the functioning is based on that of natural systems, or which have analogies with it" [7], as exemplified in Fig. 2.

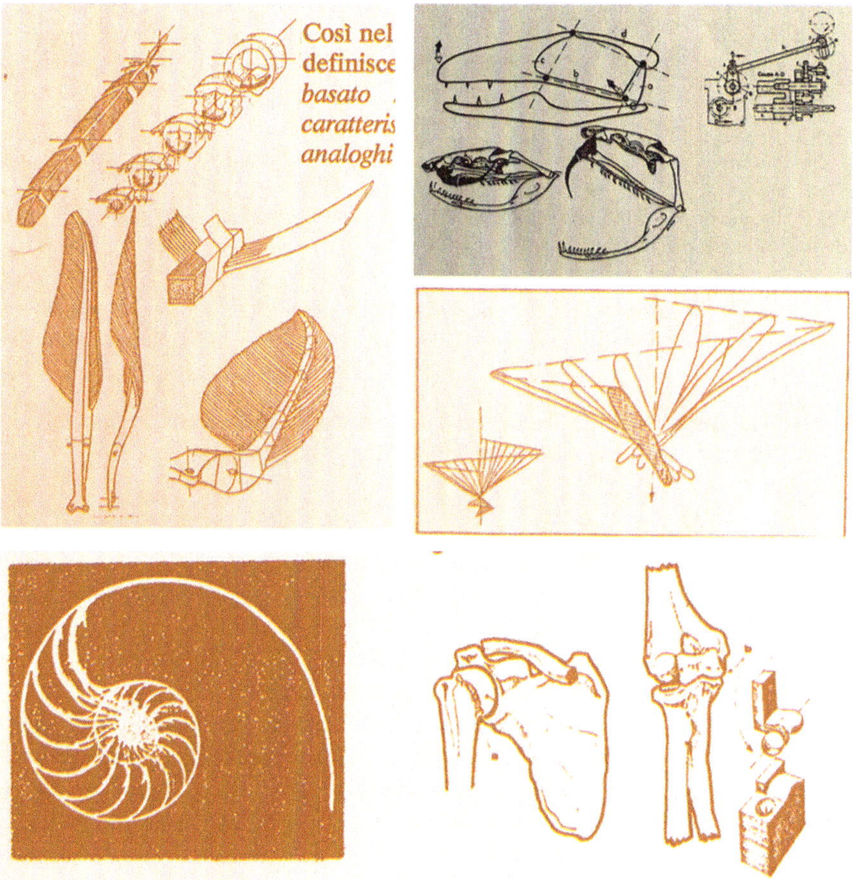

Fig. 2 Images provided from the CRSN photographic collection coordinated by Prof. Carmelo Di Bartolo, who allowed us to use them

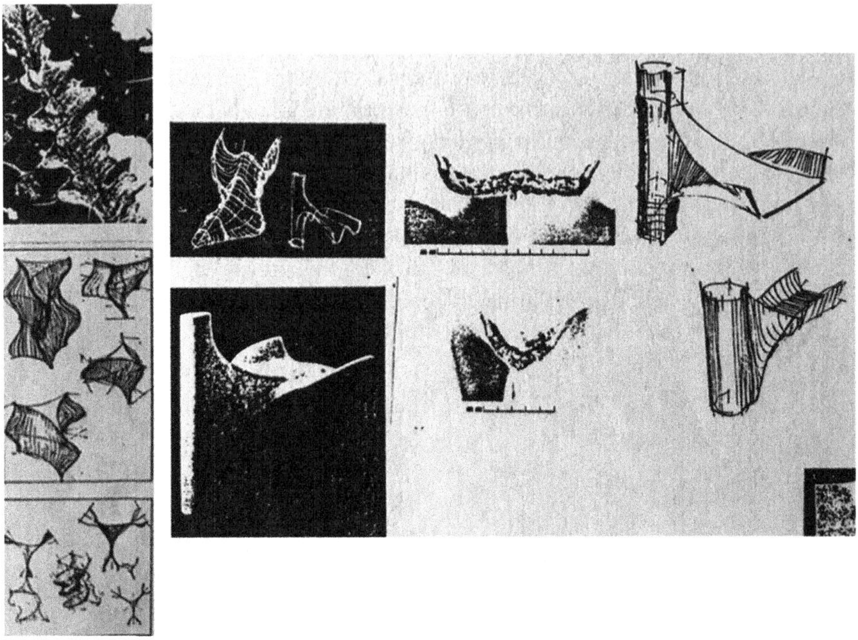

Fig. 3 Images provided from the CRSN photographic collection coordinated by Prof. Carmelo Di Bartolo, who allowed us to use them

Although the term and this science were born a few years ago, we found several authors who dealt with this subject, so it is interesting to see the different definitions (Fig. 3):

Lucien Gerardin, from the Thompson-Houston Company in Paris, defines bionics as "the art of applying knowledge about living systems to the solution of technical problems" [5]

For Victor Papanek,

Bionics is the science that studies the use of biological prototypes for the production of synthetic systems created by man. We can also say that bionics can be understood as the study of the basic principles of nature, to discover applications of principles and processes useful to human beings [6].

Bionics, according to Gui Bonsiepe, is understood as:

A study of the living system to apply its technical and mechanical principles to technology. It will be particularly adapted to stimulate the ability to perceive three-dimensional detail and the formal principles that structure it, something also to increase the capacity for transformation when examining and analyzing an analogous object [2].

This type of experimentation at HfG Ulm in the 1960s used the bionic analysis of natural phenomena and their three-dimensional transformation solely as a didactic purpose to improve the structural vision and creative interpretation of students.

Unlike those who simply understand bionics as a simple element to teach basic design principles – whether analyzing natural phenomena or as a starting point for academic design or seeking to give an exclusively morphological interpretation of nature – bionics can also be understood as a fundamental part of design activity. Prof. Fabrice Vanden Broeck from UNAM, Mexico, defines bionics, adapting it to the development of the biodesign project:

Biodesign – It is the study of natural systems and organizations, with the purpose of analyzing and recovering functional, structural and formal solutions, to be applied to the resolution of human problems, through the creation of technologies and the creation of objects or systems of objects [3].

According to the author, *biodesign* is based on two activities:

1. *Basic research and experimentation,* which start from the observation of natural phenomena and do not necessarily reach its design application. This type of attitude generates a series of innovative data that can be used later in other specific projects.
2. Another activity is to look for *solutions to a specific project through the analogy of functions* (Fig. 4). To achieve this, it is necessary to have access to a database for basic investigations and also a minimum of scientific knowledge of the principles that determine the forms in nature and a methodology to better approach natural phenomena.

As for Prof. Carmelo Di Bartolo from CRSN/IED in Milan, "The interest of bionics is directed towards design and drawing; one must study nature, in an attempt to seek information usable by man to improve his environment." Continues Di Bartolo: any natural event, structure, form, behavior, and function can be useful for a bionic type of investigation [4], as viewed in Fig. 5.

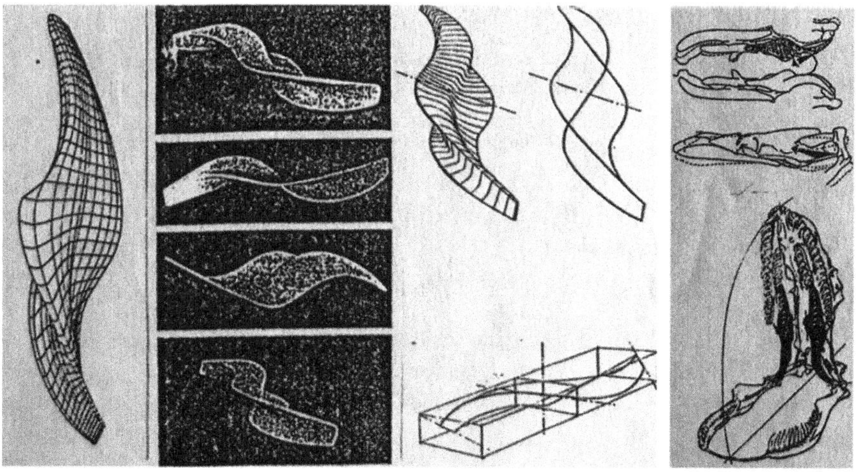

Fig. 4 Images provided from the CRSN photographic collection coordinated by Prof. Carmelo Di Bartolo, who allowed us to use them

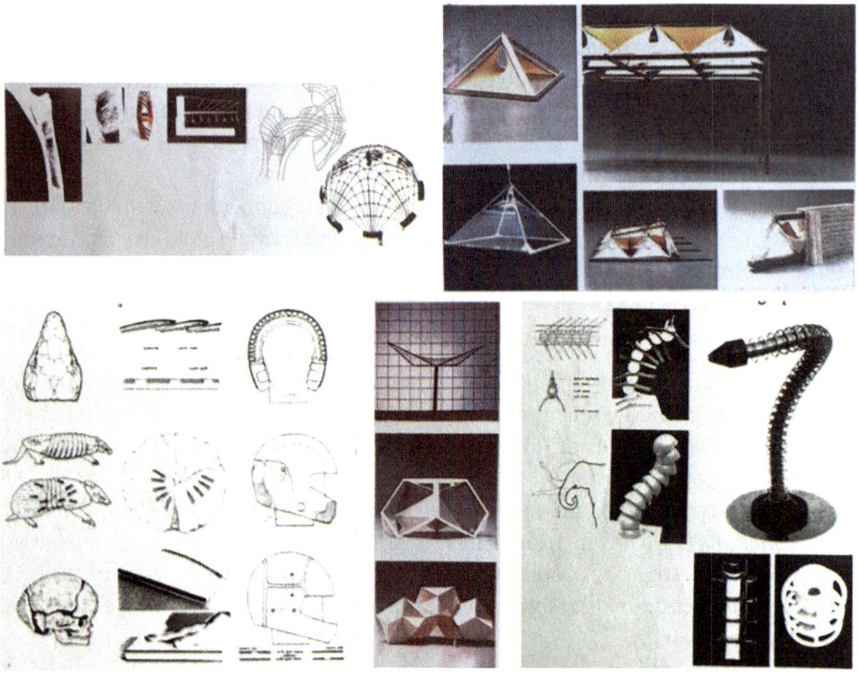

Fig. 5 Images provided from the CRSN photographic collection coordinated by Prof. Carmelo Di Bartolo, who allowed us to use them

Authors and researchers in this field consider nature a possible field of investigation for the design of artifacts, but on the other hand, they feel the lack of uniformity of procedures in facing these challenges. Despite the difficulties of uniform planning in methodological and universal terms in the bionic field, on the other hand, we feel the need to expand our knowledge in the natural/biological field, due to being able to have a more global vision and of different disciplines but very complementary on the other hand.

With that in mind, let's see what Ken Yeang says:

To carry out a bionic activity, we will need knowledge of the technical sector related to the project in question, a shallow knowledge of biology, at least that part where answers can be found, and a language that allows this communication with the biologists [8].

Life is a particular manifestation of matter that is configured in thousands of highly efficient and functional frameworks. Therefore, it is very difficult to draw a line between biology and all aspects of non-biological sciences, bearing in mind that questions have been very common for hundreds of years.

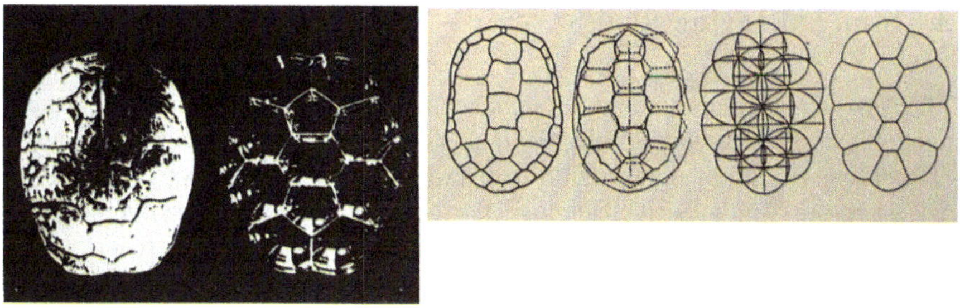

Fig. 6 Images provided from the CRSN photographic collection coordinated by Prof. Carmelo Di Bartolo, who allowed us to use them

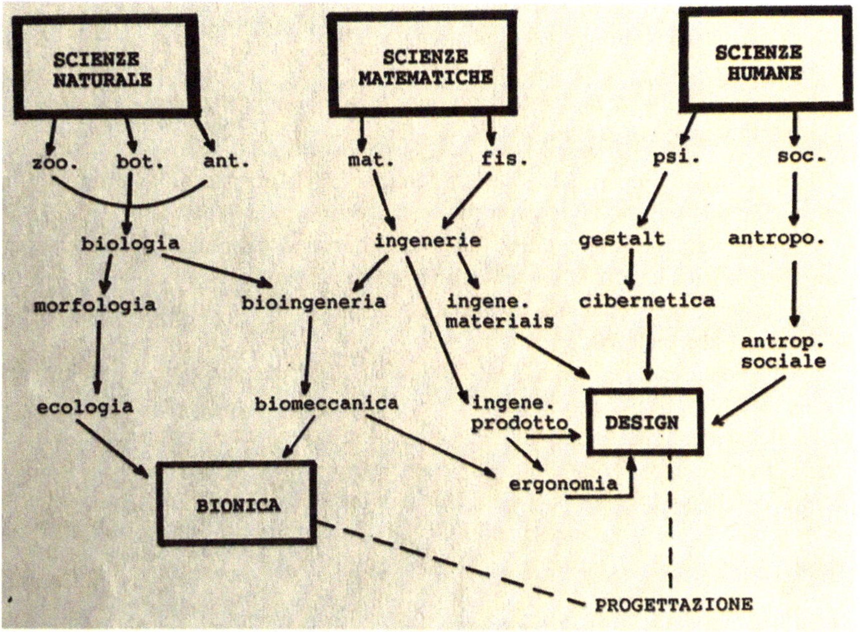

Fig. 7 Images provided from the CRSN photographic collection coordinated by Prof. Carmelo Di Bartolo, who allowed us to use them

Biology aims to analyze and understand the structures and functions in living beings, at any evolutionary level (Fig. 6), unlike bionics, which proposes the "modeling" of reality and "simulation" of it, in order to carry out various applications based on the different stimuli received (Fig. 7). This activity is based on the principle that on earth all living beings are the result of more than 2 million years of evolution and, therefore, of natural selection; current species are those that have best adapted to the environment.

3 Study of Living Beings

All living beings [plants, animals, fungi, and microscopic organisms] present a certain number of characteristics that distinguish them from the artificial world and an extraordinary unity of organization that reveals their origins to some common or closest parent. The most typical characteristic of living beings, the one that most distinguishes them from inanimate beings, is the ability to reproduce, that is, to give rise to new living beings that are the same or similar to oneself.
This property is always accompanied by other important characteristics, for example:

1. *Metabolism*, that is, the ability to transform chemical compounds
2. *Self-regulation*, the ability to establish your own rules
3. *Irritability*, the ability to react to external stimuli
4. *Ability to move*
5. *Mimicry*, the ability to camouflage one's external appearance as a measure of protection and self-defense

Some of these properties we can find in man-made machines, for example, the automobile that moves. The thermostat reacts to external temperature variations in order to maintain the temperature of the heater or refrigerator and many others constant. No man-made artifact possesses all of these properties together, and above all, none is capable of reproducing an artifact equal to itself, even if we provide it with all the necessary materials.

Living beings can be extremely different in size, shape, and way of life, but they are very uniform from a biochemical point of view. They are all made up of similar molecules with slight variations. It can even be said without fear of error that it is a single "living matter," common to all beings endowed with life. The most abundant molecule in living organisms is water; 60% of our body is made up of water, but in some fruits such as watermelon and in some marine animals, such as jellyfish, the water reaches 90%.

3.1 The Evolution

The concept of evolution probably represents biology's greatest contribution to our culture, in all its manifestations. The study of evolution allows us to better understand the characteristics of our species, our origins, and our possible future. Lamarck (1744–1829) was the first to suggest the concept of evolution for living beings. Lamarck with his theory:

...had observed that each living being undergoes transformations that make it adaptable to a particular type of life and suggested that the mechanism responsible for this specialization was the use or disuse of particular organs (for example: a deer that runs quickly to escape from a fox, it will develop muscles adaptable to running; elephants have long trunks to look for more food), and he believed that such characteristics could be transmitted to their descendants.

A classic example that has always been linked to Lamarck's name is the one relating to the giraffe:

According to your theory...in ancient times the giraffe had a short neck. But even with this inconvenience, he tried to bite the leaves from the highest branches of the trees, naturally driven by the need to find a new source of food. By force of taking it higher and higher, the neck began to grow. This longer neck than normal, obtained through various exercises, was eventually transmitted to his descendants, and in turn they continued to make it longer and longer to meet the demands of ever taller leaves... and so it continued. So, a little at a time, the giraffe got a very long neck like we are used to seeing nowadays.

The weak point of his theory was the idea that some characteristics acquired during life could be transferred to his descendants.

Talking about evolution is like quoting Charles Darwin (1808–1882), not only for the profane but also for the scientist. The Darwinian theory, based on natural selection, still constitutes the center of knowledge about evolution today, even if, over time, it has undergone some modifications.

Observing how breeders of plants and animals used natural selection to obtain particular varieties, Darwin thought that a similar type of selection that occurred in nature and interacted between populations and the environment was "natural selection." Darwinian theory supports, however, the existence of individuals of a kind of competition, in which only the most adaptable organisms manage to survive and reproduce, thus transmitting the advantageous character to their descendants; this concept has been indicated as "selective advantage." According to Darwinian evolutionary theory: "it is true that through the process of natural selection of the most advantageous characters of a single species, many other different species originated."

The physical structures of organisms can provide interesting insights into their parents. A typical example is the structural modifications that the limbs of some amphibians, reptiles, birds, and mammals have undergone based on the different functions they must perform; some of these examples are the prey capacity of monkeys; adaptation to flight in birds and bats; adaptation to swimming in seals, whales, and penguins; and the changes induced in the different ways of walking on a surface in most animals. These limbs, modified based on the functions to be performed, have different shapes, despite being made up of the same bone parts.

4 Taxonomy of Living Beings

Classifying is a way of organizing information and means grouping objects or ideas based on their similarities. The science that deals with giving a name to living beings and organizing them into a classification scheme is systematics or taxonomy, a word that derives from the Greek *taxis* = order.

Modern systematics today seeks to use a natural classification system to reconstruct the history of evolution. Living beings are in fact grouped according to a scheme that reflects order in the different groups and, therefore, the structure of their evolutionary tree. Carl Von Linneo (1707–1778), a Swedish naturalist from the eighteenth century, established a simple method to classify organisms, assigning them a binomial nomenclature characterized by a first name relating to the genus and one relating to the species. For his classification, Linneo used a technique similar to that used to communicate an address: a limited number of indications (name, postal number, street, city, country, etc.) are sufficient to ensure the individualization of a given person, among the many that populate planet Earth. Thus, all living organisms were subdivided into two broad groups: the animal kingdom and the plant kingdom. Linneo organized each kingdom into so many phyla (singular phylum) and therefore each phylum into a class, each class into an order, each order into a family, each family into a genus, and each genus into a species.

4.1 Similarity and Diversity

One of the most used criteria for individualizing affinities between different species is similarity. Two structures that were originally very different from each other can become similar in the course of evolution, adapting to perform the same function and to the same type of environment. The wings of butterflies, birds, and bats are examples of this similarity. The respective wings are similar because they respond to a certain physical law, but they belong to taxonomically very different organisms. The similarity between wings derives from an analogy of these functions so that similar environments favor the evolution of similar structures. On the contrary, structures with the same embryonic development can evolve in different ways; in this aspect, this fact is spoken of as homology.

The forelimbs of vertebrates, even though they are very different in shape, however, due to the function they perform, are all homologous, that is, they have the same embryonic origin. We can conclude by saying that similar structures can be analogous or even homologous and that homologous structures are often not similar at all. Therefore, to correctly use the criterion of morphological similarity, it is necessary to distinguish homology and analogy.

5 The Kingdoms of Living Beings

The progenitors common to all living beings were probably very similar to current prokaryotes, that is, bacteria and blue algae. Protists are the first eukaryotes (single-celled organisms) derived from prokaryotes. Multicellular organisms were derived from protists, divided into three kingdoms: the plant kingdom, the animal kingdom, and the fungi kingdom.

5.1 The Vegetable Kingdom

Plants are autotrophic organisms, and in order to nourish themselves they derive organic food substances using materials of a non-biological nature, such as water and carbon dioxide. Plants must absorb energy to enrich their own molecules.
The basic energy source is the sun, and the process of synthesizing organic substances from water and CO_2 is called photosynthesis. Another difference between animals and plants comprises the immobility of the latter. Instead of moving around in search of food, plant organisms developed true structures capable of gathering materials and energy necessary for photosynthesis.

Plants are regrouped and classified according to morphological and evolutionary criteria. The plant kingdom is subdivided into two large groups: vascular plants (which are the most evolved) and nonvascular plants, which do not have conductive tissue. Nonvascular plants are those that grow in any terrain and are always those most abandoned in humid areas. In fact, due to the lack of conductive tissue, they are not capable of transporting water or any nutrients over long distances. In this way, water moves by osmosis and other materials by diffusion. Due to the lack of supporting tissues, its stem cannot exceed more than 2 or 3 cm in height, like mosses, for example.

On the contrary, vascular plants already have perfect adaptation to terrestrial pathways, thanks to the development of specialized tissues for different functions. Among them is xylem for transporting water, salts, and nutritional substances from one part of the plant to another and also contributing to improving mechanical resistance. In general, these vascular tissue cells are elongated and have thick walls. And they are made up of two types of tissue: phloem and xylem. Phloem is the tissue that transports sugars and starch to the various parts of the plant, and xylem is specialized in transporting water and mineral salts from the root to the various parts of the plant. The various vital functions are developed in structures that are increasingly specialized.

Roots are organs specialized in absorbing water and mineral salts from the ground and are also capable of supporting plants in the soil. The trunk carries water and mineral salts from the roots to other organs. Tree trunks and branches are woody. Those plants that do not have trunks are called herbs and have a flexible trunk that cannot support heavy weights. Leaves constitute the main photosynthetic organ of the plant; they are generally flat, which increases the surface area exposed to the sun's rays.

5.2 *Animal Kingdom*

The animal kingdom is subdivided into two subkingdoms: animals and metazoa. The general characteristics of animals and metazoa in particular are the following:

1. *They are multicellular organisms.*
2. *They reproduce sexually.*
3. *They went through several stages of embryonic development.*
4. *They have the ability to move around.*

Animals are also distinguished by other important characteristics: the symmetry of the body, the presence or absence of cavities in the internal part of the body, and the presence or absence of segmentation and cephalization. The first attempt at multicellular organization was carried out by Porifera or marine sponges, which present a very primitive level of organization. But, in reality, the first multicellular organisms were the coelenterates, to which corals and jellyfish belong.

Flatworms were the first organisms with bilateral symmetry. From this point onwards, several evolutionary roads separate. One of these led to the development of parasitic worms and another that resulted in three important groups: mollusks, annelids, and arthropods. In all these groups we find many forms that ventured to conquer dry land, such as snails among mollusks, roundworms among the annelids, and most insects among the arthropods. Another important evolutionary branch of flatworms bifurcated into two notably distinct groups: the echinoderms and the chordates. Chordates are part of all those animals that have a spinal column such as fish, amphibians, reptiles, birds, and mammals.

6 Teaching/Methodological Experiment

When we talk about bionics applied to the development process of a project, one immediately thinks that the product must be something linked to robotics or even something far from our everyday world. This is because, commonly, the type of scientific information that reaches us is of an extremely mathematical or neurophysiological nature, forgetting that bionics, as a "crossroads science" [5], allows the possibility of relationships with several other sciences.

Another very particular aspect that causes the designer to be removed from the general discourse of bionics is due to its own characteristics. We do not adopt very scientific and universally known terminologies. For example, we do not know the technological process in depth – even if we have to specify engineering data – nor do we have sufficient knowledge of biology to develop a project inherent to scientific themes. Over time, it was enough for this idea to mature, and we felt the need

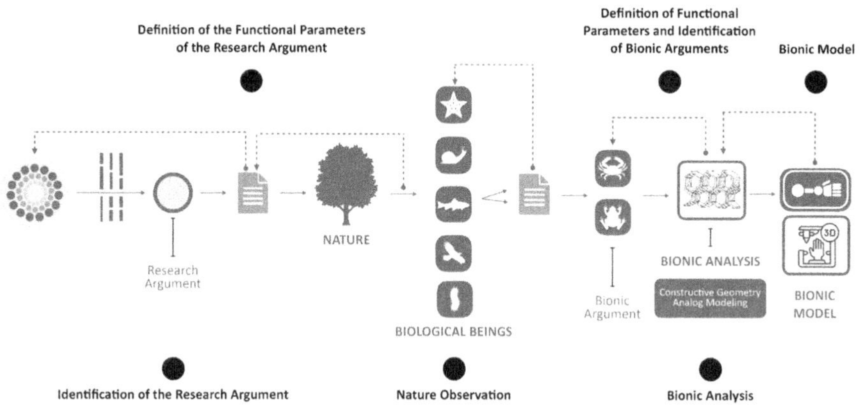

Fig. 8 Didactic/methodological proposal

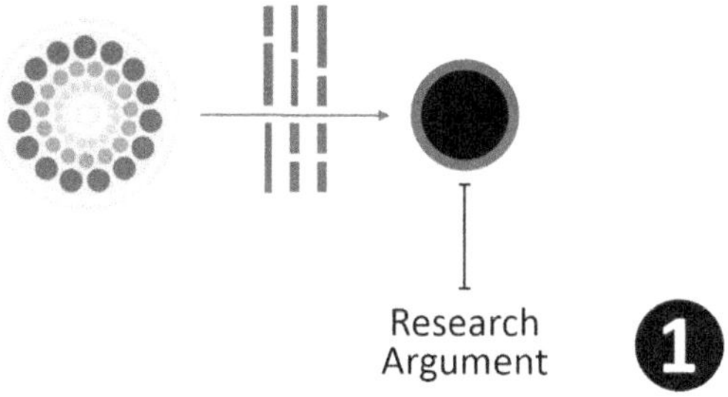

Fig. 9 Research Argument.

to fill this gap in terms of acquiring what I call a "base language" of the designer in a bionic sense, in order to allow a fruitful relationship with all scientific fields.

In relation to this research, we can state that we do not intend to create a method or even to make a huge effort on "medolatrism" (a concept adopted by G. Bonsiepe) but the intention of being able to organize a modality of approach in academic/scientific work (more directly aimed at design and architecture students), which seeks to apply what in biology is known as the *classification system of living beings*. This proposal (Fig. 8) therefore comprises using the taxonomic system applied to living organisms as a starting point for bionic research, obtaining suggestions from subjects or natural phenomena, for later application in the design of artifacts.

6.1 The Process

6.1.1 Identification of the Research Argument

The research argument (Fig. 9) must first be defined, in general terms, as an open data system. Such arguments should not be classified as a design problem, and as a natural element to be studied in its smallest details, but above all something that has a certain breadth of knowledge already available to be faced at a didactic level, in view of the fact that defining a field of research on a very specific argument excludes the possibility of a more global bionic definition. For example, we can highlight fractal structures as important and explanatory arguments for research purposes, structures in nature, the energy, the boards, natural packaging, locomotion systems, and so on. At this stage, it should encourage discussion about understanding and knowledge of the topics and about the availability and scientific potential for an investigation, analyzing the different levels of complexity.

6.1.2 Definition of the Functional Parameters of the Research Argument

After having identified the investigation argument, the initial questions are defined, that is, about the argument to be treated; it is not much about the nature of the natural phenomenon to be treated but about its function. It's not about asking "what it is" but "what it does," "how it does it," and "how it works." By defining questions in this way, we have the possibility of bringing together data and information by analogy of functions (Fig. 10).

6.1.3 Nature Observation

From the moment we establish and understand the type of function that exists in our research argument, we must begin to carry out a first check in the biological field, to identify models and natural systems that have functions and characteristics that are closer and more adaptable to our field of study (Fig. 11). To face such an operation, bibliographical research must be carried out. This is followed with a search under microscopes, preparation of photographic documentation, organization of the collected data based on files and databases, and production of a spreadsheet that

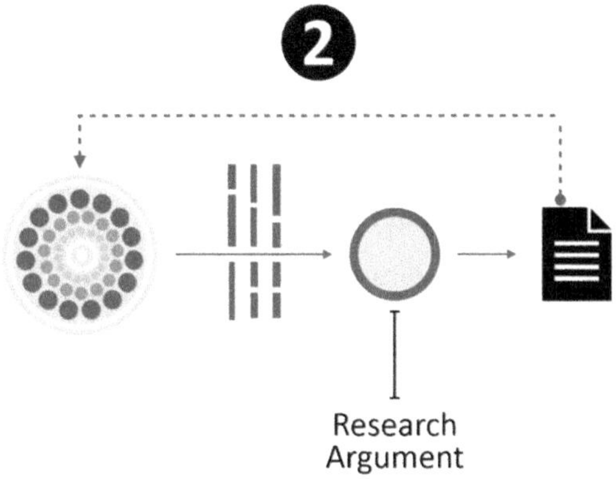

Fig. 10 Functional parameters of the research argument

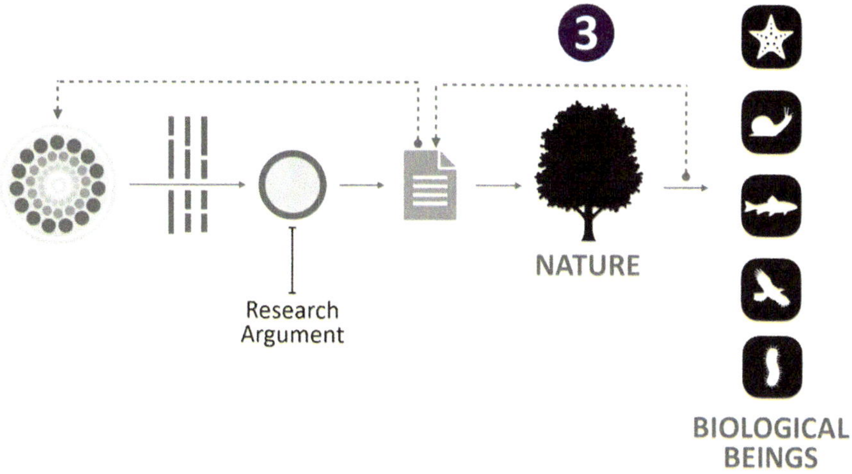

Fig. 11 Nature observation

follows the same principle as the classification of living beings. Therefore, it is necessary to ration in operational terms, seeking to specify the functions carried out by the identified organisms. These first spreadsheets/sheets have the characteristic of having a global view of the entire research, that is, comprising the maximum number of natural models analyzed. Therefore, the animal and plant kingdoms are proposed as research scopes using the following classification.

6.1.4 Definition of Functional Parameters and Identification of Bionic Arguments

The result of the previous phase comprises several sheets that contain ideas that create, at each moment, a large database on different arguments to be chosen. However, it is always good to remember that at this stage the functions and design requirements are defined and specified. Based on these parameters, we will be able to identify some natural elements that somehow better and more adequately satisfy the design problem (Fig. 12).

If, for example, the research argument identified was "natural packaging," and the established design requirements are packaging that perform the function of protecting the external environment, exchanging oxygen, defending against high temperatures, and absorbing moisture, in this case the elements that best perform these functions could be:

In the Animal Kingdom
The marine sponge

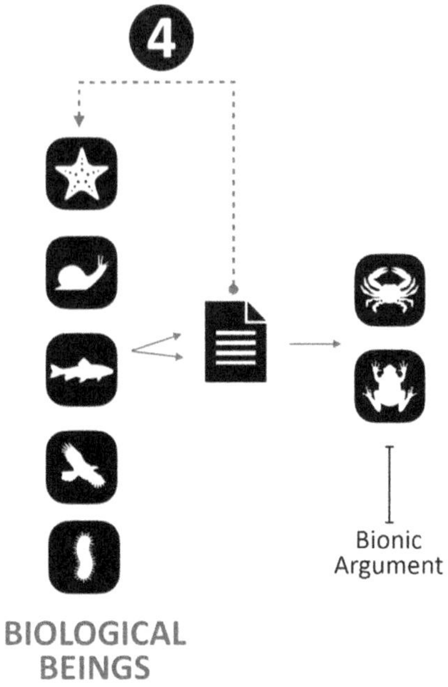

Fig. 12 Definition of functional parameters and identification of bionic arguments

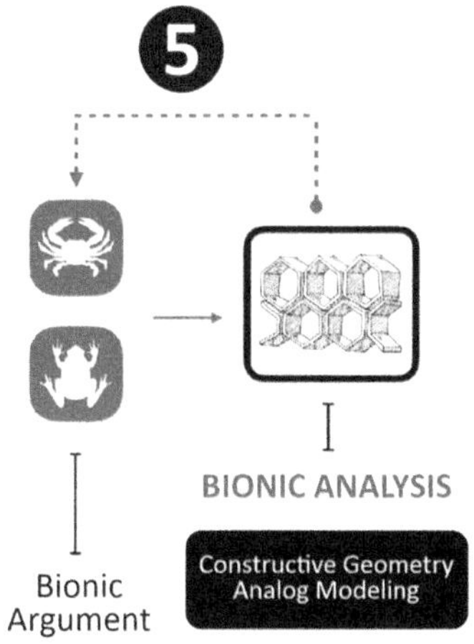

Fig. 13 Bionic analysis

The tracheal system of insects
The human pulmonary system
In the Vegetable Kingdom
The leaf transpiration system
The bean pods

These natural and biological elements that were identified are now identified in this research as "bionic argument."

6.1.5 Bionic Analysis

Analysis in this case is observed as an initial moment of knowledge, which allows us to formulate hypotheses and create connections and analogies so that we can operate in the form of a synthesis (Fig. 13). At this point, we move forward trying to understand the functioning of these systems through the analysis of "constructive geometry." Very important: it is never about copying nature but about being able to reach a synthetic (from a geometric point of view) and analogous (from a functional point of view) understanding of how we understand such mechanisms.

In a second phase, we seek to interpret what was obtained in the form of a "geometric synthesis," building 3D models and mental schemes based on the knowledge derived from the analyzed structures. It can be said that this stage of analysis corresponds exactly to bionic projection. Of the natural elements analyzed through constructive geometry and analogical modeling, results are obtained that interpret natural structures into possible artifacts (notably not finished designs from the point of view of their production) but three-dimensional artifacts where the matrix of the studied natural element is recognized. Taking advantage of and using the same example previously described, if the bionic argument applied to "natural packaging" was identified, the possible solutions develop more or less in this direction.

6.1.6 Bionic Model

The result of the previous phase will be a model that, from the point of view of industrial design, is not sufficiently finished and finalized for design, but from a bionic point of view it will be a "biological prototype" (V. Papaneck), which allows for verification, in its entirety, of the information of a bionic/biological nature acquired by understanding the natural phenomenon studied (Fig. 14).

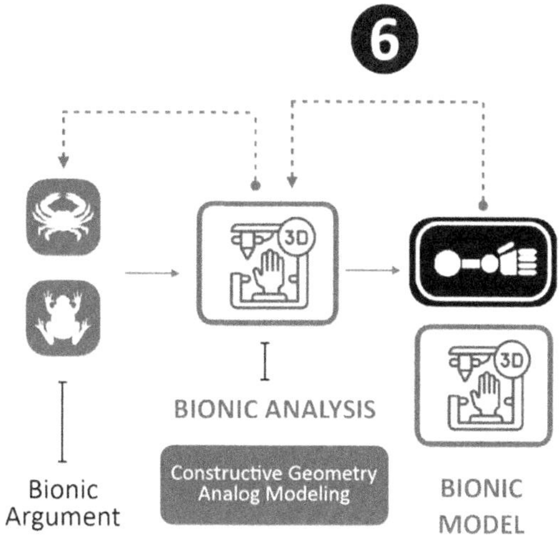

Fig. 14 Bionic model
Note. **All graphs and drawings of the methodological process were developed by the author in 1992 and adapted and redesigned in 2022 [1]**

7 Conclusion

Studying nature is undoubtedly a very complex thing; however, it is necessary to have a deeper knowledge of these aspects and their contents. When we talk about bionics, it is not enough for a designer to refer only to nature with its formal and symbolic aspects, but it is necessary to individualize a specific operational modality of research to arrive at a project.

Current living organisms are the results of continuous natural selection and adaptation that act in a particular way to provide more suitable forms and structures suitable for the particular elements found in their habitats.

According to John Too, in "*Progettare secondo la natura*": biology "is a model that reflects the most important way in which the natural world works." And according to our indications, based on all the information collected and available in this proposal, we can say without fear of error that "the Biological field tends to be a new model for projection" [1].

Analyzing this concept further, it is possible to further enhance it and state that "The Natural World could be a reference model for design in the near future" [1].

Therefore, it is now more evident that in bionics we have the possibility of observing the natural world with a new vision that unifies the research of those that are the scientific canons and seek their applicability in a more technological universe. In fact, everything we reflected on in this research proposal was to seek to propose not a complete method but didactic/design indications of the applicability of a taxonomic model of living beings identified in biology, as a reference to the world of projection. To finalize our proposal for this postdoctoral degree, we aim to:

1. *Characterize and apply data organized according to academic/scientific work, to be used as an instrument for reflection and criticism.*
2. *Establish these rudimentary methodological procedures to use natural elements (echinoderms and corals) as suggestions and ideas for better design of artificial objects.*
3. *Increasingly deepen our knowledge in the natural/biological field for adequate and safe intervention in the bionic field.*

References

1. Arruda, A. J. V. (1992). *Verso un modelo didattico metodogico*. tesi di mestrado, CRIED.
2. Bonsiepe, G. (1975). *Teoria y Práctica del Diseño Industrial* (pp.124–134). Editorial Gustavo Gilli.
3. Broeck, F. V. (1982). *O uso de analogias biológicas* (Vol.15, pp. 97–100). Revista Design e Interiores.
4. Di Bartolo, C. (1981). *Struture naturali e modelli bionici*. apostilha IED.
5. Gerardin, L. (1968). *La bionica*. il saggiatore.

6. Papanek, V. (1973). *Progettare per un mondo reale*. Arnaldo Mondadori.
7. Stelle, J. (1959). *Bionics symposium – Living Prototypes – The key to new technology* (Wadd technical report, pp. 60–600).
8. Yeang, K. (1995). *Designing with nature. The ecological basis for architectural design*. McGraw-Hill.

Amilton José Vieira Arruda, Graduated in Industrial Design Product Project at the UFPE, Master in Design and Bionics at IED in Milan, and Ph.D. in Research in Industrial Design at the University of Milan Politecnico. He finished his postdoctoral in Design and Bionics at IADE European University UNIDCOM Lisbon and postdoctoral in Bionics at the University Luigi Vanvitelli, Naples, Italy. He is currently associate professor IV at the UFPE Design Course and coordinates the BioDesign Lab.

Bionics and Design: 3D Microstructural Characterization of Cork for the Development of Conceptual Products

Melissa Pedroso Schoffen, Mariana Kuhl Cidade, and Felipe Luis Palombini

1 Introduction

There are many applications where impact dampening is required, from automobile [1] and aerospace [2] industries to sports [3] and construction [4]. Crashworthiness [5] or the ability to slowly reduce an impact by decelerating and decreasing a body's momentum is key for maximum energy absorption, particularly when safety constraints are mandatory, such as in car bumpers or bulletproof vests. Following similar characteristics, dampening is also crucial for an improved feeling of comfort and relief from uniformity in pressure points, as needed in seats, footwear, and mattresses (Fig. 1). These are distinctive features commonly found in cellular materials [6–8] as well as thin-walled structures [9–11].

Cellular solids have their properties derived from their microstructural characteristics, such as whether they have a closed or open cellular structure (e.g., a foam or a scaffold, respectively), their cellular arrangement, the orientation and aspect ratio of cells, and the cell wall thickness and shape, among others [6]. In addition to those morphological variables, the general properties of cellular material are also greatly dependent on the relative density $\left(\frac{\rho^*}{\rho_S}\right)$ – or volume fraction – of the solid [13, 14]. The relative density is a dimensionless unit representing the ratio between the

M. P. Schoffen
Industrial Design Undergraduate Course, Federal University of Santa Maria (UFSM), Santa Maria, RS, Brazil
e-mail: melissa.schoffen@acad.ufsm.br

M. K. Cidade · F. L. Palombini (✉)
NOVA Lab – Innovation and Sustainability in Design Lab – Department of Industrial Design, Graduate Program in Architecture, Urbanism, and Landscaping (PPGAUP), Federal University of Santa Maria (UFSM), Santa Maria, RS, Brazil
e-mail: mariana.cidade@ufsm.br; felipe.palombini@ufsm.br

A. J. V. Arruda, F. L. Palombini (eds.), *Biomimetics, Biodesign and Bionics*, Environmental Footprints and Eco-design of Products and Processes,
https://doi.org/10.1007/978-3-031-51311-4_2

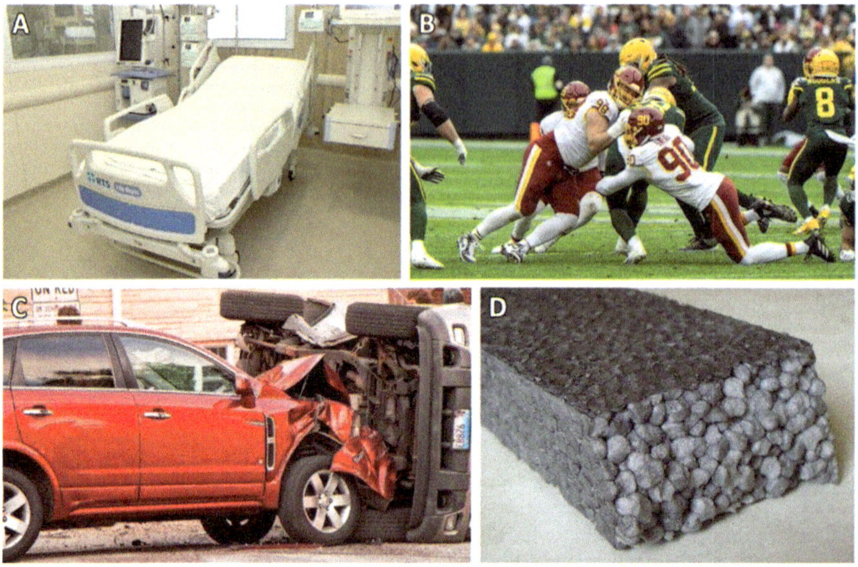

Fig. 1 Application examples of impact-dampening materials in (**a**) hospital mattresses, (**b**) football equipment, (**c**) car bumpers, and (**d**) insulating foams for construction. (Adapted from [12], license by CC-BY)

density of the whole cellular material (ρ^*) and that of the solid cell wall itself (ρ_S) [15]. In addition, the relative density can also be inversely related to the porosity (φ) of that region, where the higher the relative density, the lower the porosity, as well as a relative density of 1 is equal to a completely dense material, with no porosity. In this sense, once a cellular solid has a porous characteristic, its overall density tends to be lower than that of conventional materials, also contributing to an overall higher efficiency. Either in terms of load-to-weight ratio or the sheer amount of material employed in its manufacturing, cellular materials also gain the advantage of reducing the energy required to transport it, such as in the case of expanded polystyrene (EPS) applied on packaging.

Besides its low density, impact absorption, and superficial comfort, these features are also closely related to sound [16] and vibration [17] dampening in cellular materials. In simple terms, this happens when different mechanical frequencies cause a solid to oscillate and resonate while moving, which may either lead to undesirable consequences, such as noise and acoustic discomfort, or even severe catastrophic failures, in the case of structures. Thermal and electrical insulation are other properties that can be associated with the field of cellular solids. Due to the presence of a multiphase material, composed of a solid cell wall and some medium (which can be air, N_2, CO_2, or some other gasses), it may also present some insulating properties of interest, depending on the application.

Today, architects, engineers, and designers have a wide array of cellular solids options to choose from, for instance, MatWeb.com [18], a database of material properties, returns over 1500 instances. Depending on the project, one can look up for the reduction of the impact loads over a passenger compartment in a vehicle or improving the comfort on the midsole of a running shoe, as well as insulating walls for energy efficiency cooling. Many synthetic materials developed in the last century exhibit those characteristics, such as polymer- [19], metal- [20], and ceramic-based [21] foams; however, there is a group of materials that have gained attention due to the combination of many of the traditional cellular characteristics with new attributes. Natural materials can be described as raw materials that only require minimum processing prior to their application and can be obtained from plant, animal, or mineral sources [22, 23]. In addition to their availability and direct utilization – like most woods – natural materials may exhibit some key properties for sustainability, such as having a renewable origin if properly managed.

One of the first studied cellular materials in nature, with a variety of properties of interest, is cork. This plant's tissue was historically important during the development of imaging techniques when Robert Hooke (1665) described its microscopic structure and established the term "cell" as a way to define those repetitive, small structures [24]. Despite being found in the tree barks of many species, the cork tissue grows into a substantially thicker and more continuous layer in the cork oak (*Quercus suber* L., Fagaceae), where it is extensively explored as bottle stoppers in the Mediterranean region (Fig. 2).

Fig. 2 Cork in its various forms: (**a**) the tissue is extracted from the bark of the cork oak (*Q. suber*); (**b**) its main utilization is the manufacturing of wine bottle stoppers; (**c**) the remaining material is agglomerated and used for the production of secondary materials, like sparkling wine stoppers and board panels; and (**d**) more recently, cork has been laminated for the manufacturing of a leatherlike alternative fabric. (Adapted from [12, 25], license by CC-BY)

Cork is used to cap bottles of wines and sparkling wines not only due to its mechanical properties but also to its physical and chemical characteristics [15]. Mechanically, cork is a highly compressible material and has a near-zero Poisson coefficient: the deformation, which can be an expansion or a contraction, of a material in directions perpendicular to the original direction of loading. This means that once compressed to fit into a bottleneck, the cork won't deform in the transversal direction, facilitating the processes [26]. As for the physical and chemical properties, cork is also remarkably impermeable to liquids while allowing some oxygen exchanges which contribute to wine aging [27], besides being chemically inert, thus not negatively influencing the taste of wine [28, 29]. In addition to the main manufacturing application of bottle stoppers, as shown in Fig. 2, cork can be agglomerated into panels and many more shapes, being employed in many industries, including insulation, flooring, composites, furniture, consumer goods, leatherlike fabrics, and many more [28–32], while even being renewable. Therefore, cork can be considered not only a valuable and sustainable material for multiple applications but may have the potential of being an important source of inspiration, on which new materials and structures can be based. This chapter was developed in the NOVA Lab – Innovation and Sustainability in Design Lab, from the Federal University of Rio Grande do Sul, under the research lines of bionics, digital manufacturing, and sustainability.

2 From Characterization to Bionics

Bionics, biomimetics, biomimicry, and biodesign [33, 34] are some of the terms used to designate the field of applied sciences at which inspiration is drawn from the natural world for the development of innovative projects. Overall, the methodology of design based on the study of bionics is established upon the billions of years of evolution life has taken to surpass and adapt to the most fierce environmental conditions, by prioritizing energy efficiency while maintaining a naturally sustainable transformative cycle. Bionics can be followed via two approaches, known as bottom-up and top-down [35, 36]. Essentially the first one relies on the interpretation and understanding of natural phenomena being studied, parametrized, or abstracted and applied to a technical project. The other one begins with a technical question, where possibilities to address it are looked for in nature and then allowing it to be solved efficiently. In both cases, the natural subject must be investigated thoroughly, by means of multiple imaging and testing techniques, to allow its main principle to be fully comprehended and that it may be successfully applied to a project. Finally, this project may be related to any desired application, mainly design, engineering, and architecture but also to other fields like business, economics, medical sciences, and more.

Those are some examples where 3D technologies were used to investigate the natural world in order to promote new perceptions regarding their anatomy and physiology. Among the main techniques used, X-ray microtomography (μCT) was

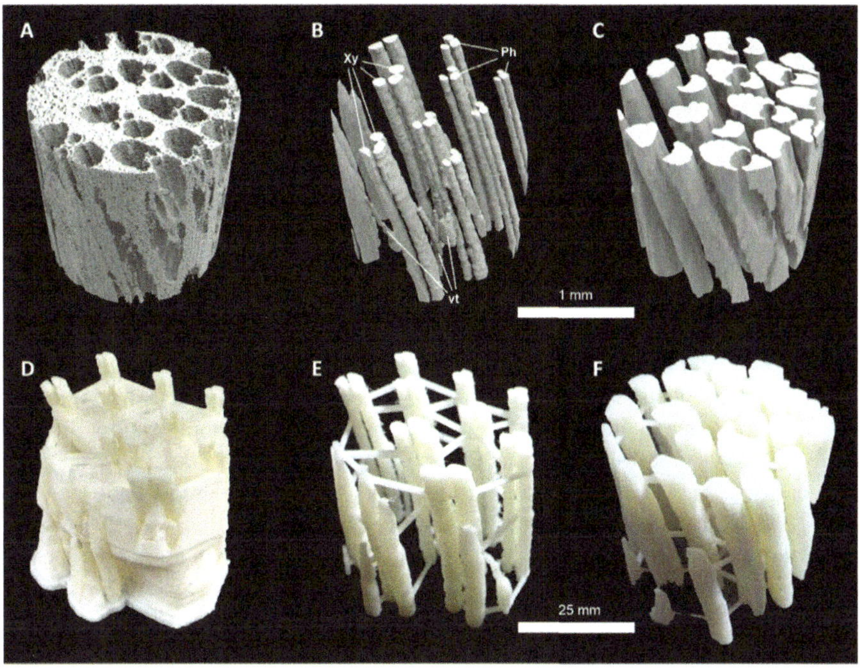

Fig. 3 X-ray microtomography (μCT) in bamboo: virtual models of the segmented regions of the (**a**) parenchyma, (**b**) conducting elements, and (**c**) sclerenchyma (fiber bundles). 25:1 scale, 3D printed models of the conducting elements (**d**) with and (**e**) without support material, and the (**f**) sclerenchyma. (Figs. **a**, **b**, and **c**: Reprinted (adapted) with permission from [14], copyright Elsevier® 2016. Figs. **d**, **e**, **f**: Reprinted (adapted) from [39], license by CC-BY)

explored, leading to the discovery of new functionalities of some already-known tissues and structures (Fig. 3). Just like its clinical counterpart, the technique is based on the quantification of the absorption of X-ray radiation by a sample, being rotated after projection. On the other hand, μCT is differentiated by the high spatial resolution of the projections, being generally in the micrometer scale. After completion, the X-ray projections are combined using an algorithm, leading to the generation of 3D volume, which can be exported as sequential images in a stack [37, 38].

One of the main advantages is that the image stack can be sectioned virtually in any plane noninvasively, thus allowing the scanned sample to be analyzed and studied more thoroughly, using quantitative and qualitative investigations. Another important benefit of working with μCT-based studies is the ability to segment different regions of interest (Fig. 3a–c) and export them as STL files with a surface mesh to be used in 3D printing (Fig. 3d–f), generating a realistic and untouched physical model of the scaled sample to be used for instruction purposes, for example [39].

Another important advantage of working with μCT is the possibility to perform virtual simulations in the digitalized high-resolution sample. The same segmented 3D model employed in 3D printing can also be discretized into a volumetric mesh, to be used in finite element analysis (FEA) [40, 41]. In a simplified way, this numerical approach uses a known finite number of elements to divide a continuum volume, investigate them separately in several types of analyses, and then combine them back to the original volume, thus allowing a representative approximation of a realistic scenario [42]. Some of the analyses include static and explicit structural analysis, thermal analysis via heat transfer, electromagnetic, fluid dynamics, and more [40, 43].

Recently, μCT-based FEA has also been of great importance to drive new insights into hitherto known properties or characteristics of the biological world (Fig. 4). For instance, in plant sciences, the parenchyma tissue in bamboo was investigated and evidenced to have an important role in the overall structural strength of bamboo [14], at when only the sclerenchyma (fiber bundles) was acknowledged to have this function in the plant (Fig. 4a). Similarly, the derivation of the vascular system of bamboo, from one main culm to three secondary branches in the nodal region, was analyzed [44] being found that in the case of a compressive load, the plant has a propensity to preserve its main axis from collapse, thus maintaining its growth system (Fig. 4b). Moreover, the tank inflorescence of bromeliad was investigated regarding the role of the accumulated water [45], which was then assessed to be part of a protection mechanism by preventing thermal shocks that would otherwise harm the development of flowers and reproduction (Fig. 4c).

Finally, once known, these insights promoted by state-of-the-art, noninvasive, high-resolution 3D technologies can be used to develop new projects by means of bionics, biomimetics, and biodesign, as shown in Fig. 5. For instance, the geometry of the cross section of the fiber bundles of bamboo was classified according to one of the types described in the literature [46, 47] and investigated regarding its mechanical role in the plant. While utilizing it as a reinforcement pattern in a thin-walled structure (Fig. 5a), the geometry proved to improve the axial and lateral

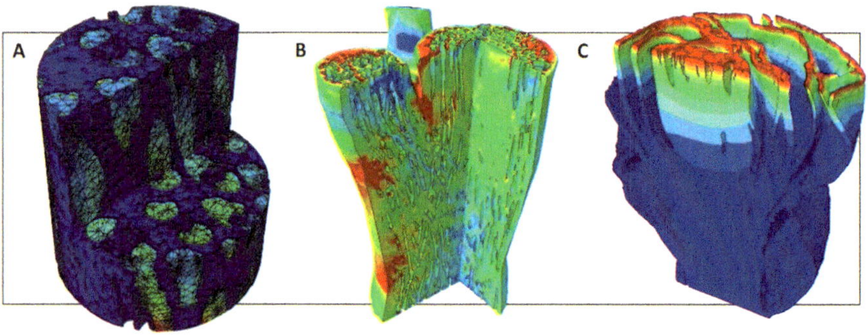

Fig. 4 New insights using μCT-based finite element analysis (FEA) in plant anatomy and its implications for plant sciences: (**a**) the mechanical role of the parenchyma in bamboo; (**b**) the preservation of the main bamboo axis during compressive load; and (**c**) the role of water as some heat protection in bromeliad's tank inflorescence. (Figs. **a**, **c**: Reprinted (adapted) with permission from [14, 45], copyright Elsevier® 2016 and 2019. Fig. **b**: Reprinted (adapted) from [44], license by CC-BY)

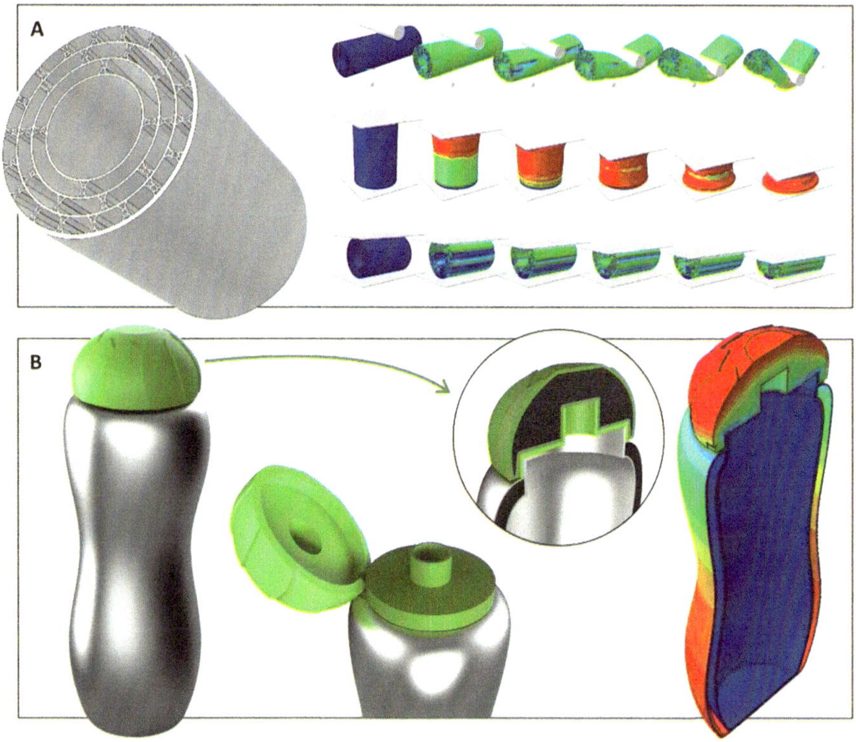

Fig. 5 Bionics and biomimetics. Example of projects founded on investigations performed in plants. (**a**) Thin-walled structure based on the geometry of the fiber bundles of bamboo and mechanical simulations. (**b**) Insulated water bottle based on the inflorescence tank of bromeliads and thermal simulation. (Fig. **a**: Reprinted (adapted) with permission from [9], copyright Elsevier® 2020. Fig. **b**: Reprinted (adapted) from [48], license by CC-BY)

compressive strength as well as the flexural strength of the model, when compared to similar other bioinspired models from the literature [9], also demonstrating that it can have a similar role in the plant. In a different approach, an insulated water bottle or flask was designed (Fig. 5b), utilizing the properties of the tank inflorescence of some bromeliads species in the inside of the outer cap, by means of acting as a thermal mass to prevent heat exchange between the inside and outside [45, 48]. In both cases, simulations were performed to assess the performance of the designed models.

3 Microstructural Characterization of Cork

As seen by the previous examples, 3D technologies are gaining significant importance both in terms of learning new insights about the anatomy and physiology of the natural world and contributing to the development of new designs following the methodology of bionics, biomimetics, and biodesign. Cork can be defined as a closed-cell foam, a natural cellular material, with elongated cells oriented in the radial direction of the plant. Therefore, it can be modeled as a honeycomb structure that has two major planes: radial and non-radial. Mechanically, it is known for having a high compression resistance, shock absorption capability, elasticity, and a near-zero Poisson effect while compressed. All that while having an exceptionally low density, allowing it to be used in many applications. However, all those characteristics are due to the overall performance of the cork tissue, known as phellem, which can be assessed and determined on a macroscopic scale.

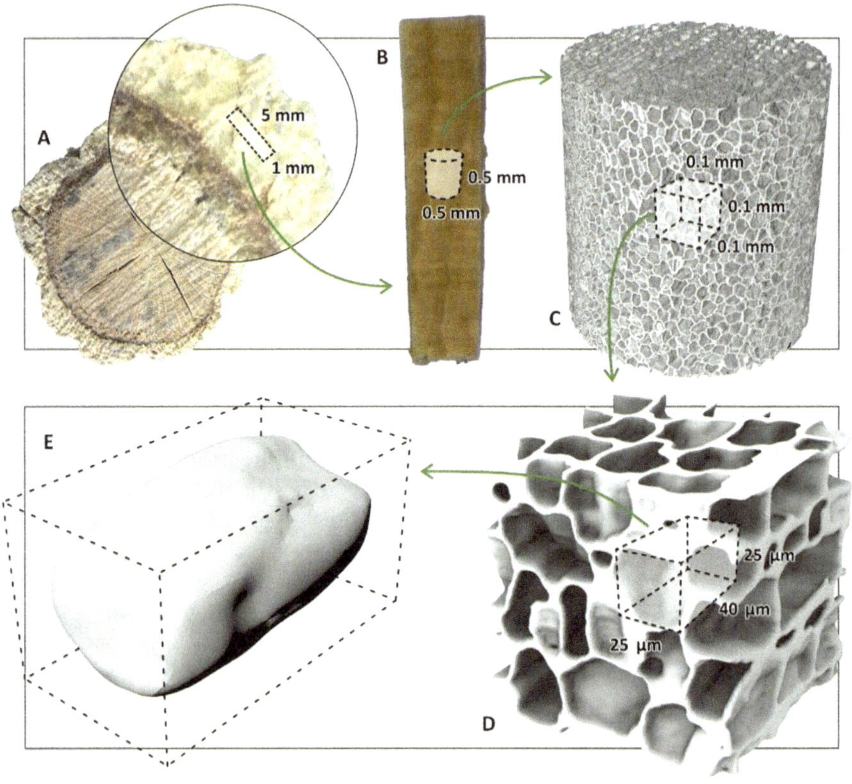

Fig. 6 3D microstructural characterization of cork (*Q. suber*): (**a**) cross section of the stem showing the rectangular sample of size 1 × 1 × 5 mm; (**b**) small rectangular sample showing a preview of the cylindrical µCT scanned region, of 0.5 mm in diameter and in height; (**c**) cubic region of interest of side 100 µm, comprising approximately 60 cells; (**d**) example of cork cell extracted from the µCT. (Reprinted (adapted) with permission from [15], copyright Springer-Nature® 2023)

In order to assess the performance of the tissue on the microscopic scale, high-resolution 3D technologies are required to allow the material to be analyzed at a cellular level (Fig. 6). A sample of cork was cut from the periphery of a section of the stem of *Q. suber*, visualized in Fig. 6a. The sample consisted of a prismatic geometry of size 1 × 1 × 5 mm, and a Xradia® 510 Versa μCT (Zeiss X-ray Microscopy, Pleasanton, CA, USA) equipment was used to scan it. An X-ray source voltage and current of 30 kV and 65 μA, respectively, were used, along with a 20*X* optical magnification, native to the equipment. A total of 1601 projections were taken, each with 24 s of exposure time, resulting in an overall acquisition time of approximately 14 h, whereas a final submicron voxel size of 0.56 μm was obtained.

The 3D volume of the reconstructed μCT model consists of a cylinder of height and diameter of approximately 530 μm, as can be seen in Fig. 6c. This obtained shape is due to the μCT acquisition method, where a series of X-ray projections (1:1 aspect ratio image with pixel size side of about 0.5 μm) are accumulated, having their vertical center in common. Each projection corresponds to one acquisition of the sample, which is vertically rotated between them in that center, thus resulting in a final combined shape of a cylinder. From this 3D reconstructed volume, a cubic region of interest (ROI) was selected, visualized in Fig. 6d, of side 100 μm. This ROI consists of around 60 cork cells, which were segmented and analyzed independently. One of the cells was extracted from the ROI and measured, being shown in Fig. 6e. Most of the cells are elongated in the radial direction of the original sample (Fig. 6a), having an average length of approximately 40 μm and lateral sides of about 25 μm. It is noteworthy that the cells are arranged in such a way that they can fill a volumetric region in spite of its volume, due to their organization as a honeycomb. Geometrically, this arrangement can be described as having a Voronoi pattern, which is based on a subdivision of given space between cells, whose boundaries are equidistant from the two neighboring seeds. Therefore, this could be explored later as a conceptual design where the cork honeycomb structure is applied to different objects.

In addition to a 3D quantitative characterization of the cork region of interest retrieved from the μCT, it is possible to analyze it numerically. Therefore, the same region was exported as a mesh and converted via Gmsh software [49], an automatic and free 3D mesh generator aimed at finite element analysis (FEA). The software was used to generate a volumetric tetrahedral mesh that could be exported into Abaqus/CAE (Dassault Systèmes Simulia® Corp., Providence, RI, EUA). Following the same boundary conditions and constitutive properties presented in our previous paper [15], an explicit (dynamic) step in FEA was conducted, in order to visualize the inner deformation of the cell walls of the material (Fig. 7).

The analysis was defined as a set of the extracted region of interest from cork and two rigid (undeformable) plates, located at the top and at the bottom. The top plate was set to be displaced downwards, up to a total deformation of 50% of the cork region. Fig. 7a–c shows different levels of deformation of the whole set, being 0%, 25%, and 50%, respectively. In the details, the cut view of the internal walls shows that they collapse by means of bending and buckling [7, 50, 51], indicating that their geometry can be explored as energy absorption structures [9, 15].

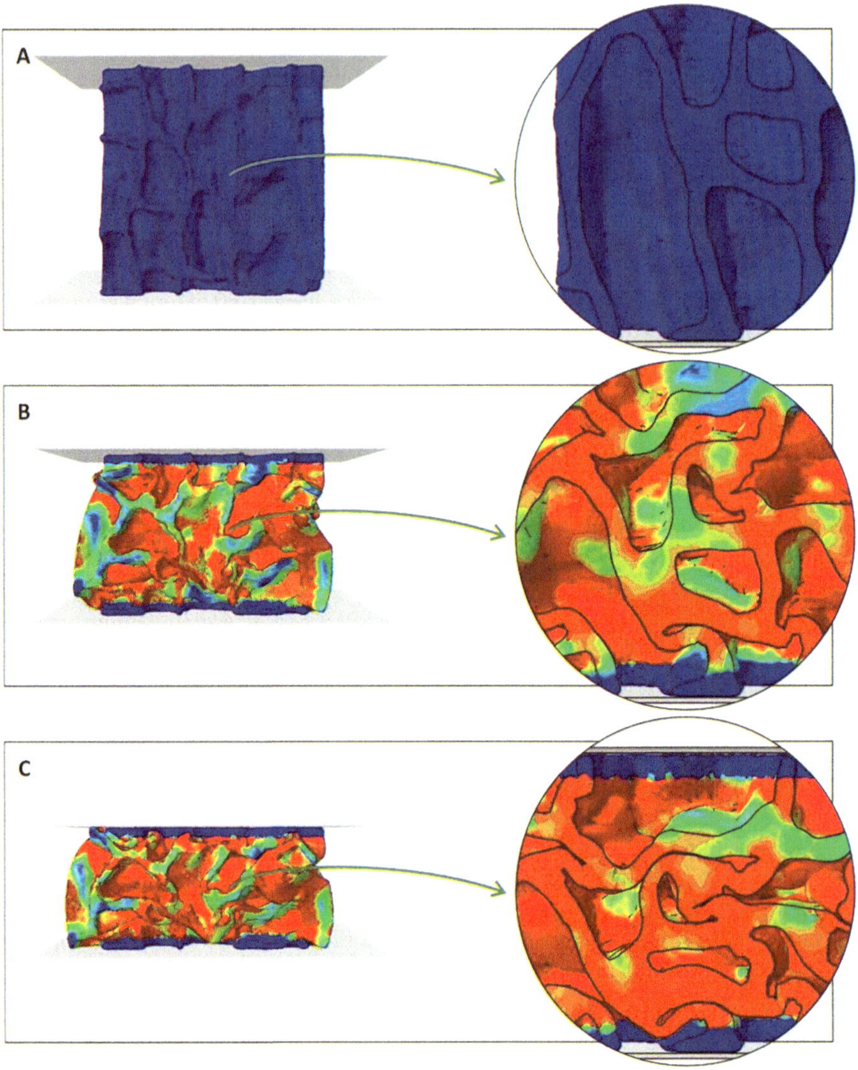

Fig. 7 Finite element analysis based on X-ray microtomography of cork (*Q. suber*) showing the collapse of the cell walls during compression: (**a**) deformation of 0% of the total height, (**b**) deformation of 25%, and (**c**) deformation of 50%

4 Conceptual Design Applications

Following the characterization of cork as a 3D honeycomb structure by means of X-ray microtomography and explicit finite element analysis, it could be observed that cork showed a maximum energy absorption efficiency of 66% [15]. This was due to the high plateau regime cork cell walls experience while being compressed.

Throughout the loading application, cell walls buckle and bend individually, thus gradually dissipating the energy applied in the load. In this regard, Fig. 8 shows examples of some conceptual designs in which a cork-inspired, honeycomb cellular structure can be applied.

There are two main applications for cellular structures inspired by the microstructure of cork considering energy absorption. The first one is aimed at protection, where the main objective is the dissipation of the energy of a certain load applied into a region of an object, such as in the case of bike helmets or shoes for impact sports, like basketball. In this case, the load can either be absorbed slowly, to reduce the velocity at which it is transferred, or quickly, aiming at the return of energy, in the case of running shoes, for instance. As for the other application, cellular structures can be associated with comfort. Cushioned products like furniture, bike saddles, and headphones, for instance, can also benefit from a cork-inspired, cellular structure, by means of a progressive deformation, such as foams made out of viscoelastic polymers.

Fig. 8 Conceptual designs of some applications of cork-inspired honeycomb cellular structures aimed at energy absorption both in terms of protection (impact absorption or propulsion) and comfort (progressive deformation)

5 Conclusions

Impact dampening is an area of study where the deceleration of an applied load is key to improving a desired characteristic or performance. There are many applications in product design at which such property is sought. One way to look for different ways to approach this issue is by means of bionics. Also known as biomimetics or biodesign, this is a research field where the design process is assisted by the natural world, where geometries, textures, colors, functions, structures, or arrangements are used as a source of inspiration. In order to apply the selected characteristics of a natural subject, one must observe and analyze them correctly, in order to better understand how it works as well as its capabilities and limitations.

Cork is a remarkable natural material, capable of absorbing energy in a macroscale, due to the morphology of its microstructure. This chapter presented a 3D characterization of the cellular structure of this plant tissue, by utilizing the state-of-the-art, noninvasive technology of finite element analysis (FEA) based on the real geometry of a sample scanned via X-ray microtomography (μCT). The μCT analysis showed that the cellular structure of cork resembles that of a honeycomb, which can be simplified in the design of a product via the application of a Voronoi pattern. By using an explicit mode in FEA, it was possible to visualize the mechanical deformation and collapse of the cork cell walls during a compressive load, leading to a maximum energy absorption efficiency of 66%. During the application of the load, the failure of the cell walls occurred due to the bending and buckling of discrete regions that, once combined, contribute to energy dissipation.

After the 3D characterization, conceptual designs were proposed in which the two main applications of the cork micromechanical behavior could be explored. First, regarding protection, a cork-inspired cellular structure can be sought due to the dissipation of the load, either by means of a slow absorption or a fast energy return. Second, there is the comfort provided by the progressive deformation of individual regions within the material, just like the effect observed in the FEA. In either case, it is evident that looking for efficient structures in nature can provide interesting insights on how to develop products with improved characteristics.

Finding appropriate methods and techniques to study a given sample is key to have in mind while defining the characteristics or problems a design project aims to address. In this research, a noninvasive, high-resolution 3D technology was vital to observe how a microscopic region deforms during loading, consequently allowing a comprehensive interpretation of the mechanism by which cork material is notable for. In the future, the realistic simulations of μCT-based FEA can be used to study different types of materials, hence supporting the development of improved and more efficient materials and products.

References

1. Abdullah, N. A. Z., Sani, M. S. M., Salwani, M. S., & Husain, N. A. (2020). A review on crashworthiness studies of crash box structure. *Thin-Walled Structures, 153*, 106795. https://doi.org/10.1016/J.TWS.2020.106795
2. Yang, X., Ma, J., Wen, D., & Yang, J. (2020). Crashworthy design and energy absorption mechanisms for helicopter structures: A systematic literature review. *Progress in Aerospace Sciences, 114*, 100618. https://doi.org/10.1016/J.PAEROSCI.2020.100618
3. Hoshizaki, T. B., Post, A., Oeur, R. A., & Brien, S. E. (2014). Current and future concepts in helmet and sports injury prevention. *Neurosurgery, 75*, S136–S148. https://doi.org/10.1227/NEU.0000000000000496
4. Ramli Sulong, N. H., Mustapa, S. A. S., & Abdul Rashid, M. K. (2019). Application of expanded polystyrene (EPS) in buildings and constructions: A review. *Journal of Applied Polymer Science, 136*, 47529. https://doi.org/10.1002/APP.47529
5. Baroutaji, A., Sajjia, M., & Olabi, A. G. (2017). On the crashworthiness performance of thin-walled energy absorbers: Recent advances and future developments. *Thin-Walled Structures, 118*, 137–163. https://doi.org/10.1016/J.TWS.2017.05.018
6. Gibson, L. J., & Ashby, M. F. (1999). *Cellular solids: Structure and properties* (2nd ed.). Cambridge University Press.
7. Gibson, L. J., Ashby, M. F., & Harley, B. A. (2010). *Cellular materials in nature and medicine*. Cambridge University Press.
8. Prabhu, S., Raja, V. K. B., & Nikhil, R. (2015). Applications of cellular materials – An overview. *Applied Mechanics and Materials, 766–767*, 511–517. https://doi.org/10.4028/www.scientific.net/AMM.766-767.511
9. Palombini, F. L., Mariath, J. E. A., & Oliveira, B. F. (2020). Bionic design of thin-walled structure based on the geometry of the vascular bundles of bamboo. *Thin-Walled Structures, 155*, 106936. https://doi.org/10.1016/j.tws.2020.106936
10. Tarlochan, F., Samer, F., Hamouda, A. M. S., et al. (2013). Design of thin wall structures for energy absorption applications: Enhancement of crashworthiness due to axial and oblique impact forces. *Thin-Walled Structures, 71*, 7–17. https://doi.org/10.1016/j.tws.2013.04.003
11. Huang, H., & Xu, S. (2019). Crashworthiness analysis and bionic design of multi-cell tubes under axial and oblique impact loads. *Thin-Walled Structures, 144*, 106333. https://doi.org/10.1016/j.tws.2019.106333
12. Wikimedia Commons. https://commons.wikimedia.org/wiki/Main_Page. Accessed 10 Sep 2023.
13. Palombini, F. L., Nogueira, F. M., de Oliveira, B. F., & de Araujo Mariath, J. E. (2022). Two-way bionics: How technological advances for bioinspired designs contribute to the study of plant anatomy and morphology. In F. L. Palombini & S. S. Muthu (Eds.), *Bionics and sustainable design* (pp. 17–44). Springer.
14. Palombini, F. L., Kindlein Junior, W., de Oliveira, B. F., & Mariath, J. E. A. (2016). Bionics and design: 3D microstructural characterization and numerical analysis of bamboo based on X-ray microtomography. *Materials Characterization, 120*, 357–368. https://doi.org/10.1016/j.matchar.2016.09.022
15. Palombini, F. L., de Oliveira, B. F., Nogueira, F. M., et al. (2023). 3D cellular characterization and finite element analysis of cork compressive behavior based on high-resolution X-ray microtomography. *Wood Science and Technology*. https://doi.org/10.1007/s00226-023-01483-5
16. Sagartzazu, X., Hervella-Nieto, L., & Pagalday, J. M. (2008). Review in sound absorbing materials. *Archives of Computational Methods in Engineering, 15*, 311–342. https://doi.org/10.1007/s11831-008-9022-1
17. Al Rifaie, M., Abdulhadi, H., & Mian, A. (2022). Advances in mechanical metamaterials for vibration isolation: A review. *Advances in Mechanical Engineering, 14*, 168781322210828. https://doi.org/10.1177/16878132221082872

18. Online Materials Information Resource – MatWeb. https://www.matweb.com/index.aspx. Accessed 10 Sep 2023.
19. Okolieocha, C., Raps, D., Subramaniam, K., & Altstädt, V. (2015). Microcellular to nanocellular polymer foams: Progress (2004–2015) and future directions – A review. *European Polymer Journal, 73*, 500–519. https://doi.org/10.1016/J.EURPOLYMJ.2015.11.001
20. Singh, S., & Bhatnagar, N. (2018). A survey of fabrication and application of metallic foams (1925–2017). *Journal of Porous Materials, 25*, 537–554. https://doi.org/10.1007/s10934-017-0467-1
21. Hammel, E. C., Ighodaro, O. L. R., & Okoli, O. I. (2014). Processing and properties of advanced porous ceramics: An application based review. *Ceramics International, 40*, 15351–15370. https://doi.org/10.1016/J.CERAMINT.2014.06.095
22. Cidade, M. K., Perini, J. T., & Palombini, F. L. (2022). Bionics for inspiration: A new look at Brazilian natural materials for application in sustainable jewelry. In F. L. Palombini & S. S. Muthu (Eds.), *Bionics and sustainable design* (pp. 195–223). Springer.
23. Ormondroyd, G. A., & Morris, A. F. (2018). *Designing with natural materials*. CRC Press.
24. Peters, W. S. (2023). The cells of Robert Hooke: Pores, fibres, diaphragms and the cell theory that wasn't. *Notes and Records: the Royal Society Journal of the History of Science*. https://doi.org/10.1098/rsnr.2022.0049
25. Free Images & Free stock photos – PxHere. https://pxhere.com/en/. Accessed 10 Sep 2023.
26. Fortes, M. A., & Teresa Nogueira, M. (1989). The poisson effect in cork. *Materials Science and Engineering: A, 122*, 227–232. https://doi.org/10.1016/0921-5093(89)90634-5
27. Echave, J., Barral, M., Fraga-Corral, M., et al. (2021). Bottle aging and storage of wines: A review. *Molecules, 26, 713*. https://doi.org/10.3390/MOLECULES26030713
28. Silva, S. P., Sabino, M. A., Fernandes, E. M., et al. (2005). Cork: Properties, capabilities and applications. *International Materials Reviews, 50*, 345–365. https://doi.org/10.1179/174328005X41168
29. Pereira, H. (2007). *Cork : Biology, production and uses*. Elsevier.
30. Gil, L. (2015). New cork-based materials and applications. *Materials, 8*, 625–637. https://doi.org/10.3390/ma8020625
31. Knapic, S., Oliveira, V., Machado, J. S., & Pereira, H. (2016). Cork as a building material: A review. *European Journal of Wood and Wood Products, 74*, 775–791. https://doi.org/10.1007/s00107-016-1076-4
32. Carvalho, L., & Williams, B. (2014). Let the cork fly: Creativity and Innovation in a Family Business. *15*, 127–133. https://doi.org/10.5367/IJEI.2014.0146.
33. Arruda, A. (2023). *Map research bionics transdisciplinary ecosystems in bionics, biodesign and biomimicry: Scientific and technological aspects for a design culture*. Editora Blucher.
34. Palombini, F. L., & Muthu, S. S. (2022). *Bionics and sustainable design*. Springer.
35. Speck, O., Speck, D., Horn, R., et al. (2017). Biomimetic bio-inspired biomorph sustainable? An attempt to classify and clarify biology-derived technical developments. *Bioinspiration & Biomimetics, 12*, 011004. https://doi.org/10.1088/1748-3190/12/1/011004
36. Kindlein Júnior, W., & Guanabara, A. S. (2005). Methodology for product design based on the study of bionics. *Materials and Design, 26*, 149–155. https://doi.org/10.1016/j.matdes.2004.05.009
37. Palombini, F. L., Cidade, M. K., Oliveira, B. F., & Mariath, J. E. A. (2021). From light microscopy to X-ray microtomography: Observation technologies in transdisciplinary approaches for bionic design and botany. *Cuadernos del Centro de Estudios en Diseño y Comunicación, 149*, 61–74. https://doi.org/10.18682/cdc.vi149.5516
38. Boyd, S. K. (2009). Micro-computed tomography. In *Advanced imaging in biology and medicine* (pp. 3–25). Springer.
39. Palombini, F. L., Kindlein Júnior, W., Silva, F. P., & Mariath, J. E. A. (2017). Design, biônica e novos paradigmas: uso de tecnologias 3D para análise e caracterização aplicadas em anatomia vegetal. *Design e Tecnologia, 7*, 46. https://doi.org/10.23972/det2017iss13pp46-56

40. Boyd, S. K. (2009). Image-based finite element analysis. In *Advanced imaging in biology and medicine* (pp. 301–318). Springer.
41. Palombini, F. L., Lautert, E. L., Mariath, J. E. A., & de Oliveira, B. F. (2020). Combining numerical models and discretizing methods in the analysis of bamboo parenchyma using finite element analysis based on X-ray microtomography. *Wood Science and Technology, 54*, 161–186. https://doi.org/10.1007/s00226-019-01146-4
42. Zienkiewicz, O. C., Taylor, R. L., & Zhu, J. Z. (2013). *The finite element method: Its basis and fundamentals* (7th ed.). Butterworth-Heinemann.
43. Kurowski, P. M. (2004). *Finite Element Analysis for design engineers*. SAE International.
44. Palombini, F. L., Nogueira, F. M., Kindlein Junior, W., et al. (2020). Biomimetic systems and design in the 3D characterization of the complex vascular system of bamboo node based on X-ray microtomography and finite element analysis. *Journal of Materials Research, 35*, 842–854. https://doi.org/10.1557/jmr.2019.117
45. Nogueira, F. M., Palombini, F. L., Kuhn, S. A., et al. (2019). Heat transfer in the tank-inflorescence of *Nidularium innocentii* (Bromeliaceae): Experimental and finite element analysis based on X-ray microtomography. *Micron, 124*, 102714. https://doi.org/10.1016/j.micron.2019.102714
46. Liese, W. (1985). *Anatomy and properties of bamboo*. In: Proceedings of the international bamboo workshop. Hangzhou.
47. Palombini, F. L., & Nogueira, F. M. (2023). *Bamboo science and technology*. Springer.
48. Palombini, F. L., Oliveira, B. F., & Mariath, J. E. A. (2021). Biônica e Seleção de Materiais : Design de garrafa térmica bioinspirada na inflorescência-tanque de bromélias. In G. G. Oliveira & G. J. Z. Núñez (Eds.), *Design em Pesquisa – Volume 4* (pp. 310–326). Marcavisual.
49. Geuzaine, C., & Remacle, J.-F. (2009). Gmsh: A 3-D finite element mesh generator with built-in pre- and post-processing facilities. *International Journal for Numerical Methods in Engineering, 79*, 1309–1331. https://doi.org/10.1002/nme.2579
50. Gibson, L. J., Easterling, K. E., & Ashby, M. F. (1981). The structure and mechanics of cork. *Proceedings of the Royal Society A: Mathematical, Physical and Engineering Sciences, 377*, 99–117. https://doi.org/10.1098/rspa.1981.0117
51. Niklas, K. J., & Spatz, H.-C. (2012). *Plant physics*. University of Chicago Press.

Melissa Pedroso Schoffen, Undergraduate student in Industrial Design at the Federal University of Santa Maria (UFSM). She is a researcher in the Innovation and Sustainability in Design Lab (NOVA Lab) at UFSM.

Mariana Kuhl Cidade, Assistant professor in the Department of Industrial Design (DDI) and in the Graduate Program in Architecture, Urbanism, and Landscaping (PPGAUP), at the Federal University of Santa Maria (UFSM). Dr. Cidade graduated in Industrial Design and received her M.Sc. and her Ph.D. in Design (PGDesign) at the Federal University of Rio Grande do Sul (UFRGS). Her research includes sustainability, contemporary jewelry, materials recycling, bionics, and product design. She is associated with the Innovation and Sustainability in Design Lab (NOVA Lab) at UFSM.

Felipe Luis Palombini, Assistant professor in the Department of Industrial Design (DDI) and in the Graduate Program in Architecture, Urbanism, and Landscaping (PPGAUP), at the Federal University of Santa Maria (UFSM). Dr. Palombini received his M.Sc. and his Ph.D. in Design (PGDesign) at the Federal University of Rio Grande do Sul (UFRGS). His main research topics are focused on bionics, sustainability, digital manufacturing, and materials characterization. He is associated with the Innovation and Sustainability in Design Lab (NOVA Lab) at UFSM and with the Design and Computer Simulation Research Group (DSC) and the Plant Anatomy Laboratory (LAVeg), at UFRGS.

Bio-digital 'Material Systems': New Hybrid Ways for Material-Driven Design Innovation

Sabrina Lucibello and Carmen Rotondi

1 Introduction

> We are nature, and we can only generate nature. So are the objects and artefacts to which we give what we call "intelligence" and "life" [1].

Nature has always inspired design, starting from Vitruvius' search for harmony, passing through Art Nouveau and developing in principle with an exclusively decorative, symbolic or semantic function. Subsequently, with bionics, biomimetics, biodesign and digital biofabrication, the reference extends to nature's processes and systems until it becomes a model, measure and mentor in the design of the artificial [2].

In recent decades, technological advancement and the introduction of increasingly sophisticated tools have made it possible to deeply understand nature and its functioning, bringing natural and artificial worlds ever closer and making matter increasingly full of performance, functions expanded and capabilities enriched. Starting from the 'imitative' method in biomimetics up to the co-creation approach of biodesign, the boundaries between nature and artifice are progressively fading, fostering an expansion of design and paving the way to post-anthropocentric and systemic development perspectives aimed at the imagination of new nature-artifice co-evolution forms. This is the field of action of bio-digital practices where computational thinking merges with biological issues, and product, process and matter meet to transfer the living's generative and performative qualities to the artifice.

S. Lucibello (✉) · C. Rotondi
Department of Planning, Design, Technology of Architecture (PDTA)_Sapienza University of Rome, Rome, Italy
e-mail: sabrina.lucibello@uniroma1.it; carmen.rotondi@uniroma1.it

A. J. V. Arruda, F. L. Palombini (eds.), *Biomimetics, Biodesign and Bionics*, Environmental Footprints and Eco-design of Products and Processes,
https://doi.org/10.1007/978-3-031-51311-4_3

2 Nature as a Source of Inspiration

Nature has always been an essential reference for humans, both from a semantic, functional and ethical point of view. In particular, the vital tension between 'nature' and '*technè*' concepts has generated some of the most exciting lines of contemporary research, which considers the existence of multiple 'natures' rather than a single 'nature'.

Since the primitive era, nature has been a formal and functional inspiration for the 'creation of new artefacts intended to compensate (and complete) the inherent shortcomings of our body' [3]. However, today, technology makes it possible to 'replicate, invade and disseminate' the body [4], expanding the field of inspiration from form to 'substance' up to the process, also generating a new bio-hybrid aesthetics, mirroring the extension to the 'plural' aspects of nature [5].

If, in fact, from a linguistic and semantic point of view, nature has always been seen in the past as the unique generator of formal, organic or structural matrices[1] underlying solid geometric rules,[2] today we can look at nature as plural in which ecological themes, archaic and traditional references, innovation on materials, utopian themes and advanced technologies intertwine or alternate, highlighting not a single method but the richness of the possible reading levels able to generate a bio-inspired aesthetic. This approach welcomes the infinite variability and richness of nature's nuances (biodiversity), fostering a new population of hybrid figures that redefine the morphological aspect of our designed environment. Patricia Urquiola, a Spanish designer who has established an interesting and fruitful working relationship with many Italian companies, speaks[3] about hybridisation as a border rich in different elements, analogies and assimilations. Urquiola also talks about a second hybridisation[4] always developed following a rigorous design process.

From a functional point of view, nature is also a reference for achieving and maintaining dynamic equilibrium, helpful for the ecosystem's survival by reproducing, for example, its laws, constraints and system logic. Nature is, therefore, no longer just a model to preserve and formally or analogically imitate but also an exemplar in terms of balance, advantage, evolution and progress [7], with significant consequences both in terms of the process (energy saving in production and in the decommissioning or recycling phase) and of the product (less energy consumption during the consumption phase and use of alternative sources). This is

[1] Formal geometric studies have united biology, applied arts and architecture since the eighteenth century (Durand, Viollet Le Duc, Greenough, Semper, Sullivan, D'Arcy W. Thompson, Steadman). An exemplification of the formal approach is, for example, the work of Violet le Duc who, using studies on organic structures such as animal skeletons or bat wings, supported by a study of forces and weights, creates rational structural systems (as in Assembly Hall project), which first influenced Art Nouveau and in particular Hector Guimard, up to Louis Sullivan and Frank LloydWright.

[2] As for Munari [6].

[3] From an interview in Repubblica, 2 March 2010.

[4] From a 2011 interview by A. Coppa: www.livinginterior.it/interno-a-patriciaurquiola/0,1254,58_ART_3868,00.html

the field of action of biomimicry,[5] i.e. the conscious study of biological and biomechanical processes, as a source of inspiration for improving human activities and technologies. From this perspective, nature can be considered a 'model', a 'measure' and a 'mentor' for the design of artefacts[6] [2].

This type of approach – based on the systematic study of 'natural prototypes' and oriented towards understanding the biomorphological, mathematical, geometric, functional, structural, mechanical and cybernetic aspects – has generated two main research orientations: on the one hand, the investigation of neurophysiological systems of living organisms to reproduce them artificially and generate innovation in the field of neuroscience, cybernetics, electronics, robotics and prosthetics (i.e. electronic retina, electronic cochlea, heart valves, orthopaedic prostheses, etc.) and on the other hand, the analysis of the biological dynamic balances between form, function and aesthetic elements or between energy and matter, in an attempt to draw useful biomechanical, structural and functional models applicable to artefacts. However, there is more because, in the post-human era,[7] nature and *technik*[8] prefigure new scenarios in which individuals will even be able to control and modify their morphology, somehow incorporating the artefacts' performance within the body, so much that it is difficult to identify where humans end and machine begins, as well as talking about cyber-prostheses or extensive-sensory prostheses appears simplistic [10]. In this regard, Gregory Stock configured the birth of a global superorganism in which humans, held together by technology and multimedia, represent the equivalent of cells in an animal organism [11]. Like the evolution history of the natural world, we can draw an evolution history of artificial species in which – even before the automaton, the robot[9] and the 'humanoids' – there is the 'replicated body' or the 'replicant' in Ridley Scott's 'Blade Runner' [12], in all respects morphologically similar to the human being but naturally more performing.

The examples so far belong to 'analogue bionics' field,[10] defined by the *Treccani* Encyclopedia (in 'bionic' session) as 'the science that studies the sensory and motor

[5] Mathieu Lehanneur is a French designer very successful in the biomimicry field who has concentrated his most recent activity on studying the interactions between the body and the environment, living systems and the scientific world, combining organic materials with digital technologies. Among his best-known projects is 'Andrea' (2008), the air purifier that works with plants.

[6] This is the field of action of 'hybrid design', i.e. the design approach that aims to transfer to the design of innovative products and services the complexity inherent in the logic and principles of the biological world as sort of 'new genetic code' [8].

[7] The term 'post-human' or 'trans-human' was born within 'cyber delicate' niches and the new age subculture to establish itself definitively thanks to the Post Human exhibition created by Jeffrey Deitch in 1992, which brings together an artistic production of the 1980s dedicated to the theme of the self and the body.

[8] Khanna and Khanna [9] talk about '*tecnik*', that is, the union of the deterministic (mechanical + scientific) and the constructivist dimensions (which deals with the effects on men and society).

[9] The Bohemian writer Karel Capek coined the term robot and etymologically means 'toil'.

[10] Analog bionics operates through a procedure which, starting from the description of the process (analysis) and passing through the translation of the biological description into a physical-mathematical scheme, arrives at the concrete realisation of the scheme (through synthesis carried out with an electronic device) in order to constitute an analogical model. On the contrary, dynamic bionics can capture nature's adaptation and development strategies to transform them into actions.

functions of living organisms to identify sophisticated solutions for technical problems and to reproduce or enhance such living organisms with electronic or other devices type'. The natural reference acts as a set of complex processes based on dynamics of 'evolution, genesis and integration of phenomena to be transferred' [13]. Another example is the all-Italian project called 'ICub'[11] about a humanoid robot created by a group of researchers from the Italian Institute of Technology (IIT) in Genoa. It is based on a very advanced IT platform, open-source hardware, and sophisticated algorithms derived from bio-inspired cognition patterns that allow ICub to learn. 104 cm tall and weighing 22 kg, ICub is like a child of about 3 years old who is estimated to be able to have 53 degrees of freedom of movement: 7 for each robotic arm, 9 for each robotic hand (3 for the thumb, 2 for the index finger, 2 for the middle finger, 1 for the ring-little finger pair, 1 for abduction/adduction), 6 for head movements, 3 for the chest and spine and 6 for each robotic leg [14]. Here, we are dealing with a copy or a reproduction in terms of form, structure and function of nature's solutions, which paves the way to the experimentation of real technologic grafts into the body and to body technology hybrids, configuring what Caronia calls the 'invaded body' [4]. Thanks to the cross-pollination of cutting-edge sectors such as information technology, biotechnology, pervasive computing, robotics, neuroscience and nanotechnology, the prefix 'bio' is increasingly replaced by the prefix 'neuro', thus prefiguring the *Homo evolutis* defined by Enriquez and Gullens [15]. We thus move from bionics to the design of living things up to biomimicry.

Therefore, the design must be fluid, experiential, intelligent, performing and, above all, self-sufficient. As in an inverse process, technological innovations bring us back to the nature of which we can imitate the form – increasingly organic thanks to complex production and IT processes able to manage complex meshes – the behaviour – increasingly naturally alive and intimately able to self-sustain – and the infinite variety, prefiguring a change in our way of observing, perceive and, why not, imagine the future. Meanwhile, the evolution of biomimicry has focused attention on nature as a technical and procedural model for design. Again, the advancement in scientific research brings us one step ahead towards the integration of biological materials into design projects. Today, what could have appeared as a metaphor in the utopias of the 1970s – i.e. the idea of an artificial world increasingly endowed with organic qualities – begins to appear simply as a mature fruit of bioengineering research. The idea of drawing on nature to transpose its technologies and processes to the artificial is bypassed by the reality of the facts. An example is the victimless leather, industrially produced in vitro starting from replicating cells and tissues. This perspective opens up a new meaning in the relationship between

[11] The IIT presented ICub at the Genoa Science Festival in 2009, after a series of increasing complexity prototypes based on research begun in 2003.

humans and nature, diverging the market for leather products from the age-old processes of breeding, animal slaughter and leather treatment.[12]

The exhibitions 'Design and the Elastic Mind' in 2008 and the more recent 'Biodesign' in 2012, held at the Museum of Modern Art in New York, demonstrate how, at an international level, the relationship between design, nature and science is rapidly evolving. These exhibitions show a design that will be increasingly able to incorporate the use of living matter within objects, collaborating with specific laboratories or proposing itself as a promoter of experiments in this sense. It is, therefore, not just a linear relationship that applies the results achieved by science to the world of objects (according to the classical scheme) but the re-evaluation of the visionary attitudes of design. The field of nature is highly ambiguous and in tension between science and myth. We can talk about imaginative, naked, alchemical nature without neglecting its ecological dimension as the now-acquired substrate of the design ideology. However, nature can also be a mythical reference and a technical and inexhaustible source for the imagination, which pushes us to explore every possibility of matter manipulation, whether organic or inorganic. An archetypal or deliberately primordial dimension and an extremely sophisticated production process can coexist in the same product. Therefore, the world of nature can be an aesthetic attractor and, at the same time, be searched with the analytical eye of the scientist. 'When chemists invented chemistry – Ettore Sottsass writes – they immediately invented new colours that do not exist on the terrestrial planet [...]. They would also be delighted to invent colours that imitated light or the colours of flowers or fish, which have within them that special 'phosphorescence' which is life' [16].

3 Bio-digital Practices: Towards a New Material-Driven Biological Paradigm

Since its origins, therefore, design culture has been based on observing nature and studying living systems, glimpsing new models for technical and expressive solutions in the harmony of their forms and the efficiency of their mechanisms. Nature, understood as a set of interacting and perfectly configured elements in a finalised way, is assimilated to a design system which, over billions of years of experience and *trial and error*, has accumulated a series of 'prototypes' useful for the design and production of new artefacts [2]. Technological sophistication and the evolution of biological knowledge have led over time to the flourishing of numerous approaches, which have, from time to time, changed the way of observing nature and – consequently – the nature of innovations.

[12] Specifically, the projects 'The Pig Wings' (Oron Catts & Ionat Zurr, 2000–2001), pig mesenchymal tissue cells used for biodegradable/bioabsorbable polymers (PGA, P4HB) and 'Victimless Leather. Prototype of Stitchless Jacket grown in a Technoscientific Body' (2004), biodegradable polymers, bone and connective tissue cells.

Nowadays, new concepts introduced by the omics sciences, the greater understanding of the biological processes of growth and self-organisation but, above all, of an informational matrix at the basis of all living things, lead design to extend the biological reference to the processes we use to shape the material world and give life to new solutions. At the same time, there is a growing awareness that our industrial system is founded on a logic opposed to the natural one, leading to ecological problems and a condition of paralysis for which, if we want to endow the artifice with living qualities, we cannot continue to use traditional manufacturing methods. So, the construction strategies nature adopts, the materials it uses and how it manipulates them can stimulate the design culture to experiment with alternative ways of conceiving and creating its products. At the same time, new bio-digital potential allows us to analyse and reproduce the generative, chemical, physical and molecular processes underlying living things [17], extending the paradigm to what nature does and how it produces. This is mainly due to an evolution of technological culture from a digital and informational perspective, which has led design to rethink processes, tools and materials according to a more nature-like logic and generative methods. The democratisation of increasingly sophisticated technologies and their independent use by designers have changed how design concepts are visualised and manipulated, their implementation, the relationships between design and other disciplines and the languages underlying the possibilities of technical, aesthetic and symbolic transformation of matter.

First of all, in the 'century of physics' in which 'we split the atom and transformed silicon into computational power' [18], digital and informational superstructures have dematerialised the areas of human intervention and have given rise to an enhancement of their transformative abilities, reaching the point of producing at the Mother Nature quantum scale and significantly changing the structures of thought and operation. Advanced reading and representing tools offered by digital technology lead design to deal with the languages and conceptual paradigms underlying biological processes, such as growth, evolution, reproduction and transformation. A case in point is generative design, which – taking advantage of the opportunities offered by information technology, new modelling software and digital manufacturing technologies – is inspired by genetics' principles and merges a product's conception and production phases as in a single biological growth process. Furthermore, the large quantity and variety of data that can be processed today allow this process to be applied to all types of artefacts, from everyday products to architecture and large infrastructures, also benefiting from the biological-like advantages that derive from it, such as the possibility of on-demand production (i.e. which occurs only at the moment of need and for a single project) or the perfect adherence of the final solutions to the conditions of the context (as in the case of architectures adapted to the site-specific atmospheric conditions or wearable products modelled on the body of a specific user). Veronika Kapsali, professor of materials and design at the London College of Fashion, states that technological development since the Digital Revolution has allowed us to look at the informational phenomena that occur in nature in a different light, inspiring a 4D biomimetic design in which both spatial and temporal elements are considered so that the properties of an object change with context and future stimuli [19]. With this definition, Kapsali seeks to introduce a vast field of biomimetic experimentation that ranges from robotics to programmable materials and which, thanks to the possibility of introducing information into human-made structures, allows the reproduction of advanced living behaviours such as self-assembly, self-repair, self-organisation and autonomous movement.

This process evolves up to the contemporary 'century of biology' [20], in which biotechnological and biosynthetic potential is the extension of digital properties to nature – which becomes comprehensible to the last detail, programmable and manipulable – and, inheriting from them design-based languages and methodologies, they bring biology closer to design than ever before [21]. In such a context, design approaches and methodologies also change by integrating systemic thinking with a set of disciplines, both humanistic and scientific, as well as broadening their range of intervention from the micro-scale (analysing and designing how things work starting from the most profound details) to the macro-scale (trying to understand and direct relationships and interconnections from a cultural, social and environmental point of view). The renewed attention to relationships, not only between disciplines and different human actors but also with other biological and artificial entities, opens up new and exciting opportunities for design to undertake cross-scale and trans-media projects, from the studio of complex dynamics underlying the living to the conditioning of the community's daily behaviour, for an interconnected system between society, nature and technology.

In particular, bio-digital creates a space of convergent knowledge in which computational thinking meets biological issues [22] and in which digital, network and advanced manufacturing aspects meet the transformative possibilities of matter, whether inert or biological, to transfer the generative and performative qualities of the living to the artifice. Thanks to new technological means that allow us to combine physical, digital and biological *computation*, the system of relationships is strengthened, allowing the control and design of the structural organisation at the material scale and transforming digital information into physical performance and functionalities, as well as physical and biological properties in new input data for the project, first of all changing the design practices and conceptual frameworks [23]. Indeed, bio-digital practices do not refer simply to manufacturing with biological or living building blocks but that design and fabrication procedures are aligned with the natural growth and information processes underlying them, producing complex structures that mimic the intelligence, specificity and active qualities of 'naturally developed' entities. Just as in nature, the typical hierarchical sequence of form, structure and material design is typically bottom-up reversed, as the material informs the structure, which informs the form [24]. Materials are designed for highly specialised functions rather than being assigned to preconceived shapes. The control and structuring of the material organisation are informed by environmental and material performance constraints, establishing seamless relationships between material, product and performance.

Moreover, a growth-based approach that starts from structuring the material according to biological paradigms and informative datasets allows the transfer of many properties such as structural heterogeneity for specific adaptations and performativity [25], responsiveness, dynamic and transformative behaviours (sensing, motion, shape change), bioreceptivity and so on.

From this perspective, we distinguish within the material design practices at the intersection with biology a specific pathway that we can define as 'bio-digital fabrication' or 'digital biofabrication' [26], which is different from other practices based on the use of living matter aligning instead with biofabrication practices born in the field of tissue engineering and their difference from other living matter-based technological platforms [27]. They are (Fig. 1) [27]:

(a) *Growing design* that matches with biomanufacturing and refers to the possibility of producing sustainable materials and biomolecules harnessing living matter as a raw material through making, crafting and tangible practices.

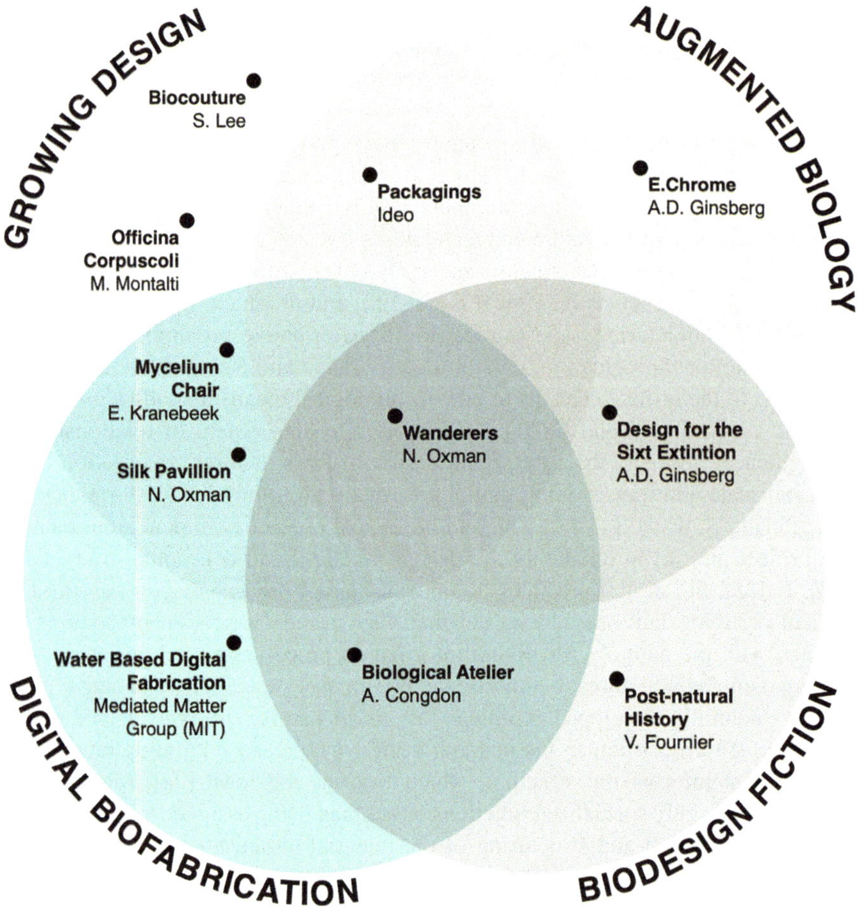

Fig. 1 Material design practices at the intersection with biology. For each area are listed some representative design projects. (Version adapted by the authors from Camere and Karana [28])

(b) *Augmented biology* that matches synthetic biology and refers to the synthesis of new biological entities and materials from genome engineering.
(c) *Bio-digital fabrication* that complies with biofabrication and stretches the possibilities of biological manipulation by adopting computational tools, digital advanced manufacturing technologies and mathematical modelling tools involving nature as the object of design processes.
(d) *Biodesign fiction* that does not properly focus on the productive paradigms of the present but imagines – often with conceptual visions – far provocative futures for advanced biotechnological developments.

Bio-digital fabrication, therefore, introduces us to a new material-driven biological paradigm that extends computational and biological principles to matter itself, becoming intrinsically sensitive, active and programmable.

First, a strong connection between digital and physical spheres translates into aesthetically and functionally augmented materialities, in which living and intelligent qualities are given by the intrinsic properties of matter itself [20]. Despite the first wave of digitality involved drawing and visualisation tools for design, digital production tools (e.g. digital manufacturing) and the derived process chains, the resulting connection of the digital with the physical – or the material – translates into informed and rewritten materials in the same way of the objects equipped with artificial intelligence, thanks to computational structures (e.g. cognitive computing, next-generation computer visualisation) and digital procedures (e.g. digital fabrication techniques, digital material representations, algorithmic shape generation methods) able to work at multiple scales (e.g. 3D microfabrication). The results are hybrid, mono-material but heterogeneous systems able to react, adapt and process data as natural ones. Examples are products and architectures with biological properties and optimised starting from the physical-chemical characteristics of the material rather than from the integration of electronic components with consequent advantages also for sustainability: from Jane Scott's experiments on 'programmable knitting' which exploit the hydromorphic properties of some fibres to shrink and unwind along specific directions [29, 30] to the 'water-based digital fabrication' developed by the Mediated Matter Group at MIT, a new platform for 3D printing of natural biopolymers (chitin, pectin, cellulose, etc.) which exploits the water-driven variability of their optical and structural properties to create material structures with gradient properties [31]. Other designers experiment with two-dimensional elements able to self-assemble into three-dimensional elements, such as Christophe Guberan who, in 'Hydro-Fold' project (2017), supports the ability of wet paper to deform for creating prints that self-assemble into solid configurations [32] or the Self-Assembly Lab at MIT which contributes to the development of soft robotics by experimenting with shape memory materials that react to external conditions (such as heat, humidity, light, electrical impulses, etc.) and are structured to reach specific deformation patterns [33]. Also part of this category are the numerous 4D printing experiments, defined as the '3D printing of multi-material structures that change shape and physical properties over time' [34], often bio-inspired by biological processes such as the movements of natural plants induced by swelling and elastic anisotropy [35]. These examples make clear how the design practice and imagination are strongly expanding, fostering a permanent confrontation between information and materialisation [36] and relying on a new form of digital mediation between real matter and virtual information, leading to new formal and functional impulses and contexts [37].

Secondly, developments in scientific disciplines such as molecular biology, genetics or synthetic biology bring the biological sphere into contact with the previous digital and physical ones, bringing the material revolution a step forward and, consequently, the design practice to explore further innovation paths. Biological matter and organisms can be used as interactive interfaces (biosensors and bioactuators) instead of synthetic ones, as in the case of 'BioLogic', a fabric that uses the mechanical behaviour of Natto bacteria deposited on portions of fabric to respond to humidity [38], of 'Organic Primitives', pH-responsive organic materials used for sensing, activation and interaction [39] or bioluminescent organisms to perform light-emitting behaviours. At the same time, the biological and biomechanical processes through which nature works and builds can be transferred to products from the material itself (like growth, repair, mutation, replication and biodegradability) and can be inspirational for intelligent behaviours. Relevant in this sense are the numerous experiments that emulate the bioprinting process used in tissue engineering for larger product and architectural scales. Essentially, the process consists of designing and printing three-dimensional 'scaffolds' that can accommodate living cells, directing their growth according to specific parameters and patterns. For example, in the 'Bio-ex-Machina" project, Maurizio Montalti and the Co-de-iT team printed a mixture based on organic waste and mycelium spores with a customised robotic arm to create new morphologically complex furniture, whose structural and aesthetic properties result from the combination of algorithms, robotic behaviours and biological growth of the mycelium [40]. The latter makes these structures extremely light, resistant and long-term biodegradable. Professor Marcos Cruz and his multidisciplinary team at the Bartlett School of Architecture (UK) have developed a bioprinting technique for large-scale, custom-printed immobilisation of microalgae, combining physical exploration of alginate-based hydrogel under printing conditions, digital exploration of algae growth pattern and printer parameters and biological understanding of algae survival ability and adsorption capacity [41]. Many innovative and sustainable scenarios arise from this research: they imagine a future in which algae ramifications cover buildings' facades, insulate, guide rainwater, purify cities' air and establish new aesthetics.

Moreover, the integration and interaction among atoms, bits and genes go beyond scaffolding strategies and bioresponsive surfaces; they encompass structures where the interplay between matter and data is not confined solely to the generative phase of the product, giving rise to fascinating exchanges between the network and biological matter. An example is the 'Fungar Project', the first 'intelligent' fungal building able to react adaptively to light, temperature and air pollution changes [42]. A novel bioelectric system has been developed, enabling the exchange of information between the fungal structure of the building, which serves as a biosensor, and computer systems that analyse this dataset and accordingly adjust the entire interconnected network of home devices.

Therefore, with bio-digital manufacturing, informed relationships between the physical, digital and biological spheres are possible thanks to sophisticated tools and technologies able to disassemble reality into its elemental components – atoms, bits and genes – and translate datasets from one dimension to another, enabling interaction, development and mutual growth between entities and disciplines belonging to them. Design practices and product conceptualisation are changing: designers are now engaging with the emerging language of *material systems*. These are not aesthetically pleasing products created through an intuitive exploration of materials or assembling distinct functional parts. Instead, they result from a novel creative process, given by a dialogue between the designer and autonomous, computerised systems, able to make decisions and adapt to their context or specific conditions. Taking biomimicry to the material level, digital biofabrication allows designers to observe the substances and logic nature employs to assemble and structure its materials, developing more sustainable and strategic ways of artefact production and replicating complex, tailored and emerging qualities. The essence of physical matter undergoes a transformative shift, hybridising with digital and natural systems and opening up new possibilities for design research, re-emphasising the tangible aspects of things and reinventing them through sophisticated technologies.

4 Bio-digital 'Material Systems'

According to Neil Gershenfeld, the MIT Center for Bits and Atoms director, we are witnessing a significant shift from 2D to 3D, transitioning from programming bits to programming atoms, approaching the ending of the Moore curve, representing the maximum point of digital evolution, where the digital world fuses seamlessly with physical matter and, ultimately, with the natural world [43]. The convergence of advances in computation and digital technologies, their extension to biological and physical matter and the intellectual influence of cybernetics and systems theory are compelling designers to engage with tangible hybrid 'materialities' able to grow, interact and transform continuously with the external environment, as well as exhibit complex biological qualities.

In recent years, many definitions of complex functional materials have been developed, from smart materials able to perceive and respond to external stimuli through reversible changes [44] to augmented materials, able to adapt to the environment through constant interaction thanks to a network of miniaturised sensors and actuators perfectly integrated into matter [45]. Furthermore, recent definitions of bio-smart materials [46, 47], bio-synergistic materials [48] and hybrid interactive systems [49] elaborate on the mechanical and biological response to stimuli of natural-based materials as a means of intelligent behaviours.

Nature, with its structures and construction laws, has always been a precious source of inspiration for the design of materials and new technologies, fostering the development of many of the previously listed categories and leading to aesthetically and functionally 'augmented' solutions through the biomimetics [7], the cooperation of digital production [50] and living matter involvement, increasingly manipulated and transformed through sophisticated technologies [51]. With bio-digital, biological processes are transformed into technological possibilities for experimentation and both the very concept of material and the conceptual frameworks for their imagination and design change, bringing nature and artifice closer than ever. This, in addition to opening up new potential for design – which are first-hand experiments thanks to the accessibility of technological tools and within multidisciplinary groups – also requires a different approach from designers to the theme of 'materiality', integrating systemic considerations more aligned with biological logic. All organisms in nature, including biological materiality, are considered open and complex systems [52] whose generation and performativity are not given simply by a material aggregate but by a combination of structural aspects of spatial organisation and functional aspects of temporal organisation, in permanent relationship with the external environment [53], to return a 'material system' – or an 'informed materiality' – in which it is almost difficult to discern between the material, its structure and its performative elements. Therefore, the transition from material to 'material systems' cannot be understood as a simple consequence of technological development but as part of a new way of understanding materiality, opening up to a different conception, imagination and production. In this sense, after the dematerialisation induced by digital, designers return to the use of matter, not understood in the traditional sense of a physical substance that materialises an idea but as an aggregate of interrelationships that we can understand, reinvent, separate and recompose, changing the game of variables and their interactions. It represents the reinstatement of an iterative and recursive process, where the designer collaborates with an agent capable of autonomous decision-making and functional adaptation to specific conditions. This collaboration unfolds across hierarchical levels, from micro to macro – defining the elements and their spatial-temporal relationships – and from macro to micro – analysing results and potentially repeating the process in a new form, considering the contextual conditions and their variability. It is a system where design action is not all-encompassing but instead adjusts to elements, properties and relationships, especially when we do not consider the involvement of only inert matter (which still possesses its agency, albeit predictable) but also of biological matter, artificial intelligence and human bodies.

Furthermore, with the transition into 'material systems' and the ability to imbue matter with thought through information and its structuring, we witness an expansion of designable qualities and their variability within a single system, resulting in an increasingly organic-like essence. The resultant hybrid concretisations encompass nearly all the degrees of freedom found in natural phenomena, exhibiting biological traits such as self-organisation, redundancy, self-generation, multifunctionality, interactivity, reactivity, adaptation and flexibility, therefore suitable for complex and changing contemporary living.

In this context, emerging forms of bio-digital production open up new conceptual and inventive paths for material transformation. Their ultimate goal is understanding the profound qualities matter can express, whether in technical, aesthetic or symbolic terms. They also inspire fresh multidisciplinary research strategies, offering advantages such as diverse forms of expression, performativity and novel aesthetics, as well as the opportunity to reimagine the paradigms of sustainable production.

To comprehensively explore this vast and intricate topic in all its facets, we present below 5 projects conducted by the authors at the Saperi&Co Centre, a research and service infrastructure of the Sapienza University of Rome, comprising 18 departments spanning various disciplines, including the humanities and sciences. Benefiting from the 'anti-disciplinary' context of Saperi&Co and the opportunity to collaborate with advanced manufacturing tools and machines, the authors have undertaken a series of material experiments over recent years within multidisciplinary teams.[13] These endeavours aimed to comprehend and promote the emerging realm of 'material systems' while simultaneously charting innovative paths in biomimicry, biodesign and bionics, spurred by contemporary cultural and technological contexts. In particular, we will elucidate how we have harnessed the principles of nature to manage the interactions between the physical, digital and biological spheres and integrate material-product-performance, merging the phases of the artefact conception and production into a single creative process.

[13] We specify that:

For the 'ScobySkin Patches' project, the research team was composed of Sabrina Lucibello, Carmen Rotondi, Chiara Del Gesso, Lorena Trebbi (Department of Planning, Design and Technology of Architecture, Sapienza University of Rome), Daniela Uccelletti, Emily Schifano (Department of Biology and Biotechnology 'C. Darwin', Sapienza University of Rome), Luciano Fattore and Riccardo Martufi (Saperi&Co).

The 'Hygroscopic Dynamism' project is part of the master's thesis work in the Master of Science in Product Design (Sapienza University of Rome) by the student Elisa Nicolia and entitled 'Responsive Hygromorph Skin for the Built Environment' (supervisor: Prof. Sabrina Lucibello; co-supervisor: PhD Rotondi Carmen).

For the 'Evolving Echinoids' project, the research team comprised Carmen Rotondi and Eugenia Maria Canepone (Department of Planning, Design and Technology of Architecture, Sapienza University of Rome). The project was for the 'Echinodesign' exhibition developed by the Hybrid Design Lab research laboratory at Città della Scienza (Naples).

The 'Bioprinting for end-of-life augmentation' project is part of the thesis work for the PhD in Planning, Design and Technology of Architecture (Product Design curriculum) of Rotondi Carmen (tutor: Prof. Sabrina Lucibello, co-tutor: Stefano Marzano – Philips Design). The experiments were carried out at the Saperi&Co center and in collaboration with the 'Bioprinting&Biofabrication' Group of the 'E. Piaggio' centre of the University of Pisa.

For the 'Continuity' project, the research team is made up of Sabrina Lucibello, Carmen Rotondi, Camilla Gironi, Paride Duello, Diana Ciufo (Department of Planning, Design and Technology of Architecture, Sapienza University of Rome), Michela Toussan (Department of Mechanical and Aerospace Engineering, Sapienza University of Rome), Teresa Rinaldi (Department of Biology and Biotechnology, Sapienza University of Rome), Luciano Fattore and Riccardo Martufi (Saperi&Co), in collaboration with Maria Diana Contemporary Jewels.

The five projects in focus are as follows:

(a) *ScobySkin Patches*: these are transdermal patches constructed from bacterial nanocellulose, featuring adaptable bioinspired auxetic textures that conform to various body surfaces.
(b) *Hygroscopic Dynamism*: this project involves creating responsive surfaces through multi-material 3D printing, drawing inspiration from the hydromorphic capabilities observed in pine cones.
(c) *Evolving Echinoids*: this initiative explores 3D printing on elastic fabric with patterns inspired by the structures found in sea urchin shells.
(d) *Bioprinting for end-of-life augmentation*: it focuses on bioprinted constructs with programmed degradability, primarily based on chitosan.
(e) *Continuity*: this project entails rendering ceramics bioreceptive using organic waste materials.

4.1 ScobySkin Patches

Typically, natural systems start with a relatively limited range of molecular components but orchestrate creating a diverse array of functional materials whose 'intelligence' goes beyond mere stimulus-response behaviour towards the ability to behave efficiently within specific contexts or situations [54]. Nature achieves this not by inventing entirely new materials but by enhancing the properties of those already available through qualitative structuring.

With this premise in mind, our project had two main objectives. Firstly, it sought to explore novel applications of bacterial nanocellulose, considering its chemical-physical and aesthetic-perceptive attributes. Secondly, it aimed to assess its compatibility and adaptability with emerging processes and tools specific to digital manufacturing.

Bacterial nanocellulose, or Scoby, is a biofabricated material generated through a symbiotic culture of bacteria and yeast fermentation. The production of this organic material relies on biological processes. So the manufacturing process harnesses the natural growth and reproductive mechanisms of microorganisms which, in the case of nanocellulose, unfold in two phases: first, yeast converts sugar into ethanol, and then bacteria transform ethanol into acetic acid, resulting in the release of nanocellulose fibrils as a by-product. These nanofibrils float in the culture medium, forming a three-dimensional biofilm on the surface, whose characteristics make it particularly suitable for replicating leather qualities from an aesthetic, tactile, visual and olfactory perspective.

Additionally, Scoby possesses hydrophilic properties, allowing it to absorb water and active ingredients. While this property might have limited relevance in leather-based applications, it proves immensely beneficial in the biomedical sector. For these reasons, the project envisions utilising bacterial nanocellulose as a second skin for creating transdermal patches or transdermal therapeutic systems (TTS) designed for drug delivery through skin absorption. The TTS offers numerous advantages, including immediate visual confirmation of drug administration and precise control over the absorption rate, thus maintaining optimal blood concentrations for specified durations. Furthermore, the material's hydrophilic nature enables it to absorb fluids and moisture, which is particularly useful for wounds, burns and skin protection while allowing easy wound monitoring. The project thus aimed to apply bacterial nanocellulose in this specialised biomedical context and experimented with 'subtractive' technology using CNC lasers. This approach aimed to enhance its application potential by creating auxetic bio-inspired textures, making transdermal patches adaptable to various body surfaces, shapes and sizes (Fig. 2).

Design tools and methods typical of bio-digital fabrication have been implemented: the biological aspect has mainly involved the production process of the material; from a physical point of view, there was an exploration of the material characteristics to identify the suitable application fields for their exaltation; with digital, the properties beneficial for the application context have been augmented, fostering a qualitative improvement of the overall material system.

4.2 *Hygroscopic Dynamism*

In nature, energy utilisation is meticulously planned to minimise the effort required for its production. Nature strives to find solutions with a favourable cost-benefit ratio, emphasising comprehensive design at all levels and employing extended timeframes in processes [55]. From a material perspective, this means starting with the molecular-level properties of the material to structure, inform and form it to be intrinsically performative, sensitive, responsive and dynamic.

Fig. 2 Transdermal therapeutic systems (TTS) realised with bacterial nanocellulose and 'augmented' with bio-inspired auxetic textures applied successively with CNC laser-cutting machine. (Copyright: Sabrina Lucibello and Carmen Rotondi, 2019)

The direct relationship among variable elements (so much that when one varies, the entire system connected to it varies) is paramount, as even a minor alteration with minimal energy expenditure can lead to significant, sometimes vital, changes throughout the interconnected system. This logic and meticulous attention to detail are exemplified in the passive behaviours of specific living structures, which, harnessing material reactions to external stimuli, vary their structure to achieve specific movement patterns.

Drawing inspiration from the hydromorphic capabilities of pine cones, the 'Hygroscopic Dynamism' project explored the intricate interplay among physical, digital and biological aspects, particularly concerning passive behaviours and how these dynamics could be integrated into material systems to minimise energy impact in the built environment. In pine cones, individual bracts close in response to humidity and open in dry conditions, regulating the dispersal of seeds. This behaviour, first comprehensively studied by Colin Dawson at the original Center for Biomimetics at the University of Reading in the 1990s, is remarkable because it operates without external energy. Instead, it relies on a reversible mechanism driven by two fabrics with distinct hygroscopic properties: one fabric is more porous and swells in humidity conditions, while the other is rigid, hydrophobic and anisotropic, with fibres oriented in a specific direction. When exposed to humidity, the first fabric thickens, causing the more rigid one to bend perpendicular to the fibre direction, resulting in the opening of the cone.

This natural phenomenon has inspired various projects, such as Sphere Macro (2006), a performance fabric developed by Nike, featuring carved textures similar to bracts that open in response to humidity to keep athletes dry. In the HygroSkin Meteorosensitive Pavilion (2013), the pine cone's hygroscopic principle was transferred to a composite material by layering woods with varying swelling properties and applied for the pavilion's openings, enabling them to open and close in response to changing weather conditions. The opening dynamics were carefully engineered and controlled by orienting the wood fibres along desired curvature axes.

Similarly, 'Hygroscopic Dynamism', in alignment with research conducted by MIT researchers [56], placed significant emphasis on this aspect, and the main objective was to investigate novel transformation patterns for 3D-printed hydromorphic structures. In particular, it involved multi-material printing experiments, combining two filament types to replicate the two fabrics found in pine cone bracts: a hydrophilic filament that swells in humidity (Laywood®, an environmentally friendly FDM filament composed of approximately 40% recycled wood cellulose and a binding polymer) and a hydrophobic filament (PLA or Nylon). The latter was the rigid and anisotropic component, bending along the orientation of 3D-printed 'fibres', e.g. the infill's printing directions.

From a physical point of view, we meticulously considered the distinct characteristics of the two materials – rigidity, printability, deformability, etc. – for printing active layers (using Laywood®) and passive layers (employing PLA or Nylon). The printing parameters (layer height, print orientation, material flow, etc.) were studied digitally to program specific deformations (curling, folding, etc.). These deformations were carefully directed by orienting the printing direction perpendicular to the

Fig. 3 The final prototype of a hydromorphic structure for the built environment: each component is designed to acquire specific movement patterns. The prototype was firstly full water submersed and then air-dried. (Adapted with permission from 'Responsive hygromorph skin for the built environment', thesis work of Elisa Nicolia (Supervisor: Prof. Sabrina Lucibello; co-supervisor: PhD Carmen Rotondi), 2021)

desired movement and adjusted in terms of intensity, total or partial (Fig. 3). From a biological perspective, we closely examined the patterns and physical principles underpinning the pine cone's biological behaviour to optimise our outcomes.

The envisioned application for these material systems lies in the future city, where the built environment embodies intrinsic intelligence and human-made structures engage in a continuous and harmonious dialogue with the natural ecosystem they inhabit. These hydromorphic panels have been primarily applied to self-sustaining architectural structures, including solar tracking systems for architecture [57], architectural skin systems [58] and shading systems, which, in addition to eliminating the need for electromechanical support systems, promote a circular economy and the recycling of waste cellulosic materials. We developed three distinct concepts tailored to diverse application contexts. Firstly, a symbolic dome-shaped pavilion served as an explanatory showcase of the overall dynamic behaviour of hydromorphic panels. Secondly, a more expansive canopy composed of multiple components was envisioned for potential urban shelters (Fig. 4). Lastly, we envisioned a futuristic portrayal of tomorrow's cities if we replace actual buildings with dynamic, ever-evolving structures intricately synchronised with the planet's transformations. Across all three solutions, we integrated adaptive systems that

Fig. 4 Concept of futuristic urban shelter composed of multiple components and characterised by a hydro-responsive and hydromorphic roof. (Adapted with permission from 'Responsive hygromorph skin for the built environment', thesis work of Elisa Nicolia (supervisor: Prof. Sabrina Lucibello; co-supervisor: PhD Carmen Rotondi), 2021)

transformed the role of the skin or envelope, shifting it from a rigid barrier to a delicate mediator in the built environment.

4.3 *Evolving Echinoids*

In nature, growth occurs simultaneously across all hierarchical levels of organised information – ranging from cellular and tissue levels to organs, organisms, populations, ecosystems and biosphere evolution. Therefore, in biological materiality, every element can influence another element across any level, regardless of time and position, potentially leading to the evolution of the whole system and the emergence of what we call 'emergent properties'. In this context, the focus shifts away from the individual parts that constitute a phenomenon to the understanding of the intricate relationships among them, following a 'hologrammatic principle' according to which a part exists within the whole but, conversely, the whole also exists within the part. Hence, we cannot comprehend a complex system through the sole dichotomy of reductionism – attempting to understand the whole solely from the qualities of its parts – or holism, a simplifying approach that disregards the parts to grasp the whole [59].

Transferring and harnessing the hologrammatic principle, even in artificial contexts, can prove exceptionally effective in grappling with complexity and taking advantage of it. On the one hand, it yields significant energy savings, as the system's properties can significantly alter with the slightest change. On the other hand, the meticulous establishment of a hierarchy of relationships enables work on multiple levels of abstraction, amplifying the range of potential solutions and enhancing the adaptability of constructed systems to their intended purposes. In this context, variety accentuates differentiation and engenders complexity, leading to a multiplication of interaction possibilities and potential responses to evolving conditions [53].

The 'Evolving Echinoids' project delves into the hologrammatic principle as a source of innovation in bio-digital manufacturing. This approach, bolstered by computational thinking and advanced manufacturing tools that process extensive datasets and on-demand materialisation of complex structures, empowers designers to operate across multiple scales and design multiple qualities within a system. Bio-inspired by the highly complex geometric and structural configuration of the echinoid test, the project embodies its hierarchical construction adaptations across all scales, meticulously designed for optimal support and load distribution [60]. These geometric principles were meticulously examined across micro-, meso- and macro-scales and subsequently applied following a hologrammatic logic to create 'programmable prestressed textiles' (PPTs), programmable and three-dimensional fabrics crafted through 3D printing on stretched fabric. Once released from tension, the fabric exerts both tensile and compressive forces on the 3D-printed components, thus generating the final shape. It is important to note that the application of biological principles is inverted: the objective is not to create a shape capable of withstanding internal and external loads but rather a morphology that, much like sea urchins, 'grows' and adapts its form in response to the forces acting upon it.

From a physical perspective, we studied the material composition of the echinoderm skeleton, characterised by a rigid component consisting of a 3D network of biocalcite trabeculae (stereom) and an organic component made up of a cellular and extracellular matrix, including collagen fibres (stroma) [61]. We analogised the latter with stretchable and elastic fabrics, providing support, continuity and compactness to the rigid part represented by the 3D-printed plastic pattern. Moreover, this elastic component functions as an energy absorption system, a crucial feature in echinoids for preventing the propagation of fractures and in PPTs for material deformation (acting as a distorting element). From a digital standpoint, our efforts concentrated on several aspects, including the geometry of the 3D-printed rigid layer, which drew inspiration from the micro-scale's porosity and structural variability observed in stereomes [62]. We also considered the modularity of the dome-shaped structure, characterised by variable hexagonal plates arranged according to the principles of 'pure plate structures' at the meso-scale [63]. Finally, we examined the overall geometric configuration, incorporating double-curvature surfaces, differential and from a point growth of the plates at the macro-scale [61, 64, 65]. Specifically, the micro-level bio-inspiration allowed for variations in the density, texture and printing direction of the 3D printing patterns on the stretched fabric, enabling customisable folding intensities and directions (Fig. 5). The principles at the meso-level facilitated the definition of support (strength) and tension (deformation) lines, while the study of geometric laws at the macro-scale led to the exploration of 2D Gaussian (intrinsic) curvature. This last principle enabled the creation of specific three-dimensional double-curved structures based on two-dimensional geometric patterns, allowing us to define 3D printing geometries a priori for precise morphological responses, further enhanced by computational and generative design

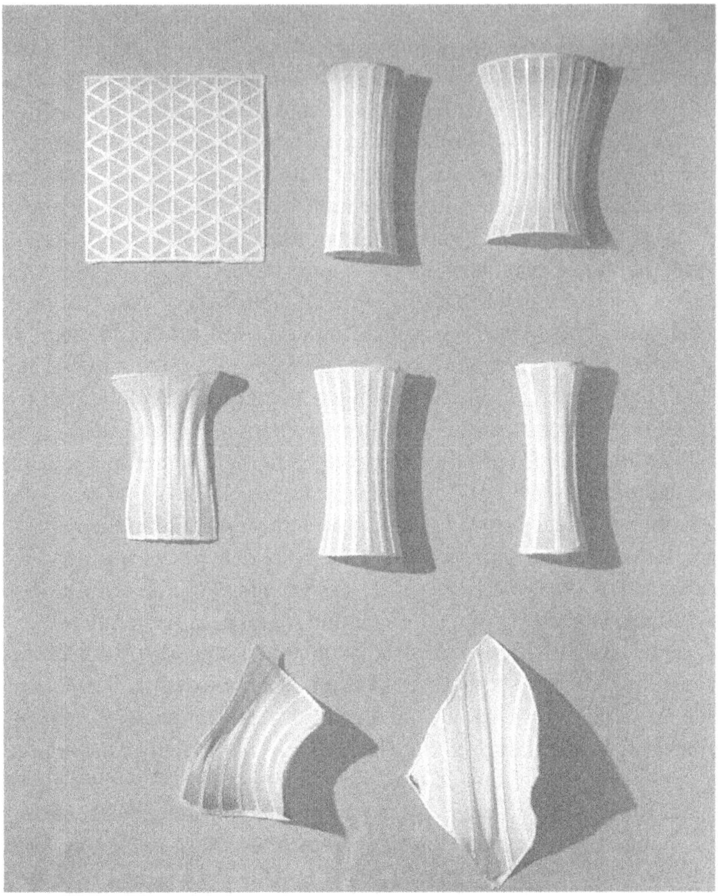

Fig. 5 Samples of trials made printing PLA filament on pre-stretched fabric, varying the printing pattern (lines and reticular), the direction (90°, 0°, 45°) and the printing density (80% and 60%). (Copyright: Carmen Rotondi and Eugenia Maria Canepone, 2022)

tools – gradually integrated into this project. The final results achieved are complex morphologies tailored for the lighting design sector and programmed to meet aesthetic requirements and deliver differential and context-specific lighting effects (Fig. 6).

4.4 *Bioprinting for End-of-Life Augmentation*

Nature's creativity is not only expressed in producing sophisticated material systems or articulated growth and adaptation strategies but also at the end-of-life phase, where nature ensures that essential components vital for the development of new

Fig. 6 Final result of 'Evolving Echinoids' project: a prototype of a lamp realised following micro–/meso–/macro bio-inspired 3D-printed textures on a prestressed textile. (Copyright: Carmen Rotondi and Eugenia Maria Canepone, 2022)

life forms return to the internal ecosystem in a simple form, easily reusable by subsequent generations [53]. This aspect represents the most significant departure between our linear production logic and nature's circular one. Many designers are focused on this aspect, both conceptually and practically. The translation of concepts like reuse, recycling, decomposition and monomateriality from nature, which are now paramount in the context of the circular economy, has been a central theme in design culture over recent decades. This shift has been driven by environmental, social and cultural imperatives, seeking alternative meanings beyond our human society's relentless drive for production. However, although the circularity of production processes and efficient resource utilisation are fundamental for environmental sustainability, there are numerous other end-of-life aspects we can learn from nature to make our projects eco-efficient but especially 'eco-innovative', namely, able to translate innovations into forms of awareness and stimulus for more responsible behaviours. In nature, everything is programmed to return to circulation according to predefined schedules aligned with the specific needs of the biocoenosis. Even patterns of degradability are controlled to achieve particular functional and expressive qualities beneficial to the entire ecosystem. A noteworthy example is leaf senescence, which represents a response to the intricate integration of various environmental signals and leaf age information and involves a complex and highly regulated process that coordinates multiple pathways, ensuring the orderly disassembly of macromolecules for transfer and facilitating communication between different aspects or species within the ecosystem [66].

The 'Bioprinting for end-of-life augmentation' project contemplates the significance of programmability even at the end of an artefact's life cycle and explores the potential for expressive and functional innovation from such an approach. Furthermore, a second objective of the project was to delve into the design and production possibilities that emerge when employing bioprinting, a hybrid process originating from the field of tissue engineering that combines the physical exploration of the base material (or 'scaffold's material') with the digital design of supportive architectures (known as 'scaffolds') and the directed growth of living cells, guided by the geometry and composition of the underlying scaffold.

Bioprinting follows processes and principles completely aligned with bio-digital manufacturing. It amalgamates the techniques and tools of additive manufacturing – ranging from 3D modelling software to bioprinters – with material engineering experiments, particularly concerning the structural and functional optimisation of biomaterial scaffolds, and biotechnology protocols, that manipulate biological matter under precise conditions. Beyond the myriad factors that designers must now consider (including rheology, biocompatibility, sterility, and time), bioprinting continually expands the realm of possibilities to create constructs that mimic the complexity and heterogeneity of biological tissues. Designers can fabricate optimised materials with spatial gradients in composition (material variability) and structure (geometry variability) that can also influence dynamic behaviours over time [67, 68].

'Gradient features' hold considerable potential for design as they allow for creating and replicating specific, dynamic biological-like properties within bioprinted materials. This concept is particularly relevant given the abundance of bio-derived and circular materials, often in the form of extrudable pastes, which may not inherently possess the properties needed to replace their synthetic counterparts and can be enhanced or tailored for new and specific applications by incorporating gradient features. Starting from these assumptions, the project can be outlined in two primary phases:

(a) The development of a robotic dispensing system designed to transform a conventional 3D printer into a bioprinter. This system can handle visco-elastic fluids with solid-like properties and is equipped with a double extruder, allowing for the use and combination of different and increased material quantity for printing (Fig. 7).
(b) The experimentation with controlled degradability constructs, starting with bio-derived and circular chitosan-based pastes (Fig. 8).

In particular, the robotic dispensing system offers three modes of operation: with a single extruder for printing homogeneous constructs with a single material; with double extruder, alternating between two extruders to create layers from two

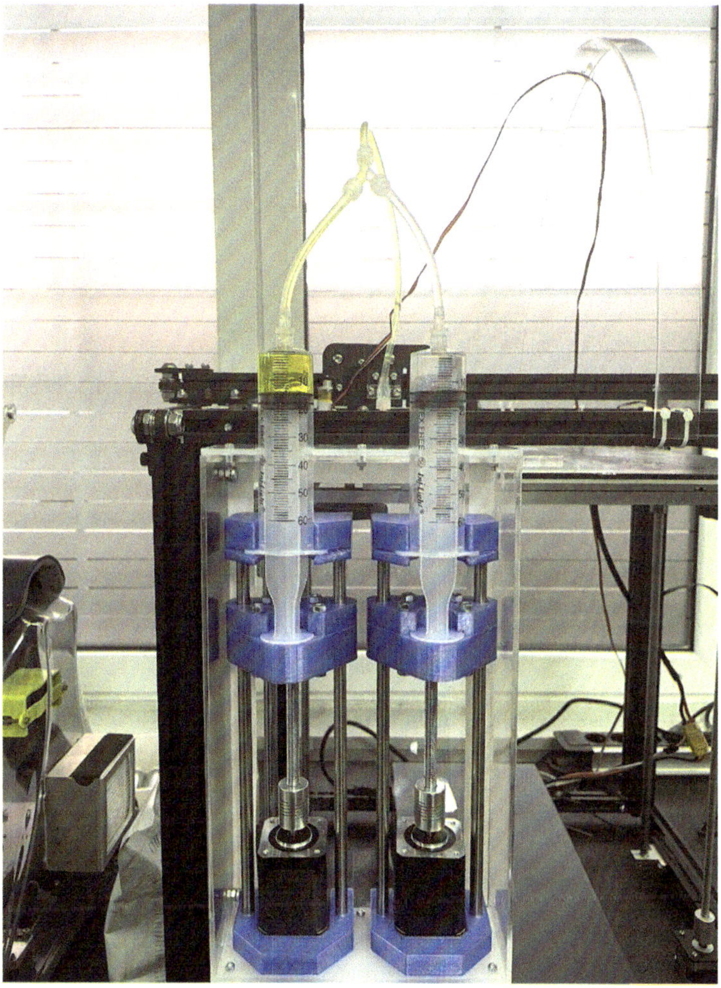

Fig. 7 Piston robotic dispensing system with a dual extruder for bioprinter implementation. Both the hardware and the software of the system were developed and adapted to visco-elastic fluids with solid-like properties. (Copyright: Carmen Rotondi, 2023)

different materials (e.g. base material and support material, biomaterial and crosslinker,[14] etc.); and with a mixing modality, enabling the simultaneous extrusion of two distinct materials, with layer-by-layer control over the percentage of material extruded from each nozzle. This last method, designed for creating functionally

[14] Biomaterials in the form of pastes often require the action of cross-linking agents during the printing process, which solidify the material layer by layer and make the constructs mechanically more resistant and less subject to deformation or loss of shape.

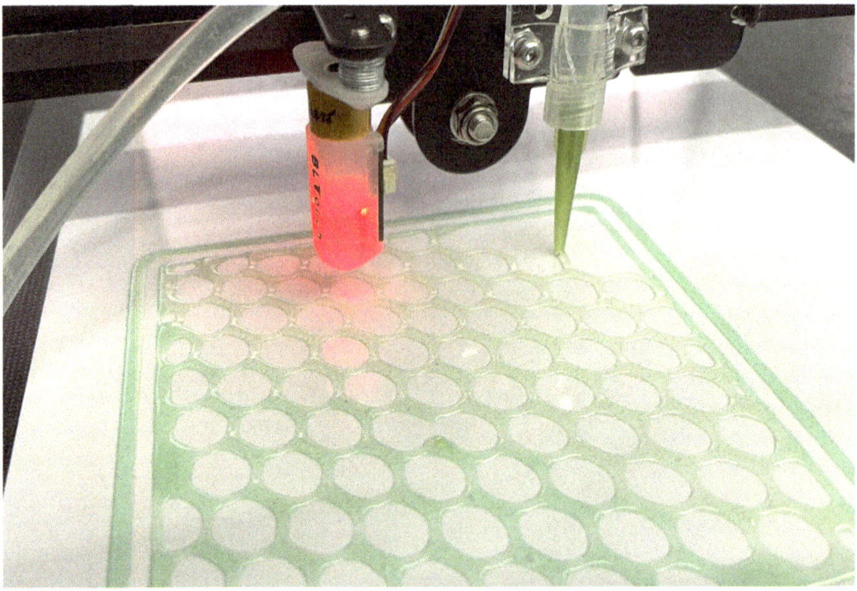

Fig. 8 Printing process of a functionally graded material system made of chitosan-based bioink. (Copyright: Carmen Rotondi, 2023)

graded systems, has been the foundation for experimenting with controlled degradability constructs.

As regards the chitosan-based pastes, they have been modified following different optical, mechanical, rheological and degradability properties. Various material compositions have been explored by adjusting the base ingredients, mainly chitosan and acetic acid, and then printed in 2D constructs with gradient material composition (mixing pastes with different properties) and micro–/meso–/macro-geometries. These geometries, much like the base material, played a role in influencing the degradability patterns of the final constructs.

The outcome of this research produced demonstrative samples illustrating the possibility of creating material systems with controlled and programmed degradability, following defined timelines and patterns, offering functional and expressive potential for design. Several long-term application possibilities emerged, including:

(a) The potential for creating scaffolds that, during degradation, release catalysts or biochemical signals to promote the cellular growth of living species. This application could transform the surfaces of products and architectural elements from inert to bioreceptive.
(b) The creation of expressive transdermal therapeutic systems (TTS) that release nutritional elements during degradation while indicating the quantity of release.
(c) The concept of a 'seasonal fashion' system that is not dictated by the timing of the economy and fast fashion trends, leading to increased waste. Instead, it operates based on the material's degradation timeline. When a material reaches

the end of its life, it disappears, returning new resources to the planet for sustainable living.

As we can imagine, experiential and communicative aspects are very relevant for collective sensibilisation and responsible behaviour improvement, especially in the last two examples. As in nature, we can employ aesthetical features and visual languages to foster involvement and interaction, stimulating sustainability and symbiotic coevolution values.

4.5 *Continuity: Bioreceptive Surfaces*

Nature's system is inherently regenerative, with each element designed to embrace, nurture and sustain new life. This regenerative quality plays a fundamental role in the cyclic processes found in nature, such as biodegradation or biodeterioration facilitated by catalysing microorganisms that induce chemical-physical transformations of materials, extracting the nutrients necessary for developing new life forms.

This regenerative capacity is also essential for maintaining ecosystems and fostering mutual benefits among all the organisms they host through symbiosis, interaction and permeability between different parts of ecosystems, resulting in new emergent functions and conditions. Bioreceptive surfaces exemplify this concept. These surfaces are receptive to colonisation by one or more groups of microorganisms without necessarily undergoing biodeterioration [69]. For example, the bark of trees, due to its structure and material properties, serves as a host for various microorganisms, including cryptogams (plants without visible reproductive organs like ferns, mosses, lichens, liverworts, sphagnum moss, fungi) and even more complex organisms. It also acts as a permeable membrane mediating between internal and external environmental conditions. The concept of 'bioreceptivity' has gained prominence in recent decades in the fields of construction, material sciences and ecology. It has led to numerous research efforts which, overcoming the negative connotation of biological colonisation, transform built environments' surfaces into 'ecological augmentation' devices without needing external technical support [69].

The 'Continuity' project, still in progress, is part of this extensive field of research and aims to explore how to add value to waste materials in a circular perspective, moving beyond their mere utilisation as raw materials or fillers for creating new substances. Specifically, the objective is to transform ceramics – typically an inert material – into bioreceptive surfaces by incorporating organic waste materials (such as coffee and cardboard), which, when subjected to high-temperature firing, create both porosity and microporosity (superficial and structural) that facilitate controlled biological growth. Porosity is critical in bioreceptivity, as it facilitates water and nutrient intake from the external environment while ensuring long-term retention within the pores, which also serve as protective havens for the original living cells. To explore the impact of different porosity levels on bioreceptivity, we created material samples with varying micro-, meso-, and macro-porosities (Fig. 9).

Fig. 9 Sample of bioreceptive ceramic. The micro-porosity is given by coffee wastes (20%), and the meso-porosity is given by cardboard wastes mashed on the surface. (Copyright: Sabrina Lucibello, Carmen Rotondi, Diana Ciufo, Camilla Gironi, Paride Duello, Michela Toussan, Maria Diana Concept Jewellery, 2023)

Micro-porosity is intrinsic to the material and depends on its composition (specifically, the percentage of waste to the ceramic quantity), *meso-porosity* is shaped by the structural geometry of the material system, while *macro-porosity* refers to the overall system geometry, which, like the others, contributes to the material's overall bioreceptivity.

Similar to the approach in 'Evolving Echinoids', the 'hologrammatic principle' and the ability to work across multiple scales to shape numerous system qualities are highly emphasised and enabled by various bio-digital tools. Concretely, at the physical level, we experimented with different concentrations of waste mixed with ceramic (ranging from 5% to 40%) and explored various micro-porosity levels. We then integrated these micro-porosities with meso-porosities achieved through 3D printing in ceramic to create porous structures or by adding extra waste on the surface, serving as 'random generators' for porous textures. Conversely, the overall geometry gives the macro-porosity, which can take on flat, concave, convex, hemispherical or cylindrical forms, influencing the material's capacity to retain water, shield cells from environmental elements, and more.

Regarding the biological aspect, extensive testing is underway to verify the actual bioreceptivity of the developed systems. In nature, bioreceptive surfaces are inhabited by stress-resistant species from various domains of organisms, including bacteria, algae, fungi, lichens and mosses, which collectively form a biofilm. However, numerous conditions must be met for these organisms to survive and thrive, including the proper nutrient levels, humidity, pH and more. Our focus is

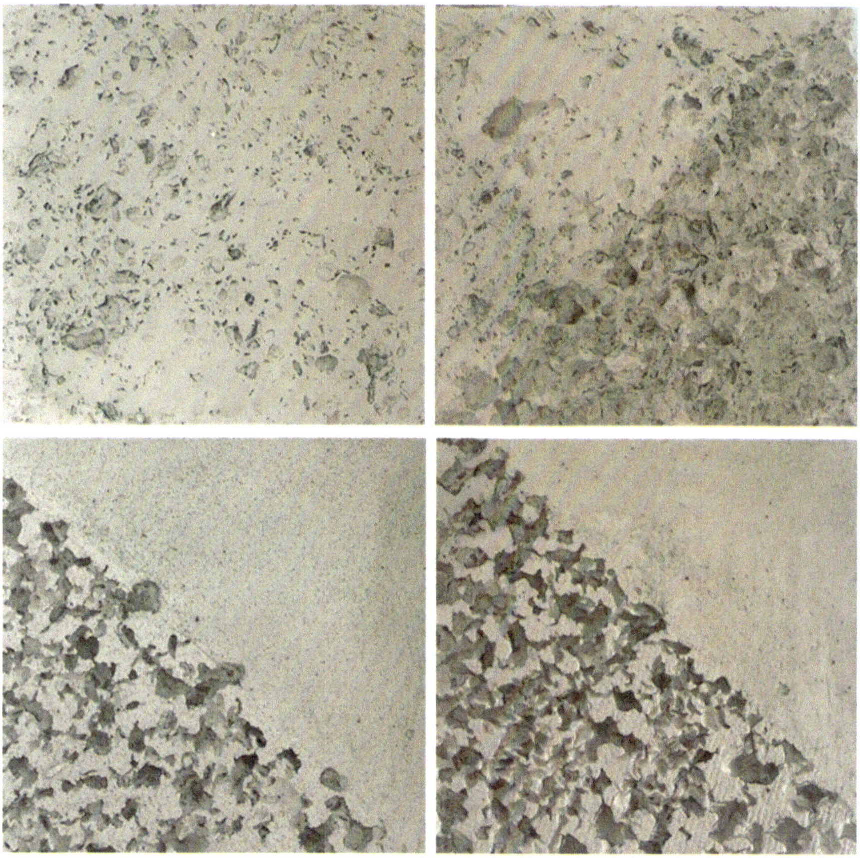

Fig. 10 First-stage growth of cyanobacteria-based biofilm. Micro-porosity and especially the surface meso-porosity have proven to be influential on the growth patterns of living matter. (Copyright: Sabrina Lucibello, Carmen Rotondi, Diana Ciufo, Camilla Gironi, Paride Duello, Michela Toussan, Maria Diana Concept Jewellery, 2023)

mainly on biofilms composed of cyanobacteria, aquatic photosynthetic microorganisms (often mistakenly associated with algae and microalgae) and those commonly found in anthropic environments on stone or cement surfaces. Furthermore, these tests provide an opportunity to illustrate and gain a deeper understanding of the structural or surface characteristics that promote biological growth (Fig. 10). This knowledge will be invaluable for designing future material systems where biological growth is intentionally directed to achieve specific patterns, functionality and colour variations.

Lastly, the digital component plays a crucial role in both the sample creation phase – supporting the modelling and the 3D printing of intricate morphologies and structural designs – and in the testing phase, enabling the collection and comparison of growth data, facilitating the establishment of guidelines for enhancing ceramics' bioreceptivity.

The primary application field for these experiments is undoubtedly in architecture, promoting green initiatives and urban regeneration by developing bioreceptive 'skins'. One of the long-term design concepts is establishing an open-air 'green museum' along the banks of the Tiber River in Rome. This vision entails a series of ceramic panels covering the riverbanks and serving as hosts for living organisms. These bioreceptive panels are meticulously designed to emulate biological works of art, fostering a connection between the city centre and the periphery while harmonising human-made structures with the natural environment for a multi-level continuity.

5 Conclusions

In the 'century of biology', a novel form of 'bio-digital industry' is emerging in which, thanks to increasingly sophisticated and digitised technologies enabling engineering and production at a biological quantum scale, it is possible to analyse and reproduce the generative, chemical, physical and molecular processes underlying natural mechanisms. Inheriting methodologies and technologies from biological fabrication, bio-digital practices foster a new material-based biological paradigm that, bringing biomimicry to a material level, allows designers to observe substances and logic used by nature for assembling and structuring its materials, developing more sustainable and strategic ways for artifice manufacturing, as well as replicating complex, tailored and emergent biological qualities.

The paper focuses on these emerging technological platforms, which are changing the investigation, simulation and testing of natural features, leading designers to interpret them through design and envisioning skills to foster the sustainable and digital transition. In particular, the focus is on informed relations between physical, digital and biological dimensions, allowing interaction, development and mutual empowerment between entities and disciplines belonging to them. Such a correlative strategy can help design to apply systemic thinking, from the scale of the material to that of the product and the process, paving the way to sustainable scenarios, not simply to reduce the human impact on the ecosystem but to enhance nature through original cooperation and integration forms between humans, biology and machines. A new way of beholding nature is rising, passing from its exploitation as a simple resource – a legacy of the previous Industrial Revolutions – to its conception as an organism with which to dialogue and experiment collaborative ways of transformation and improvement [70–72].

Additionally, aesthetics take on a different dimension when discussing material systems that grow rather than being assembled. As Frei Otto states, aesthetics in nature is a measure of ecological intelligence; it manifests through visual languages, which express the state and material transformations of the organism over time [73]. Natural objects narrate and allow us to read the phenomena that have occurred over time and the functional adaptation to the environment. This can be a point of interest for the designer who wants to design with material systems since a powerful aesthetic – which enhances relationships and makes the transformative processes deriving from them transparently through the metalanguage of art – can encourage involvement and interaction stimulating values of sustainability and symbiotic coevolution with other organisms. Overall, bio-digital practices help design to build a socially responsible society in which social and ecological issues are strongly interconnected both in the production paradigms and in the resulting aesthetics, so much so that we no longer need to ask ourselves if there is a border between nature and artifice.

References

1. Capucci, P. L. (2006). L'intelligenza del corpo, la sua evoluzione e la sua eredità. Tecnologia del vivente. In M. Pireddu & A. Tursi (Eds.), *Post-umano. Relazioni tra uomo e tecnologia nella società delle reti*. Guerini e Associati.
2. Benyus, J. M. (1997). *Biomimicry: Innovation inspired by nature*. Morrow.
3. Maldonado, T. (1997). *Critica della ragione informatica* (p. 141). Feltrinelli.
4. Caronia, A. (2006). Corpi e informazioni. Il post-human da Weiner a Gibson. In M. Pireddu & A. Tursi (Eds.), *Post-umano. Relazioni tra uomo e tecnologia nella società delle reti*. Guerini e Associati.
5. Lucibello, S., & La Rocca, F. (2015). *Innovazione e utopia nel design italiano* (p. 10). Rdesignpress.
6. Munari, B. (1966). *Arte come mestiere*. Laterza.
7. Salvia, G., Rognoli, V., & Levi, M. (2009). *Il progetto della natura. Gli strumenti della biomimesi per il design*. Franco Angeli.
8. Langella, C. (2007). *Hybrid design. Progettare tra tecnologia e natura*. Franco Angeli.
9. Khanna, A., & Khanna, P. (2013). *L'eta ibrida. Il potere della tecnologia nella competizione globale*. Codice edizioni.
10. Marchesini, R. (2002). *Post-human. Verso nuovi modelli di esistenza*. Bollati Boringhieri.
11. Stock, G. (1993). *METAMAN—The merging of humans and machines into a global superorganism*. Simon & Schuster.
12. "Bionica", in Enciclopedia Treccani, Roma: Istituto dell'Enciclopedia Italiana.
13. Langella, C. (2008, 8 December). Design biomimetico per l'innovazione sostenibile. *Digimag*. Accessed 1 september 2023.
14. Segantini, E. (2010, 27 aprile). Robotica e nanotecnologie: L'IIT alza il livello della sfida. *Corriere della Sera*. Accessed 1 september 2023.
15. Enriquez, J., & Gullens, S. (2011). *Homo Evolutis*. TED Books.
16. Sottsass, E. (1997). *Erotik design* (p. 509). Stampa Alternativa anno edizione.
17. Estèvez, A., & Navarro, D. (2017). Biomanufacturing the future: Biodigital architecture & genetics. *Procedia Manufacturing*, *12*, 7–16. Available from: https://doi.org/10.1016/j.promfg.2017.08.002.
18. Isaacson, W. (1999, 22 March). The Biotech Century. *Time.*. Accessed 1 September 2023.
19. Kapsali, V. (2016). *Biomimetics for designers. Applying nature's processes and materials in the real world*. Thames&Hudson.
20. Venter, C., & Cohen, D. (2004). The century of biology. *New Perspective Quarterly, 21* (4), 73–77. Available from: https://doi.org/10.1111/j.1540-5842.2004.00701.x.
21. Rotondi, C. (2019). Bio-visioni del futuro. Tra sparizione dei confini e bio-intelligenza [Bio-visions of the future. Between fading borders and bio-intelligence]. *DIID Disegno Industriale | Industrial Design, 62–63*(19), 119–125.

22. Lockton, D., & Ranner, V. (2017). Plans and speculated actions: Design, behavior and complexity in sustainable futures. In J. Chapman (Ed.), *The Routledge handbook of sustainable product design*. Routledge.
23. Rotondi, C. (2022). Bio-augmented materiality. Towards the next biomimicry. In T. Ahram, W. Karwowski, P. Di Bucchianico, R. Taiar, L. Casarotto, & P. Costa (Eds.), *Intelligent human systems integration (IHSI 2022): Integrating people and intelligent systems* (Vol. 22). AHFE Open Access.
24. Vincent, J. F. V. (1982). *Structural biomaterials*. Macmillan.
25. Oxman, N. (2011). Variable property rapid prototyping. Virtual and Physical Prototyping, 6(1), 3–31. Available from: https://doi.org/10.1080/17452759.2011.558588
26. Rotondi, C. (2023). How the informed relations between physical, digital and biological dimensions are changing the design practice, as well as the sustainability paradigm. *Frontiers in Bioengineering and Biotechnology, 11* (2023). Available from: https://doi.org/10.3389/vioe.2023.1193353
27. Mironov, V., Trusk, T., Kasyanov, V., Little, S., Swaja, R., & Markwald, R. (2009). Biofabrication: A 21st century manufacturing paradigm. *Biofabrication, 1*(1). Available from: https://doi.org/10.1088/1758-5082/1/2/022001
28. Camere, S., & Karana, E. (2017). Growing materials for product design. In *Proceedings of International Conference of the Design Research Society Special Interest Group on Experiential Knowledge (EKSIG)* (pp. 101–115).
29. Scott, J. (2016). Programmable knitting. In K. Velikov, S. Ahlquist, & M. Del Campo (Eds.), *Acadia 2016 Posthuman Frontiers: Data, Designers, and Cognitive Machines. Proceedings of the 36th Annual Conference of the Association for Computer Aided Design in Architecture*. Acadia Publishing Company.
30. Scott, J. (2018). Responsive knit: The evolution of a programmable material system. In C. Storni, K. Leahy, M. McMahon, P. Lloyd, & E. Bohemia (Eds.), *Proceedings of DRS2018. Design Research Society Conference: Design as a catalyst for change*. Design Research Society.
31. Mogas-Soldevila, L., & Oxman, N. (2015). Water-based engineering & fabrication: Large-scale additive manufacturing of biomaterials. *MRS Online Proceedings Library (OPL),* 1800. Available from: https://doi.org/10.1557/opl.2015.659
32. Guberan, C. (2017). Hydro-fold. In S. Tibbits (Ed.), *Active matter*. The MIT Press.
33. Tibbits, S. (2017). *Self-assembly lab: Experiments in programming matter*. Routledge.
34. Tibbits, S. (2014). 4D printing: Multi-material shape change. *Architectural Design, 84*, 116–121. Available from: https://doi.org/10.1002/ad.1710.
35. Gladman, S., Matsumoto, E., Mahadevan, L., & Lewis, J. A. (2017). Biomimetic 4D printing. In S. Tibbits (Ed.), *Active matter*. The MIT Press.
36. Johns, R. L. (2014). Augmented materiality: Modelling with material indeterminacy. In F. Gramazio, M. Koheler, & S. Langenberg (Eds.), *Fabricate 2014*. GtA Verlag.
37. Holzbach, M. (2014). MaterialDenken – Materiality and (their) design. In C. Leopold (Ed.), *On form and structure – Geometry in design processes*. Springer.
38. Yao, L., Ou, J., Cheng, C., Steiner, H., Wang, W., Wang, G., & Ishii, H. (2015). BioLogic: Natto cells as nanoactuators for shape changing interfaces. In B. Begole, K. Jinwoo, K. Inkpen, & W. Woontack (Eds.), *CHI'15: 33rd annual CHI conference on human factors in computing systems*. Association for Computing Machinery.
39. Kan, V., Vargo, E., Machover, N., Ishii, H., Pan, S., Chen, W., & Kakehi, Y. (2017). Organic primitives: Synthesis and design of pH-reactive materials using molecular I/O for sensing, actuation, and interaction. In G. Mark, S. Fussell, C. Lampe, M. C. Schraefel, J. P. Hourcade, C. Appert, & D. Wigdor (Eds.), *CHI 2017: CHI conference on human factors in computing systems*. Association for Computing Machinery.
40. Montalti, M., & Co-de-it. (2016). *BIO EX-MACHINA. Biological meets digital computing & robotics*, Officina Corpuscoli website. Accessed 15 September 2023.
41. Malik, S., Hagopian, J., Mohite, S., Lintong, C., Stoffels, L., Giannakopoulos, S., et al. (2019). Robotic extrusion of algae-laden hydrogels for large-scale applications. *Global Challenges, 4*. Available from: https://doi.org/10.1002/gch2.201900064.

42. Adamatzky, A., Ayres, P., Belotti, G., & Wosten, H. (2021). Adaptive fungal architectures. *LINKs-series, 5–6* (2021), 66–77. Available from: https://doi.org/10.48550/arXiv.1912.13262
43. Gershenfield, N. (2006, 30 October). Unleash your creativity in a FabLab [video]. *TED Talk,* YouTube. Accessed 15 September 2023.
44. Ritter, A. (2006). *Smart materials in architecture, interior architecture and design*. Birkhäuser.
45. Razzaque, M. A., Dobson, S., & Delaney, K. (2013). Augmented materials: Spatially embodied sensor networks. *International Journal of Communication Networks and Distributed Systems, 11*, 453–447. Available from: https://doi.org/10.1504/IJCNDS.2013.057721.
46. Lucibello, S., Ferrara, M., Langella, C., Cecchini, C., & Carullo, R. (2018). Bio-smart materials: The binomial of the future. In W. Karwowski & T. Ahram (Eds.), *Intelligent human systems integration 2019*. Springer.
47. Ferrara, M., Rognoli, V., Arquilla, V., & Parisi, S. (2018). Interactive, Connected, Smart materials: ICS materiality. In W. Karwowski & T. Ahram (Eds.), *Intelligent human systems integration. IHSI 2018. Advances in intelligent systems and computing*. Springer.
48. Parisi, S., & Shetty, S. (2020). Alive, provocative, surprising: Emotional dimensions of biosynergistic materials for socially meaningful design. *Diseña,* 17, 128–159. Available from: https://doi.org/10.7764/disena.17.128-159.
49. Parisi, S., Holzbach, M., & Rognoli, V. (2020). Paving the way to post-digital smart materials. Experiments on human perceptions of a bio-inspired cellulose-based responsive interface. In L. Di Lucchio, L. Imbesi, A. Giambettista, & V. Malakuczi (Eds.), *Culture(s). Cumulus conference proceedings Roma 2021. Volume #2*. Cumulus Association.
50. Langella, C., & Santulli, C. (2017). Processi di crescita biologica e Design Parametrico. *MD Journal, 3*, 14–27. Available from: https://materialdesign.it/media/formato2/allegati_6232.pdf
51. Myers, W. (2012). *Bio design: Nature, science, creativity*. Themes&Hudson.
52. Von Bertalanffy, L. (1968). *General system theory: Foundations, development*. George Braziller.
53. De Rosnay, J. (1977). *Il Macroscopio. Verso una visione globale*. Edizioni Dedalo.
54. Schrödinger, E. (2012). *What is life?* Cambridge University Press.
55. Vincent, J. F. V., Bogatyreva, O. A., Bogatyrev, N. R., Bowyer, A., & Pahl, A. K. (2006). Biomimetics: Its practice and theory Journal od the Royal Society Interface, 3, 471–482. Available from: https://doi.org/10.1098/rsif.2006.0127.
56. Correa, D., Papadopoulou, A., Guberan, C., Jhaveri, N., Reichert, S., Menges, A., et al. (2015). 3D-printed wood: Programming hygroscopic material transformations. *3D Printed andAdditive Manufacturing, 2*(3), 106–116. Available from: https://doi.org/10.1089/3dp.2015.0022
57. Rüggeberg, M., & Burgert, I. (2015). Bio-inspired wooden actuators for large scale applications. *PLoS One, 10*(4). Available from: https://doi.org/10.1371/journal.pone.0120718
58. Vazquez, E., Randall, C., & Duarte, J. (2019). Shape- changing architectural skins a review on materials, design and fabrication strategies and performance analysis. *Journal of Facade Design and Engineering, 7*(2), 93–114. Available from: https://doi.org/10.7480/jfde.2019.2.3877.
59. Morin, E. (2007). Le vie della complessità. In G. Bocchi & M. Ceruti (Eds.), *La sfida della complessità*. Mondadori.
60. Grun, T. B., & Nebelsick, J. H. (2018). Structural design of the echinoid's trabecular system. *PLoS One, 13*(9). Available from: https://doi.org/10.1371/journal.pone.0204432.
61. Perricone, V., Grun, T. B., Marmo, F., Langella, C., & Carnevali, M. D. (2020). Constructional design of echinoid endoskeleton: Main structural components and their potential for biomimetic applications. *Bioinspiration & Biomimetics, 16*(1). Available from: https://doi.org/10.1088/1748-3190/abb86b.
62. Smith, A. B. (1980). The structure, function and evolution of tube feet and ambulacral pores in irregular echinoids. *Palaeontology, 23*, 39–83.
63. Wester, T. (1984). *Structural order in space: The plate-lattice dualism*. Plate Laboratory, Royal Academy of Arts, School of Architecture.
64. Ellers, O., & Telford, M. (1992). Causes and consequences of fluctuating coelomic pressure in sea urchins. *The Biological Bulletin, 182*(3), 424–434. Available from: https://doi.org/10.2307/1542262.

65. Dafni, J., & Erez, J. (1982). Differential growth in Tripneustes gratilla (Echinoidea). In J. Tampa Bay & M. Lawrance (Eds.), *International echinoderms conference*. A.A. Balkema.
66. Hye, R. W., Hyo, J. K., Pyung, O. L. & Hong, G. M. (2019). Leaf senescence: Systems and dynamics aspects. *Annual Review of Plant Biology, 70*(1), 347–376. Available from: https://doi.org/10.1146/annurev-arplant-050718-095859.
67. Cheah, C., Chua, C., et al., (2003). Development of a tissue engineering scaffold structure library for rapid prototyping. Part 2: Parametric library and assembly program. *The International Journal of Advanced Manufacturing Technology, 21* (4), 302–312. Available from: https://doi.org/10.1007/s001700300035.
68. Yeong, W., Sudarmadji, N., et al., (2010). Porous polycaprolactone scaffold for cardiac tissue engineering fabricated by selective laser sintering. *Acta Biomaterialia, 6*(6), 2028–2034. Available from: https://doi.org/10.1016/j.actbio.2009.12.033.
69. Guillitte, O. (1995). Bioreceptivity: A new concept for building ecology studies. *Science of the Total Environment, 167*(1–3), 215–220. Available from: https://doi.org/10.1016/0048-9697(95)04582-L.
70. Oxman, N. (2016, 17 January). Towards a material ecology, *World Economic Forum*. Accessed 1 September 2023.
71. Natalio, F. (2018). Future perspectives on biological fabrication and material farming *Small Methods, 3*, 1. Available from: https://doi.org/10.1002/smtd.201800136.
72. Ramsgaard Thomsen, M., & Tamke, M. (2009). Narratives of making: Thinking practice led research in architecture. In J. Verbeke & A. Jakimowicz (Eds.), *Communicating (by) design*. Drukkerij Sintjoris Ghent.
73. Otto, F. (2008). *Occupying and connecting – Thoughts on territories and sphere of influence with particular reference to human settlement*. Axel Menges.

Sabrina Lucibello, Architect and Ph.D. She is Full Professor at the Department of Planning, Design and Technology of Architecture of the Sapienza University of Rome. She is President of the Degree in Design and Director of the Saperi&Co. research centre. She carries out research in industrial design (ICAR/13) on the subject of product innovation through materials for design.

Carmen Rotondi, Product Designer and researcher at Sapienza University of Rome. She conducts research in industrial design (ICAR/13) on biodesign and biofabrication and, more generally, on the hybridisation between design and science, nature and artifice. Her Ph.D. thesis focused on the new concept of "bio-augmented materiality" fostered by bio-digital practices and aimed at innovating through the contemporary eco-entangled context.

Echinodesign: A New Model for Facilitating the Dissemination and Effectiveness of the Biomimetic Design Culture

Valentina Perricone, Carla Langella, Gabriele Pontillo, and Roberta Angari

1 Introduction

Echinoids (Echinodermata: Echinoidea), known as sea urchins, are marine invertebrates populating the seas since the Middle Ordovician, about 450 million years ago. During their evolution, they underwent relevant adaptive radiation leading to a variety of specialised forms and lifestyles [1]. The evolutionary success of echinoids can be also accredited to the strategic employment of their endoskeletons adapted to resist both biotic and abiotic stresses related to marine environments [2]. These structures entail a number of functional details representing ideal models for transferring functional bio-inspired solutions to develop novel design applications [3, 4]. This approach that combines the understanding of natural structures, systems, and processes with their abstraction and translation into technological applications is known as "bioinspiration", "biomimetics", or "biomimicry" [5–7].

V. Perricone (✉)
Department of Research Infrastructures for Marine Biological Resources, Stazione Zoologica Anthon Dhorn of Naples, Naples, Italy
e-mail: vperrico@uci.edu

C. Langella
Department of Architecture, University of Naples Federico II, Naples, Italy
e-mail: carla.langella@unina.it

G. Pontillo
Department of Design, University of Study of Florence, Florence, Calenzano, Italy
e-mail: gabriele.pontillo@unifi.it

R. Angari
Department of Architecture and Industrial Design, University of Campania "Luigi Vanvitelli", Aversa, Italy
e-mail: roberta.angari@unicampania.it

A. J. V. Arruda, F. L. Palombini (eds.), *Biomimetics, Biodesign and Bionics*, Environmental Footprints and Eco-design of Products and Processes, https://doi.org/10.1007/978-3-031-51311-4_4

In this framework, the present chapter describes an interdisciplinary biomimetic hybrid research on echinoid skeletal structures between biology, engineering, and design, called Echinodesign. In particular, starting from basic research on the constitutional characteristics of echinoids, the study led to new biological knowledge that was subsequently transferred to different product design fields. The results archived were subsequently spread outside the academia to a wider community of designers, artists, and scientists from around the world through a dedicated exhibition.

2 Hybrid Investigation of Echinoid Constructional Design

In the course of time, the unique constructional design of the echinoid endoskeleton has attracted the attention of researchers from different scientific fields due to its unique morphology, structure, and material properties. Echinoids are highly adapted to different marine environments: the extant groups of regular and irregular sea urchins are widely distributed from the equator to the poles, from the bright shallow waters to the dark and cold deep sea. The evolutionary success of echinoids is certainly due to the massive strategic employment of their endoskeleton displaying a high number of functional details.

In particular, the endoskeleton consists of three main structural components: test, dental apparatus, and accessory appendages. All parts of the echinoid skeleton consist of the same basic high-magnesium calcite material; however, their microstructure displays great potential in meeting several mechanical needs according to a direct and clear structure-function relationship. This versatility has allowed the echinoid skeleton to adapt to different activities such as structural support, defence, feeding, burrowing, and cleaning. Although constrained by energy and available resources, many of the structures found in the echinoid skeleton are optimised in terms of functional performances and, consequently, can be used as role models for bio-inspired solutions in various industrial sectors.

In this context, Echinodesign research was developed, i.e. a hybrid experimental investigation on the morphological and mechanical characteristics of echinoid skeletal structure aimed to acquire new biological insights, as well as identify adaptive solutions possibly useful for biomimetic transfer applications. Through the combination of advanced biological investigation techniques with digital engineering and design techniques, the research was intended to study the morphological-functional characteristics of the skeletal system of the regular echinoid *Paracentrotus lividus*; identify the motivations and biological principles underlying the different details; and select and analyse individual details that can be reinterpreted (according to the analogies with design problems) and transferred to industrial products in the form of intelligent solutions.

2.1 The Hybrid Process

The hybrid process can be summarised in different key phases: (1) *Contextualisation.* The biological model should be contextualised in its phylogenetic frame, biological function, fabrication process, and environment. Organisms are often defined as "elegant and perfect projects" to live in specific environments; however, there are several constraints that must be taken into consideration. In morphodynamics, four different constraining factors are identified: phylogenetic, functional, morphogenetic, and environmental [8]. Paraphrasing Stephen J. Gould [9], these constraints are not only negative limits associated with structural restrictions that prevent natural selection from modifying traits in a more efficient way; they can also be seen as important factors able to guide organisms to adequate evolutive changes using only available sources. (2) *Deconstruction survey.* The living organism is divided into its main components as a sort of "reverse engineering" process. Particularly, it is schematised into different functional components (shapes, structures, patterns, textures, dynamics, behaviours, etc.) at different dimensional scales (macro-, meso-, micro-, and nano-scale) in order to reveal its constructional design and working principles [10]. During this phase, the organism is converted into 2D/3D models and analysed based on the available literature. Specific organismal components can be acquired using digital microscopy or scanned using different technologies such as photogrammetry, laser scanners, and computed tomography scanning. This survey is aimed to identify different systems and details in an attempt to answer "how and why" questions about their form and function based on the available knowledge. Subsequently, a series of details of interest emerge as well as the need to investigate unanswered questions. Hence, it is a creative phase in which all the information acquired is used to identify single elements of interest and main gaps in knowledge that need to be further explored. (3) *Investigation.* Once identified one or more elements that have high potentiality for biomimetic transfer and are useful to increase the species-specific knowledge of the organism, a project plan can be elaborated. In this experimental phase, organisms are subjected to in-depth analysis integrated with 3D CAD modelling techniques and digital simulation software, chosen according to the type of information desired. These analyses and the subsequent interpretation of data are carried out according to strict scientific methods in order to achieve reliable, verifiable, and shareable knowledge. (4) *Abstraction.* Once shapes and structures corresponding to adaptive solutions and functional principles have been recognised, it is possible to abstract and transfer them into the design of different technical products, according to function similarities and innovation needs expressed by the market. The abstraction process involves the translation of biological strategies into design strategies. The main aim of the abstraction is the description of how the biological strategy works without relying on the selected biological organism. Quick and simple drawings or diagrams showing the understanding of features and mechanisms involved in the biological model can support the abstraction, and both graphics and writing allow visualisation and verbalisation. They can also ensure a more accurate understanding of the mechanisms behind the function and help to fill gaps in knowledge. The abstracted models can be then re-elaborated in function of specific design concepts derived from the designer's intuition in the observation of organismal functional characteristics based on an analogy of function with emerging unsolved needs of contemporary society. In this case, the abstraction process is directed to the project intention of designers and contextualised in the final design application. (5) *Transfer.* The abstracted models are contextualised and transferred to the designing of new man-made artefacts that solve design problems. It is not a mere copy of natural inventions, rather it is an exploratory process intended to capture a "recipe" from natural examples and modelled as a new design concept [5]. During this part of the process, creativity emerges to reconcile what has been learned in the last five steps, remaining coherent to the hybrid process; an open mind is important at this stage. Indeed, it is another divergent creative thinking process, in which abstracted principles can acquire the form of different solutions. Lastly, a series of design-oriented concepts are elaborated where constraints and limits of application need to be considered. Thus, it is necessary to discern among the different concepts, which have the most potential and which lead to dead ends. (6) *Development and realisation.* The final phase of the hybrid process consists of the development and assessment of the design concepts considering how well they suit or solve the design problems, as well as how feasible they are. This involves complex activities, such as creating models and testing technologies. At the end of this phase, simple prototypes or other visualisations (graphics, storyboards, etc.) of the design project are needed focusing on the critical elements and solving possible technical problems. After a series of tests and validations, the bio-inspired product can be eventually patented and proposed for the market.

This hybrid research differs from the typical *bottom-up* and *top-down* biomimetic approaches [11, 12] in which the disciplines involved transfer notions and concepts, although remaining divided and confined to their own disciplinary areas. This approach can limit the creative process and advantages rising from a real integration of knowledge. Even though differing, both approaches converge in a technical application of biological knowledge generating one or several bio-inspired concepts; hybrid research integrates these approaches. As in the case of the bottom-up investigation, also in the hybrid approach the technological development of the product follows a preliminary biological analysis of "model organisms". However, a top-down component is added to the hybrid approach in which the design participates in the initial phase of the biological investigation, directing it towards the functional study of organisms responding to product needs [13]. Furthermore, an engineering component is also involved in the investigation of biological functional details up to the final product realisation.

2.1.1 Biological Model Description and Contextualisation

Echinoids are members of the phylum Echinodermata, which also includes sea stars, brittle stars, sea cucumbers, and crinoids (Fig. 1). Echinoids are traditionally subdivided into two groups, Regularia and Irregularia, mainly identifiable based on

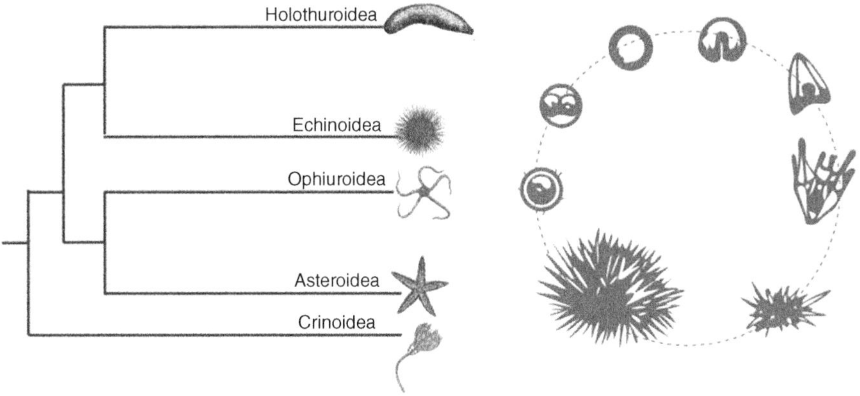

Fig. 1 Echinodermata phylogeny. Simplified Echinodermata phylogeny scheme based on Reich et al. [14]. (Adapted from Hendricks et al. [15])

skeletal morphology and lifestyle. Regular echinoids have a pentameric globular test and are omnivorous and epifaunal (i.e. living on the ocean bottom surface) [16]. In contrast, irregular echinoids are typically aboral-orally flattened and elongated or heart shaped with bilateral symmetry and are generally deposit feeders and infaunal (i.e. eating organic particles in sediments and living buried in the ocean bottom) [16]. Ancient echinoids (Palaeozoic) were characterised by a less rigid test with plates that did not interlock as strongly as modern taxa. Modern taxa include quite old groups such as Cidaroids (e.g. pencil urchins) that first evolved in the Permian, whereas other groups, such as Clypeasteroids (e.g. sand dollars) recently evolved during the Palaeocene) [16]. Regular echinoids commonly live on hard substrates, such as reefs and rocky shores. Many members are herbivorous playing a key role in the ecosystem due to their grazing activities. In exposed shallow waters and/or intertidal rock pools, they can use their mouth and spines to burrow in substrates (e.g. sandstone, limestone, granite, basalt; not hard slate), creating perfect fitting cavities providing shelters from predators. In places where urchins are numerous, the substrate can be completely honeycombed. The skeleton displays significant phenotypic plasticity in which relevant variations in environmental conditions determine graded morphological changes. For example, decreases in available food cause a reduction of the test size and an increase in Aristotle's lantern mass and demi-pyramid length [17], presumably enabling echinoids to feed more efficiently [18].

2.1.2 Deconstruction Survey and Investigation

The echinoid endoskeleton was schematised and divided into different functional components at different dimensional scales to reveal its design and architecture, gaining useful knowledge for biomimetic transfer applications.

The regular echinoid endoskeleton can be divided into five main functional components: test, Aristotle's lantern, spines, pedicellariae, and tube feet [3, 4, 16] (Fig. 2). All these components are described identifying the following aspects: (i) adaptive function, which is the role of components in the internal and external environment of the organism, ensuring survival and reproduction, and (ii) functional details and transferable properties, which are the list of details with functional properties that can be abstracted and transferred into the design of different technical products. It refers to the abstraction process that describes how the biological system works.

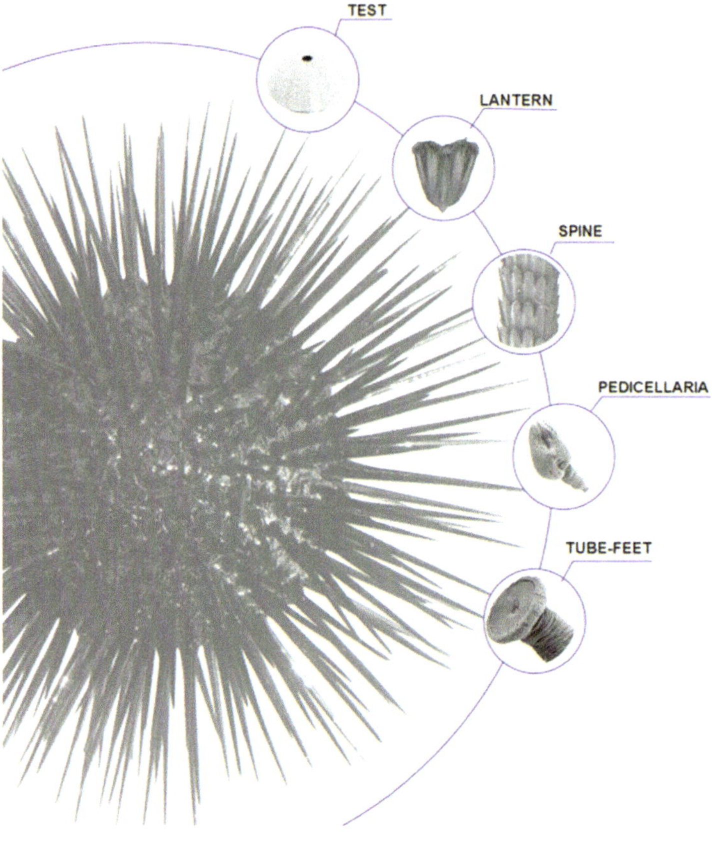

Fig. 2 Echinoid deconstruction. Regular echinoid endoskeleton can be divided into five main functional components: test, Aristotle's lantern, spines, pedicellariae, and tube feet

2.1.2.1 Test

Echinoids are characterised by a calcareous skeleton made up of interlocking plates, called test (Fig. 3). Regular sea urchins have a pentaradially symmetrical test divided into several polygonal plates organised in five interambulacral and five ambulacral columns. The plates of the ambulacral zones are typically pierced by pores for the tube feet emergence. These pore pairs are located and aligned along the outer (adambulacral) margins of the plates. Since all plates bear movable spines, the outer surface of the plate displays suitable rounded tubercles on which the spines are articulated.

Biological Function Visceral protection and mechanical support.

Shape (Fig. 3b) The test is oblate shaped assuming the form of a flattened sphere and resembling an architectural dome [19]. During growth, it acquires this shape presumably resulting in a stable geometry funicular to self-weight and coelomic pressure [20]. The internal pressure balances the surrounding water pressure and was also found to fluctuate rhythmically of about −8 Pa [20] providing the test with compressive compactness. Even though the ideal form of the test should be that of a sphere, downward forces from calcite underwater weight and tube feet pull are most probably the cause of polar flattening [20, 21].

Division into Plates (Fig. 3c) The echinoid test is a modular structure divided into plates, each with a different shape and function, providing organisms with great flexibility during growth, greater structural resistance, and damage limitation [19, 22]. Piercing loads, such as those caused by predator actions, are transferred through all skeletal plates rather than affecting one. For better comprehension, two types of hollow glass balls can be used as an example: one ball consisting of a single piece and another of many fitted together. Piercing the single-piece ball will result in cracking and in its probable destruction. Conversely, piercing the multiple-pieced ball will cause the cracking of only a single or few pieces, but the overall ball will remain intact. Hence, echinoids prevent test cracking and breaking by distributing impact forces throughout their different test plates [23].

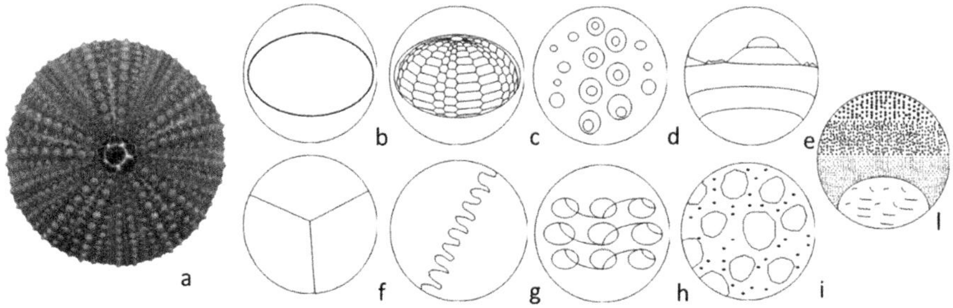

Fig. 3 Echinoid test functional details

Ornamentation (Fig. 3d) The plates are characterised by a series of species-specific tubercles and fine-relief ornaments. Tubercles serve as multiple articulation surfaces for spines (ball and socket joint) and pedicellariae allowing these appendages to exploit the wide range of movements related to their diverse functional roles [24]. The plate surface ornamentation can indicate habitats and growth patterns of extant and fossil echinoids [25].

Plate Curvature (Fig. 3e) The plates are visibly curved following the flattened sphere of the test [19]. Curved hinges avoid rotation mechanisms even with only one line of support per plate, preventing bending action transfer.

Trivalent Vertex Principle (Fig. 3f) In the echinoid test, the plates are arranged following the trivalent vertex principle, in which three plates meet at one point [26]. This is a stable configuration and a structural pattern characteristic of pure plate structures, the echinoid test being an outstanding example. As described by Wester a plate structure consists of plates and lines of support. The forces generated between the plates are only shear forces. Considering a horizontal plate in a three-dimensional space, a plate hinged along one or two lines of support is movable for forces not directed through the intersection points; conversely, a plate hinged along three lines of support is not movable by any direction of forces, resulting in a stabilised plate.

Knob-Like Joints (Fig. 3g) Knob-like spherical protrusions allow interlocking between the complementary plates preventing shearing movements [27, 28]. Under loading, the small plates of the sea urchin test interlock creating a stronger shell structure.

Collagen Fibres (Fig. 3h) Flexible fibres joining the plates dissipate normal, shear, and flexural forces, increasing the stability of the structure [22, 29]. The collagen fibres may have an important role in keeping the whole structure assembled by pre-compressing the joints, as well as in sustaining and transmitting stresses, particularly during the repair period from eventual breakage of one or more plates [27]. Additionally, unlike other shells, tests are growing structures composed of articulated hard elements. Thus, collagen-swathed sutures permit growth and local deformation reducing impact loading and therefore offsetting the eventual hazards of a thin shell [19].

Functional Porosity (Fig. 3i) Porosity is highly variable within the skeleton according to local functional needs. From a mechanical perspective, internal porosity ensures lightness and saves material without compromising the overall skeletal resistance [30, 31], as well as reduces cracking eventuality [23]. Furthermore, open pores provide an attachment site for muscles and ligaments, as well as permeability and connections to the external environment. Examples of the latter function are the madreporite, a modified highly porous plate used to filter and balance the water vascular system, and gonopores.

Structural Material Differentiation (Fig. 3l) The stereom microstructure is highly variable in density and architecture and is filled with cells and organic material (stroma). This architecture is extremely complex, largely varying not only from species to species but also within the individual echinoid test plates. The microstructure is characterised by variable oriented trabeculae of different basic types (e.g. imperforate, galleried, labyrinthic, fascicular, laminar) associated in very different combinations, creating species-specific 3D structural patterns [32]. The differentiation depends on three main factors: associated soft tissues, mechanical needs, and growth rate [32]. It optimises the use of material, whereas the structural specialisation increases the mechanical resistance, ensuring lightness and a high level of robustness.

2.1.2.2 Aristotle's Lantern

Aristotle's lantern is characterised by a *pentaradial symmetry* and represents an integrated mechanical system, formed by several articulated sub-components (Fig. 4). It is composed of ten semi-pyramids welded in correspondence of the interambulacral zones, forming five pyramids. The pyramids are connected by muscle fibres and calcareous bones (rotulae); above them other thin, bilobed, and slightly curved bones (compasses) are found. Each semi-pyramid is composed of an epiphysis articulated laterally to the patella. In total, the articulated sub-components of the lantern system can be summarised in the following 40 elements: 5 teeth, 10 semi-pyramids, 10 epiphysis, 5 rotulae, and 5 double compasses [33].

Biological Function Food grabbing, substrate scraping and digging, and support during walking.

Muscular System (Fig. 4b) The antagonistic protractor and retractor muscles allow the lantern to protrude or retract inside or outside the test, while the interpyramidal muscles determine its opening and closing. However, the retractors also play an important role in the processing of food and in the opening mechanism of the lantern. The lantern can move widely or tilt along its vertical axis, while the peristomial membrane limits rotation movement. The relationship between strength and muscle length affects lantern mobility. Increased mobility allows greater adaptability to substrate and is particularly useful for algal grazing [34].

Subdivision into Articulated Sub-components (Fig. 4c) The lantern is subdivided into several complementary components linked by bivalent joints that increase both the strength of the structure and its capacity for deformation and adaptation. This can be identified as a strategic adaptation of the structure in response to excessive mechanical stresses [33].

Structural Material Differentiation (Fig. 4d) Microstructure is highly variable in density and architecture and is filled with stroma. This architecture is extremely

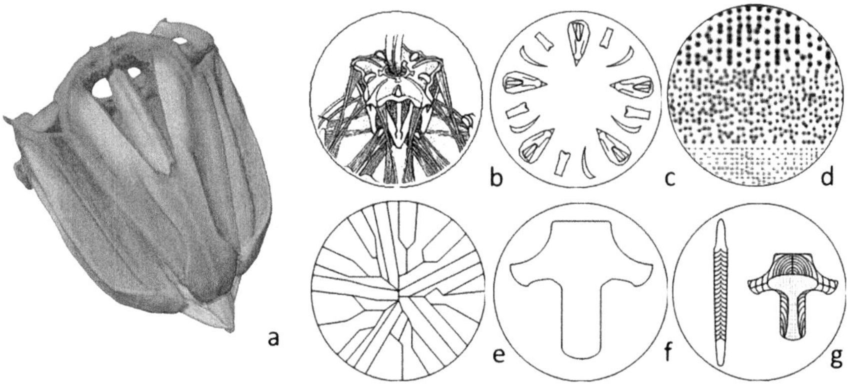

Fig. 4 Aristotle's lantern functional details

complex and variable within the different ossicles of the lantern [33]. Differentiation optimises the use of materials, whereas structural specialisation increases mechanical resistance [33]. Crystallographic alignment and composition variability along the structure also improve the strength and toughening of these components [35].

Teeth Closure (Fig. 4e) Aristotle's lantern acts as a five-toothed mechanical grab, and its particular form of closure has the ability of grabbing and cutting objects using various degrees of precision.

T-Shaped Teeth (Fig. 4f) The teeth are curved and T shaped similar to the keel of a ship. Together with the materials used, this shape results as a functional strategy to withstand normal and shear forces [35].

Teeth Material Differentiation (Fig. 4g) The teeth are continuous growing structures made of composite calcitic materials (similar to artificial ones) and consist of plates, fibres, and polycrystalline matrix. The lamellar components (primary plates) are distributed in the maximum compressive stress areas, while the fibrous elements (prisms) are subjected to traction [36]. Teeth plates are continuously replaced through abrasion determining a self-sharpening mechanism keeping the tooth tip sharp and efficient [37]. The tooth shows functional crystal orientation and Mg distribution to increase the mechanical performance [38].

2.1.2.3 Spines

The spines of the echinoids (Fig. 5) vastly differ from species to species. Some are long, hollow, thin, and pointed similar to a needle; others are thick, cylindrical or flattened, long or short, and striated or variously ornate. Spines absolve different functions, among them are defence, burrowing, wave energy reduction, and

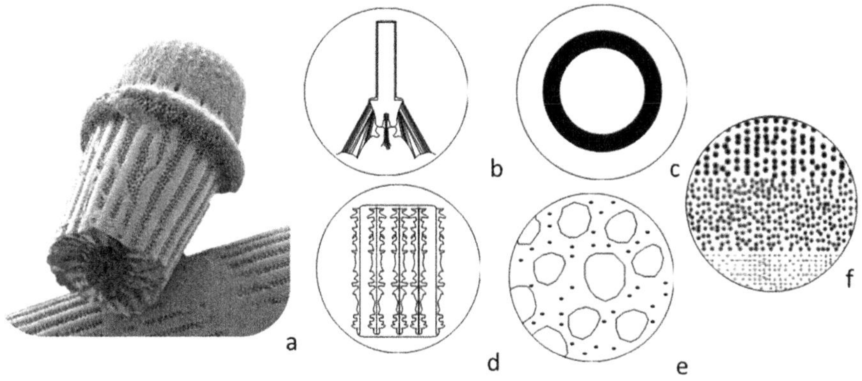

Fig. 5 Spine functional details

locomotion. Spine morphology can be smooth or ornamented by thorns, granules, and ribs. Spines are attached to the tubercles of the echinoid test by a ball-and-socket joint through an inner ring of collagenous fibres, called catch apparatus, and an outer ring of smooth muscle fibres. Muscle fibres are attached to the stereom of the areole through connective tissue ligaments. Lastly, a fluid-filled space separates the muscle fibres from the catch apparatus [39]. The catch apparatus allows spines to lock in a specific position, useful in predatory attacks.

Ball-and-Socket Joint (Fig. 5b) Special types of sliding ball joints, present in the spine-bar attachment system, allow movement between two articulated surfaces. The spherical shape of the joint consents a wide degree of freedom in movement, regulated by the muscle cone and collagen fibres [40].

Shape (Fig. 5c) Spines consist of a hollow or highly perforated central core surrounded by a high-density circular area ending in a series of radial wedges. The cylindrical shape (very common in nature, e.g. bones, spines, and stems) provides a good mechanical efficiency keeping a minimal weight: bending forces concentrate on the margins, and spine material can be spared at the central area [40].

Ornamentation (Fig. 5d) The outer surface has a distinct pattern of smoothed ribs (as in *P. lividus*) or pointed elements facing the tip of the spine. Bridges between the adjacent wedges follow a helical pattern around the longitudinal axis of the spine. The wedges and bridges that rotate around the spine axis concentrate forces in the external area of the spine structure, leaving the central cylinder intact. Spaces between wedges prevent fracture propagation throughout the structure increasing spine strength [41]. For example, compared to a monolithic cylindrical calcite structure, cracks propagate separately in each wedge rather than across the entire cross-section.

Functional Porosity (Fig. 5e) The porosity is also highly variable throughout the entire spine leading to a functionally graded organisation that results in a graceful failure behaviours [42, 43].

Structural Material Differentiation (Fig. 5f) Microstructure is also highly variable in density and architecture throughout the spine and is filled with stroma [44]. The structural patterns are able to efficiently absorb impact energy [43–45]. Crystallinity and material composition are also functionally variable [46].

2.1.2.4 Pedicellariae

Pedicellariae (Fig. 6) are small muscular appendages with increased rigidness due to calcareous spicules and fixed on small tubercles. In echinoids, there are four different types of pedicellariae: globiferous, ophicephalous, triphyllous and tridentate. All include a head, composed of two five valves and a stalk partially supported by a skeletal ossicle. Each valve has a distal blade (rim) often denticulated and a proximal articulation surface with insertion regions for the adductor muscles. The valves are made of calcitic stereom having a strong but lightweight construction covered by a pigmented epithelium often containing spicules (for a review see [47]).

Biological Function Gripping, defence, and cleaning.

Shape (Fig. 6b) Pedicellariae can be categorised into globiferous, ophicephalous, triphyllous, and tridentate [48], each one specialising in different functions varying from species to species in number, combination, and distribution on the epithelium. Often all four can be found on the same individual, with clear spatial distribution differences, suggesting that their form and functions are closely related. Each morphology specialises in efficiently keeping the organism's surface clear from algae, encrusting organisms, and debris and to deter pests and predators [47].

Closure (Fig. 6c) Pedicellariae are reactive and gripping systems providing the echinoid with an epithelium sensitive and reactive to biotic and abiotic environmental stimuli. They appear as gripping elements with variable selectivity in scale (from small debris and algae to large debris particles, parasites, and predators). The pedicellariae jaws open and close after direct stimulation through an inbuilt reflex arc or by stimulation of the test surface through the under-skin nerves. The rapid jaw closure has been recorded at speeds of about 100 ms [49].

Muscular System (Fig. 6d) It consists of antagonistic abductor and adductor muscles allowing the pedicellariae to open and close. The muscle size can be variable based on the pedicellariae type. The larger the muscle system, the greater the gripping strength of the jaw, allowing objects to be held for longer periods before fatigue. The gripping power of the pedicellariae is also supported by collagen fibres [47].

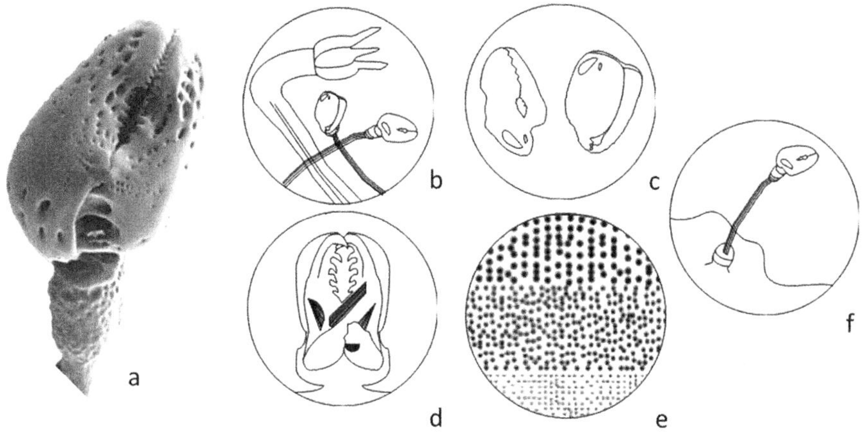

Fig. 6 Pedicellariae functional details

Material Differentiation (Fig. 6e) As for the other skeletal components, the microstructure is filled with stroma.

Articulation (Fig. 6f) Pedicellariae are attached by a long and movable stalk supported by a skeletal rod [47].

2.1.2.5 Tube Feet

Tube feet are extensions that branch out from the echinoid aquifer system (Fig. 7). They are shaped like small tubes with suction cups used for walking on seabed as well as for food handling and feeding. They also have respiratory and sensory functions. The specific tube feet attachment/detachment mechanism is related to the disc secretion of adhesive and de-adhesive substances, while the tube feet stem in extension, flexion, and retraction is regulated by hydraulic mechanisms.

Biological Function Attachment, movement, perception, covering, and food transport.

Functional Disc (Fig. 7b) The specific tube foot attachment/detachment mechanisms are related to disc adhesive and de-adhesive secretions composed of inorganic material, proteins, neutral carbohydrates, and lipids. The adhesive strength (force per unit area) ranges from 0.09 to 0.54 MPa [50].

Reinforcement (Fig. 7c) Small calcareous ossicles (rosette), composed of four elements, are used to reinforce and support the tube foot disc. The peristomal tube feet rosette differs from the coronal tube feet one. In particular, the coronal tube feet have a radially symmetrical rosette, whereas the peristomal tube feet one is oval and almost twice as broad and long [51].

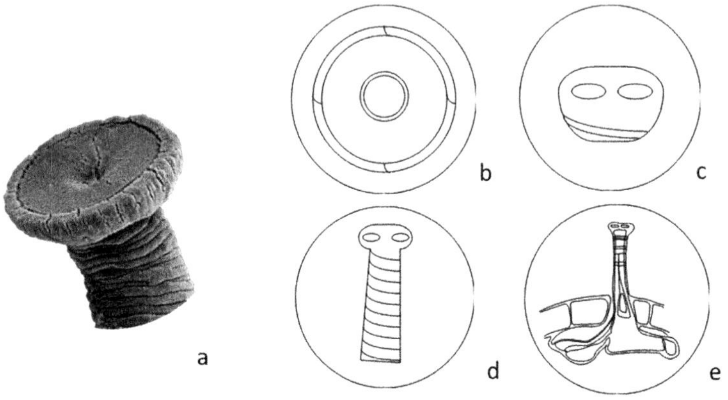

Fig. 7 Tube feet functional details

Flexible Tube (Fig. 7d) Each tube foot consists of an extensible hollow cylindrical tube that can extend, flex, and retract. It functions through a hydraulic pressure, bending and stiffening when the connected ampoules fill or empty with water, causing their movement on the surface. It is useful for covering and conveying food to the mouth. In addition, by adhering to the surface through suction cups and bending, tube feet allow the slow creeping movement of echinoderms [51].

Hydraulic System (Fig. 7e) The tube foot consists of two main parts, i.e. the ampulla and the podium. The ampulla is a seawater-filled sac surrounded by circular and longitudinal muscle. The podium is a tubular structure with longitudinal muscles. When the surrounding ampulla muscles contract, the water moves from the ampulla into the podium determining its elongation; conversely, when the podium muscles contract, the water returns to the ampulla [52].

3 Biomimetic Design Transfer

The deconstructive survey of the echinoid endoskeleton supported by literature and the specific analyses carried out on the echinoid test led to the classification and identification of shapes and structures corresponding to different functional principles. Once structure-function relationships were identified, a series of biological solutions could be abstracted and transferred into the design of different technical products, according to function similarities and innovation needs expressed by the market and society. In particular, the abstraction involved the translation of biological strategies into design schemes from which a series of design-oriented concepts can be elaborated, considering technical constraints and limits of application [53].

Following these lines, different design concepts and products were developed in the Hybrid Design Lab, i.e. a didactic and research lab founded in 2006 in Naples dedicated to experimentation of the intersections between design and biosciences. The hybrid methodology applied in HDL can be defined as biunivocal, since designers, through their representative and interpretative skills, assist scientists in gaining new knowledge; at the same time, scientists help designers to draw inspiration from their knowledge, in order to design innovative and sustainable products.

Following the hybrid process, the data obtained from the instrumental analyses, e.g. SEM and micro-CT scans, were re-elaborated to create two- and three-dimensional digital models. These were carried out by using specific software for graphic representation, in order to render the complexity of morphologies, structures, and functional details. Designers and engineers supported the biologists and worked together to interpret the data, using their modelling skills and their knowledge about the objects from and function and helping them to deduce the relationship between the echinoid structures, the generative processes, and their adaptive functional purposes.

To encourage interdisciplinary collaboration, methodological tools such as collaborative science sketching were used, in which biologists and designers draw together nature modelling and working principles to better understand organismal details and functionality. In this regard, drawing becomes a useful communication tool for both designers, who became more comfortable and open at understanding complex concepts, and biologists, who are encouraged to simplify explanatory descriptions through graphic communication while also making them concrete.

Digital biological models were also 3D printed to create physical replicas in enlarged scale, which are particularly useful to biologists, engineers, and designers to identify hidden structural details.

The participation of designers in the biological research phase enabled them to directly foresee design transfer hypotheses useful for the abstraction phase, creating "application-oriented nature models" related to the previewed application fields and the related innovation needs. For example, the processes by which the Aristotle's lantern and pedicellariae capture and hold food and particles were modelled abstracting shapes and movements that were subsequently applied into design concepts for food grabbing, holding, and locating (see *Achinos* and *Feel and Sunder* projects).

These abstraction processes and conceptualisations led to a deeper knowledge of the relationship between morphologies and adaptive functions as well as to a biological re-interpretation into innovative concepts and design prototypes for different applicative fields such as furniture, design for food, light design, biomedical design, and jewellery. Many developed products emerged as answers to unresolved or emerging needs linked to current and future lifestyles and technological scenarios.

3.1 The HDL Echinoid-Inspired Design Products

The echinoid test architecture and functional-related strategies were abstracted and applied to the design of new lightweight and resistant sportive protective gear components destined for, for example, bikes, rollerblades, skates, skateboards, as well as electrically powered scooters, specifically, *Helmest*, a sportive protective helmet (Fig. 8), and *Biomeduni*, for knee protection (Fig. 9). These echinoid-inspired protectors are characterised by a modular structure divided into perforated and curved hexagonal modules designed to ensure breathability, lightness, and material reduction. The modules follow the trivalent vertex arrangement and are particularly shaped with geometrically angled edges filled with flexible silicon materials intended to dissipate stress, guaranteeing high structural performance and function of the protectors easily realised using 3D printing. Helmets and protectors are created in 3D printing using a variation of materials between hexagonal modules and edges; the first is characterised by 20% flexibility and 80% stiffness while the second by 80% flexibility and 20% stiffness. The use of 3D printing accelerates the product development process and is usually cheaper than other fabrication techniques. Another major advantage is the possibility of creating complex customisable structures based on user size and preferences. 3D printing has been in use since 1986 as a prototyping technique; however, it is presently possible to use additive manufacturing of finished and high-performance products through an increase in their structural strength. Moreover, presently, 3D printing material quality is continuously improving, allowing the production of parts similar to those of traditional manufacture.

The echinoid shell structure with multiple spines inspired the *Polar Star*, a device aimed to help heavy PC user with guided micropauses (Fig. 10). In particular, the device is able to guide the users with specific exercises to relieve eye fatigue, through a laser light, and hands fatigue, through the soft spines that can massage the hands and reactivate the surface microcirculation relaxing musculoskeletal tensions.

The *Urcare* is a system of accessories inspired by the echinoid test morphologies for therapeutic massage with the aim of relieving temporomandibular muscle pain (Fig. 11). Each accessory can support a range of motion and postural strengthening exercises that can reduce pressure on the jaw and relax it while also controlling pain and promoting healing.

The Aristotle's lantern and its unique shape, symmetry, modularity, and dynamism inspired a modular seat for children (ages 4–9), called *Ekihmon* (Fig. 12). It consists of a base on which reversible seat modules can be placed, detached, rotated, or even moved elsewhere. It can also be tipped over to be transformed into a table. The reversibility of this furniture allows children to configure their own play and relational space, developing self-organising and creative skills.

Another product inspired by Aristotle's lantern is *Achinos*, a bio-inspired gripping device for agriculture (Fig. 13). In particular, the following echinoid lantern characteristics were imitated: (1) pentametric shape ensuring structural stability and resistance; (2) subdivision into articulated sub-components linked by bivalent joints

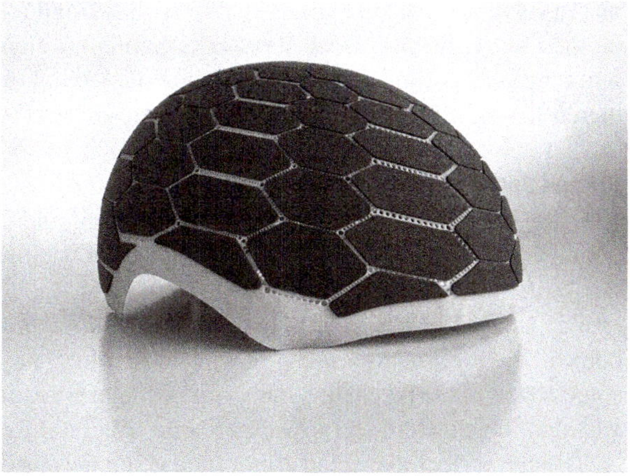

Fig. 8 Helmest. (Design: Giorgio Carotenuto, Vincenzo Giorgione, Natialia Janquin, Alessia Masiello, and Mario Matarazzo; research group: Carla Langella, Gabriele Pontillo, and Valentina Perricone)

Fig. 9 Biomeduni. (Design: Terry Celentano; research group: Carla Langella, Gabriele Pontillo, and Valentina Perricone)

Fig. 10 Polar star. (Design: Giuseppe Sorvillo; research group: Carla Langella, Gabriele Pontillo, and Valentina Perricone)

increasing both structure strength, deformation, and adaptation capacity; (3) a five-toothed mechanical grab closure and a particular closure form with the ability of grabbing and cutting objects in various degrees of precision; (4) T-shaped teeth increasing structural strength, i.e. a functional strategy withstanding normal and shear forces; and (5) a material differentiation of the teeth together with a self-sharpening mechanism, which is able to keep the blade tip sharp and efficient. This complex bio-inspired system enhanced gripping and cutting efficiency and operational safety, reducing time, producing greater yields, and increasing operational flexibility.

The echinoid spine led to the design of *e-bookshelf*, an analogical and digital bookshelf. In particular, the following echinoid characteristics were imitated: (1) spine shape and ribs ensure shape and structural resistance, and (2) central cavity and hierarchical stereom porosity provide structural lightness and functional space.

The cylindrical shape can accommodate numerous books without a large encumbrance. The user can interact with the bookcase from any angle, and the series of LEDs, positioned inside the central cylinder, facilitate night-time viewing. The projector panel can be extracted from the cylindrical structure to project e-books or multimedia products (e.g. movies) adding to the usual bookcase storage a new function as a multimedia transmitter. In this way, the E-bookshelf connects two generations, analogue and digital, allowing reading to become a moment of sharing: reading a book together with a partner or guiding children in learning (Fig. 14).

The echinoid spines also inspired the designing of *Pure*, a new bio-inspired nasal filter able to prevent inhalation of allergens, pollutants, and irritants (Fig. 15). In particular, the following echinoid characteristics were imitated: (1) spine shape and ribs ensure the ideal fitting to different nose shapes and structural resistance, and (2) hierarchical stereom porosity provides mechanical breathing filtration of pollution, allergens, and particles. The nasal filter is characterised by two cylindrical-shaped structures with longitudinal ribs. In each cylinder, three perforated layers are positioned at the centre characterised by three different pore sizes, which provide a variable degree of mechanical filtration regarding the different particles. The cylinders are linked together by a bridge facilitating the insertion and removal of the device. Additionally, variable charms are 3D printed to the bridge resembling a "septum piercing", which makes the product attractive and wearable to young people. Pura is fabricated by additive manufacturing using antibacterial materials used by the medical sector.

Inspired by and created with the echinoid spine, a coffee table *Ekino* was also developed (Fig. 16). In particular, the echinoid spine fragments from food waste were embedded into a biodegradable resin. The applied upcycling process provides educational value and sustainability awareness to the product. The incorporation of biological fragments into the resin transforms the table into a speculative object, a kind of natural fossil that showcases and enhances the exquisite details of spines and test fragments while also conveying the significance of repurposing waste by elevating its prestige.

The pedicellariae structures were taken into consideration for the development of a new kind of nut forks designed not to break the kernel altering the taste. The fork is part of the *Feel and Sunder* kit, a system of accessories designed to enhance multi-sensory perception in the nut-tasting experience and to favour the consumption, which is beneficial for brain health (Fig. 17). Similarly, a salad cutlery was also proposed (Fig. 18).

Finally, the echinoid functional details inspired the *Aerial Garden*, a plant plot designed to make the food experience healthier and more enjoyable (Fig. 19). It consists of a plant pot inspired by the attachment mechanism of the echinoid tube feet to the surfaces. The pot can be attached to windows by means of a suction cup to contain small aromatic plants. This gives the possibility to turn the kitchen window into a vertical aromatic garden.

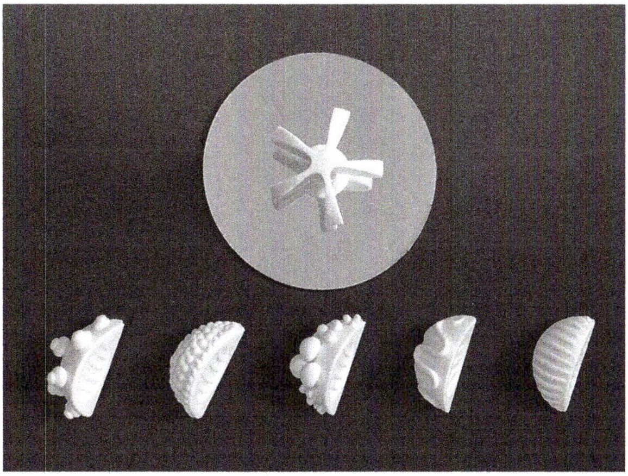

Fig. 11 Urcare. (Design: Simone Martucci, Alessia Schettino, Lucio Junior Prisco, and Maria Saputo; research group: Francesca Lenci, Carla Langella, and Gabriele Pontillo(

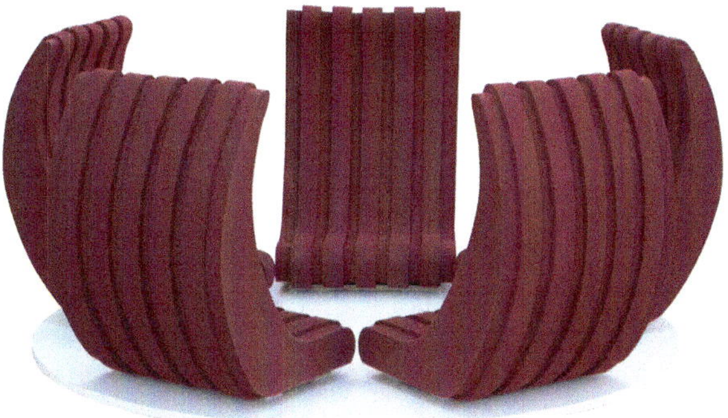

Fig. 12 Ekihmon. (Design: Melania Spatresano, Raffaela Cozzolino, Daria Cermola, Enya De Rosa, and Margot Tura; research group: Carla Langella and Valentina Perricone)

Fig. 13 Achinos. (Design: Claudio Catalano; research group: Carla Langella, Valentina Perricone, and Gabriele Pontillo)

Fig. 14 E-Bookshelf. (Design: Camilla Amato; research group: Carla Langella and Valentina Perricone)

Fig. 15 Pura. (Design: Valeria Campaniello; research group: Carla Langella, Gabriele Pontillo, and Valentina Perricone)

Fig. 16 Ekino. (Design: Giovanna Bava, Martina Buonomo, Emmanuela Murolo, and Gabriella Palomba; research group: RICCIcliamo team, Carla Langella, and Valentina Perricone)

Fig. 17 Feel and Sunder. (Design: Cinzia Gervasio, Claudia Improda, Simone Martucci, Vincenza Pellegrino, and Alessia Schettino; research group: Carla Langella, Gabriele Pontillo, Valentina Perricone, and Roberta Angari)

Fig. 18 Pedicellaria. (Design: Teresa Iavarone, Gabriele Pontillo, Carla Langella, Roberta Angari, and Valentina Perricone)

Fig. 19 Aerial Garden. (Design: Marisa Orlo, Ileana Rosy Pepe, Ilaria Troiano, Alfonso Trombetta, and Angela Zaccariello; partnership: D.re.a.m. FabLab, and Città della Scienza; research group: Carla Langella, Valentina Perricone, and Gabriele Pontillo)

4 Echinodesign: Dissemination and Interpretation of Echinoid Details Between Design, Art, and Science

Based on the previously described hybrid research and the resulting design products, a final exhibition was organised to disseminate the scientific results outside the academia, involving worldwide artists, designers, architects, and scientists. The exhibition, called Echinodesign, was organised at Città della Scienza from 24 February to 24 April 2022 (Fig. 20).

Starting from 15 April 2021, a call was opened inviting worldwide designers, artists, and scientists to submit their echinoid-inspired projects.

Candidates had the opportunity to consult a "Learning from Echinoids" guide, i.e. a document with a graphical representation and explanation of the echinoid deconstruction survey. The scientific knowledge acquired during the research was here filtered through a meta-design language and visual representation appropriate to the creative contexts to which they were addressed.

Forty Italian and international designers, artists, and scientists took part in the exhibition providing their creative interpretation of the echinoid world in multiple sectors, such as furniture, medical, jewellery, electronic devices, interaction design, new materials, graphics, visual arts, speculative design, lighting design, architecture, and building constructions. The participants were supported in the development of their projects by the exhibition curators, who were committed to verifying their coherence with the interpreted scientific knowledge and to encouraging interdisciplinary encounters in their individual concepts. Their proposals were also evaluated and validated by the scientific committee composed of designers and scientists of international relevance.

Some of the proposed products were realised with the partnership of companies interested in the innovative potential of echinoid research to increase the competitiveness of their products. Diverse products have been integrated in the catalogues

Fig. 20 Echinodesign exhibition, Città della Scienza, Naples, Italy

of the companies and craftsmen who have contributed to their development. The projects have been catalogued in different sections, according to their areas of application, namely, upcycling experiments, decorative products and games, scientific and technological products, biomedical, critical design, design for food, jewellery, lamps, and innovative interior furnishings. The exhibition was conceived as an experience rich in aesthetic and sensory stimuli to allow visitors to appreciate, in an experiential way, the multiple qualities of echinoids, understanding their environmental value and their biomimetic potential for the development of new products and technological innovations.

Accordingly, the Echinodesign exhibition was organised into different following sections, namely:

1. *RICCIcliamo*, a section dedicated to a research project developed by the University of Milan, University of Padua, and University of Genoa oriented to the reuse of the echinoid skeleton as food waste in an innovative way from a circular economy perspective.[1] The project was carried out with a multidisciplinary approach, including the collection of food waste from sea urchins; the extraction of specific molecules and substances, e.g. collagen and antioxidants (Fig. 21); their reprocessing to create products for use in regenerative medicine and breeding; the testing of feed products; and the economic and environmental sustainability analysis of the supply chain. Some of these products by echinoid upcycling are exhibited including extracted collagen, biomedical patches, and animal feed. In this section, the helmet *Galea* (L. Borrello, V. De Simone, S. Devi

[1] https://www.riccicliamo.it/en/the-second-life-of-sea-urchins/

Dasi Migliaccio, N. Mungivera, S. Garcia Jurado) was developed by using new bio-based materials obtained by food waste of echinoid skeletons and latex.

2. *Biophilic section*, dedicated to artistic, communicative, and interior design products, such as inspired ceramics and educational games directed to valorise the echinoid skeletal characteristics and learn their functional aspects. The *Aerial Garden* pot was included in this section together with *Equinoxadrez* (T. Laila, M. Vicente), a chess set inspired by the shapes of echinoid skeletal components with biophilic and scientific dissemination purposes; *Gioechino* (F. Rosetti, F. Tartaglia), a modular game tool designed for children's creative educational activities, inspired by the trivalent vertex arrangement and the joint system of the echinoid plates (Fig. 22); and *Echi* (F. Serafini, V. Picili), a tableware set inspired by the inter-specific variability of shapes, colours, and patterns of the echinoid test, with the purpose to disseminate the RICCIcliamo project.
3. *Scientific section*, dedicated to projects with high scientific and technological content. *Helmest* and *Achinos* were herein included, as well as *Tassellated velaroid vault* (F. Marmo), a scale model of an echinoid-inspired funicular vault whose shape enables equilibrium between its weight and the compressive contact forces between blocks; *Moveup* (F. Dell'Aglio, C. Scarpitti, E. Migliore), a digital application inspired by the echinoid movements through water; *The magic moment* (L. Santella, C. Caccavalle), scanning electron microscope images showing the interaction of the fertilising sperm with the surface of a sea urchin egg and its fusion with the plasma membrane and incorporation of the sea star sperm into the egg (Fig. 23); and *Echino-Patterns* (M. Deb), self-organising textures based on the echinoid test microstructure and realised with artificial neural networks.
4. *Design-for-health section*, displaying all the innovative inspired solutions for the biomedical sector. All the products were designed by ameliorating the functionality but also by making them more comfortable and aesthetically appreciable to ensure greater acceptability, compliance, and consistency in healthcare treatment. They include *Biomeduni*, *Pura*, *Polar Star*, and the *Urcare* kit together with *Echino-forceps* (A.R.M. de Oliveira, P.C.S. da Silva), dental extraction forceps inspired by the prehensile shape function variability of pedicellariae (Fig. 24).
5. *Interior design*, a sector dedicated to innovative and unusual interior furnishings such as the hybrid bookcase *E-bookshelf*; the reversible table-chairs systems *Ekihmon* (Fig. 25); *Spillo: The Microcosm of Fantasy* (C. Francolini Bianchi), a modular pavilion designed to support creative, laboratorial, and educational activities for children inspired by the echinoid test morphology; and *Echino Layers* (Sixim Lab: A. De Falco, M. Parisi), an artistic installation made using experimental textile techniques representing shape and colour variability of the echinoid skeletons.
6. *Evocative design section*, addressing emerging issues coming from contemporary society such as interpersonal communication and information systems. These include *a spine in the light* lamp (C. Langella, L. Verdolotti, G. Lama) inspired by the echinoid spines and made with ceramic foam based on diatomite

and silica, where the fragility of the material filtering the light alludes to the weakness of the current energy system based on fossil fuels; *Seventy fossil* (C. Langella, L. Verdolotti, G. Lama), a lamp made with a recycled diffuser from the 1970s updated with ceramic foam containing aluminium and silica referring to the fossilisation processes as an extension of the object life (Fig. 26); and *Messages from the abyss* (Dotfog: R. Cianniello, M. Firpo, G. Grizzuti), an evocative story based on echinoids, in which their forms and geometries generate signs, "meaningful echinography" interpreting the sense of "getting in touch", which even unintentionally means to communicate.

7. *Light design section*, proposing new forms of relationship between user and light, with ample space for customisation, innovation of materials and technologies, and environment consciousness. These include *Evolving Echinoids* (E. Canepone, C. Rotondi), 3D programmable fabrics inspired by the functional details of the echinoid test at macro-, meso-, and microscale with applications in lighting design; *Curly Lamp* (V. Golia, A. Fabozzi, M. Mazzocca, G. Vivaccia), a customisable lamp made of echinoid food waste and inspired by their generative geometries, in which the engraving on the base allows to shape the diffuser in different ways; and *AQI-NOS* (M. Massera), a lamp device inspired by the echinoid sensory system integrating sensors able to inform through light codes about the concentration of pollutants and air parameters including carbon dioxide and VOC (Fig. 27).
8. *Jewellery section*, inspired by the echinoid shapes, including *Maut* (L. Al-Jazzar, C. Scarpitti) jewels inspired by and made with the skeleton of echinoid test in collaboration with the IED Milano Jewelry Department and Partnership Swarovski.
9. *Design for food*, a section proposing new concepts of food cutlery facilitating healthy eating styles inspired by the echinoid pedicellariae. *Feel and Sunder* and *Pedicellaria* were here included.

5 The Potential of Digital in Echinoid Design: Approaches and Techniques

In recent years, the rapid evolution and increasing diffusion of digital design approaches and techniques have redefined not only design and production processes but also the way people interact with artefacts, triggering what some define as the third industrial revolution [54, 55].

In particular, the field of experimentation outlined by the digital and its potential uses has led to an exponential innovation in design and production processes in various sectors, making possible the creation of complex structures, both in aesthetic and functional terms, through short and sustainable supply chains, and that still guarantee high precision of the products made.

Fig. 21 RICCIcliamo. Antioxidant pigments. Echinodesign, Città della Scienza, Naples, Italy

Fig. 22 Gioechino. (Design: F. Rosetti and F. Tartaglia. Echinodesign, Città della Scienza, Naples, Italy)

Fig. 23 The magic moment. (Design: L. Santella and C. Caccavalle. Echinodesign, Città della Scienza, Naples, Italy)

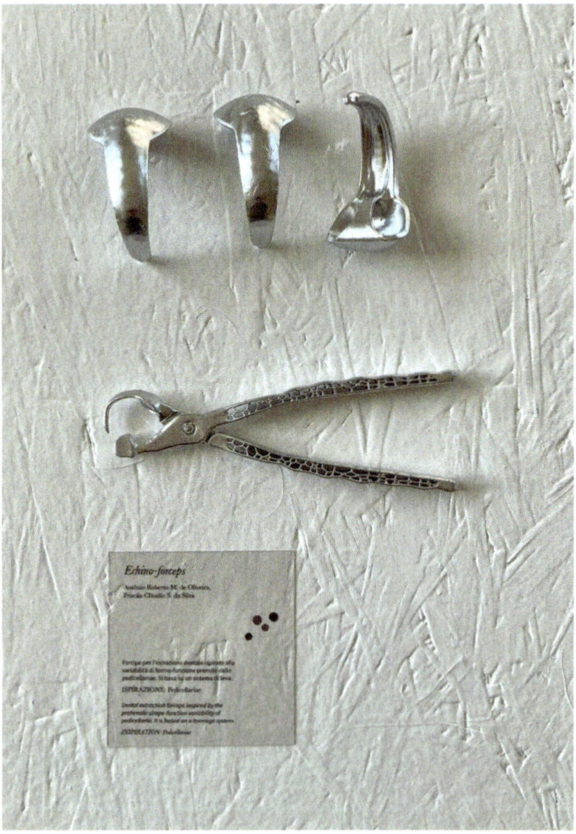

Fig. 24 Echino-forceps. (Design: A.R.M. de Oliveira and P.C.S. da Silva. Echinodesign, Città della Scienza, Naples, Italy)

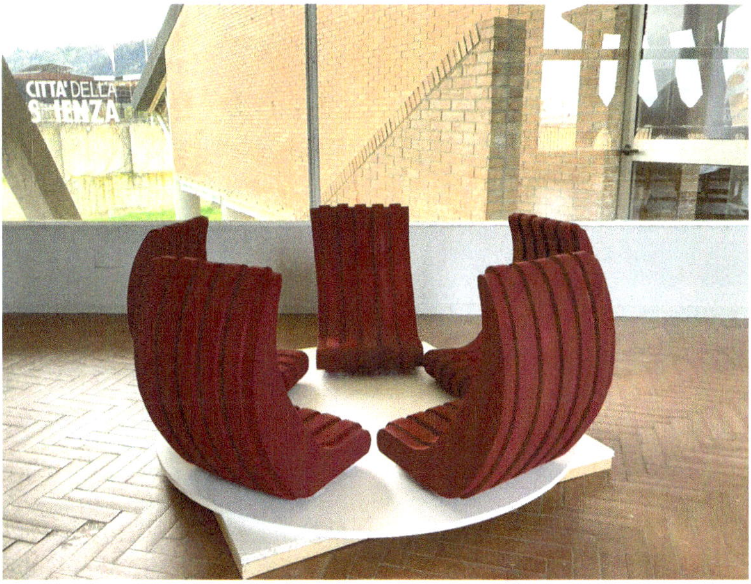

Fig. 25 Ekihmon. Echinodesign, Città della Scienza, Naples, Italy

Fig. 26 Seventy fossil. (Design: C. Langella, L. Verdolotti, and G. Lama. Echinodesign, Città della Scienza, Naples, Italy)

Fig. 27 AQI-NOS. (Design: M. Massera. Echinodesign, Città della Scienza, Naples, Italy)

For this reason, as part of the Echinodesign exhibition, digital modelling and additive manufacturing through 3D printing were used to create objects capable of satisfying the need to create complex structures such as in echinoids, which were considered the reference body for the inspiration of the projects on display, both from an aesthetic and functional point of view, giving shape to medical devices, wearables, but also highly functional objects such as work tools or augmented thanks to sensors and/or production techniques and innovative qualities.

Therefore, the approach used began with the study of the echinoid elements and functions to draw inspiration for the project. The relevance of this phase, in addition to the analysis of the organism, consists in the study of complex structures, in which the hierarchical order provides unique characteristics and mechanical properties to the designed new objects. Hence, to mimic these organisms, structural elements were analysed at various scales appropriate to the transfer of functional aspects to the designed model [56]. From this analysis, concepts of objects have emerged in which the identified function, through hierarchical structures and differentiated in performance, allows to identify innovative solutions or improvement of pre-existing objects in terms of efficiency as well as use. To do this, during the modelling phase, aspects related to the design methods have been taken into account, in particular, the observation of the needs of users as well as ergonomics, to reach not only the confirmation of geometries inspired by the reference organisms but that could also meet the needs detected in specific areas of reference, such as gestures and grips, postures, type of interaction, wearability, etc. In the modelling, therefore, expedients have been used both for what concerns the optimisation of the structures and the parts that mostly had to perform mechanical functions, such as joints or joints, and for the aspects more closely related to the surfaces of the objects, returning those characters that aesthetically recall in morphology the echinoids but that guarantee a better performance of the designed object.

From this point of view, some of the objects that best represent these concepts are the Pedicellaria vegetable cutlery and the *Urcare* massager that have been defined starting from the analysis carried out on the user and the context of use to then go to determine the functions and possible applications of the designed object, to which the complexity of echinoids has been transferred both in functional and aesthetic terms.

Subsequently to the phase of three-dimensional modelling, as in the case of the examples cited previously, it is then passed to the phase of production of the plans through the aid of the technologies of additive manufacturing.

In this regard, it is emphasised that there is growing interest in the use of these technologies, given their potential use, diversity – both in terms of technology and material – and performance. The processes of additive manufacturing, or 3D printing, exist today in fact in different technologies and materials, and, although initially they were considered only as techniques suitable for prototyping, they are increasingly evolving investing those sectors also related to the production of art prototypes and small series [57] according to the type of technology and/or material used. Nevertheless, it is emphasised that it is necessary to carry out a careful analysis on the object and the performance that you want to characterise it, mainly since the need to transfer primarily aesthetic values rather than mechanical may determine the use of certain technology and/or certain material. FDM printers, for example, which are known to be used more widely in the field of prototyping, in many cases guarantee the production of objects characterised by mechanical strength but may have a lower surface yield than other types of technologies, although it should be noted that even in this case should be carried out in-depth studies on the conformation of geometries and characteristics of printers for the additive production of models.

Once the technologies and materials considered most appropriate for a specific application have been identified, the prototypes have been printed and in some cases post-produced to ensure better aesthetics of the objects made.

In conclusion, a relevant aspect of what has been described in this section is the freedom of design offered by digital modelling and manufacturing through additive technologies, which allow the designing of complex products and components that meet functional requirements, minimise material volumes, and also ensure to reach innovative solutions that improve the design and manufacturing processes [58].

6 The Visual Identity System in Echinodesign

According to design aspects, technological developments have also increased the potential of visual communication, which in the context of identity design results in a variety of new tools and functions useful for translating the symbolic elements of identity into visual form [59].

In particular, visual identity design can consist of several elements that together contribute to the construction of the visual identity system (VIS), which several authors identify with branding, typography, and colours [60, 61]. Thus, the VIS can make a given project uniform, consistent, and recognisable. Nevertheless, the current phenomena related to the convergence between graphic design practice and the new technologies available have given rise to the design of more organic identities, promoting the ability to combine printed artefacts with those enjoyed through digital devices [60], changing the way identity projects are realised, and becoming increasingly variable and dynamic. For this reason, the elements that make up the VIS are also undergoing mutations, as the use of values and parameters useful to modify the individual visualisations of identity is increasing. However, such processes cannot forget what is the primary function of identity, which is to ensure the identification, distinction, and recognition of a project [62, 63].

This living system, which is identity, would therefore be capable according to Kapferer of giving meaning and purpose to a product, service, or entity, which is not only represented but also educates the user and influences their emotions and overall experience [64–66].

The analysis of the scholarly literature and the state of the art of reference highlights how both in academic and professional fields there is a growing interest of designers in dynamic and evolving identities investigated simultaneously in different places and through different perspectives [59], probably also because the mutation is itself a means of visualising identity elements that connote a single project.

In particular, for the Echinodesign project, the objective of the identity project was to identify in the first phase the biological reference useful for the realisation of the brand. In this regard, the idea was to find an image that could be translated into a graphic form, without losing value and intrinsic meanings. In this perspective, it is necessary to emphasise how to visually represent organisms and/or scientific phenomena, it is necessary to start from the reference image, and then through the techniques of computer graphics [67] and the level of abstraction [68], it is possible to derive a graphic sign that meets the requirements from the point of view of form and content. The image taken as a reference was that of the echinoid larva, an organism that forms a collectivity through the aggregation of individuals.

From the images shown (Figs. 28 and 29), it is possible to observe how, starting from the biological reference, the punctiform graphic elements have been deduced, which have been used in the realisation of the brand of the exhibition, a useful form which aims to graphically translate both the formal aspects of the organism from which the brand is inspired and those related to the construction of an identity image that conveys the idea of togetherness and collectively (Fig. 30).

Subsequently, a typographic and chromatic survey was carried out, useful to identify those elements that make up the VIS and that have given uniformity and recognisability to the identity project.

In this regard, an interesting aspect is related to the fact that each component of the VIS had a fundamental role in the realisation of a project that integrates different disciplines and different areas, such as design, science, biology, and art, and therefore balances and creates a balance between the aspects examined, directing the results of the analysis and investigations carried out initially by the research group.

As far as the chromatic aspects are concerned, a chromatic chart has been outlined that integrates tones typical of the seabed, as well as of some elements that constitute the echinoids and that has then been declined, through superimpositions and filters added to the biological images of the details of the sea urchins (Fig. 31).

From Fig. 32, it is also possible to observe how, for the visualisation through digital devices, and in line with what has been reported in the introductory part of this section, the brand has been dynamically declined, based on the variable conformation of the point elements that constitute it. GIFs were created for visualisation through social networks, in which the dynamism of the brand, together with typographic elements or biological images, was intended to gradually reveal the contents of the single artefact.

Another fundamental aspect related to the graphic design of the visual identity was addressed through the design of the call, as well as the editorial artefacts related to it. In particular, the call was structured to make visible the information related to

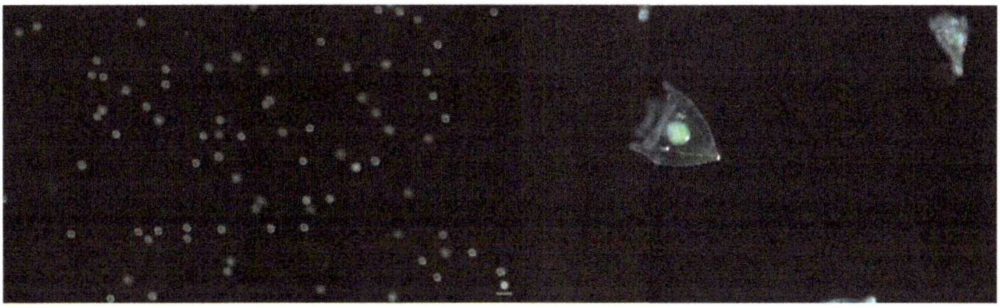

Fig. 28 Echinoid Larvae and aggregation

Fig. 29 Graphical representation of the single Echinoid larvae and aggregation

Fig. 30 Echinodesign logo. The project of the logo has been made within the course of Design for Science Visualization, held by Carla Langella at the Master's Degree in Design for Innovation, Department of Architecture and Industrial Design, University of Campania "Luigi Vanvitelli". The group components are Antonella Rosmino, Sara Coscione, and Ciro Esposito

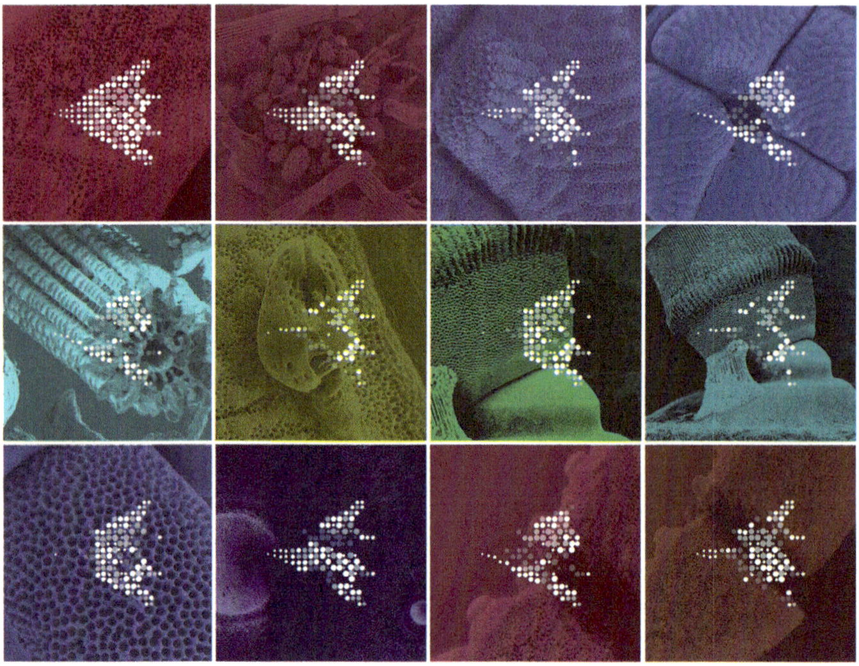

Fig. 31 Image of colour biological image chart

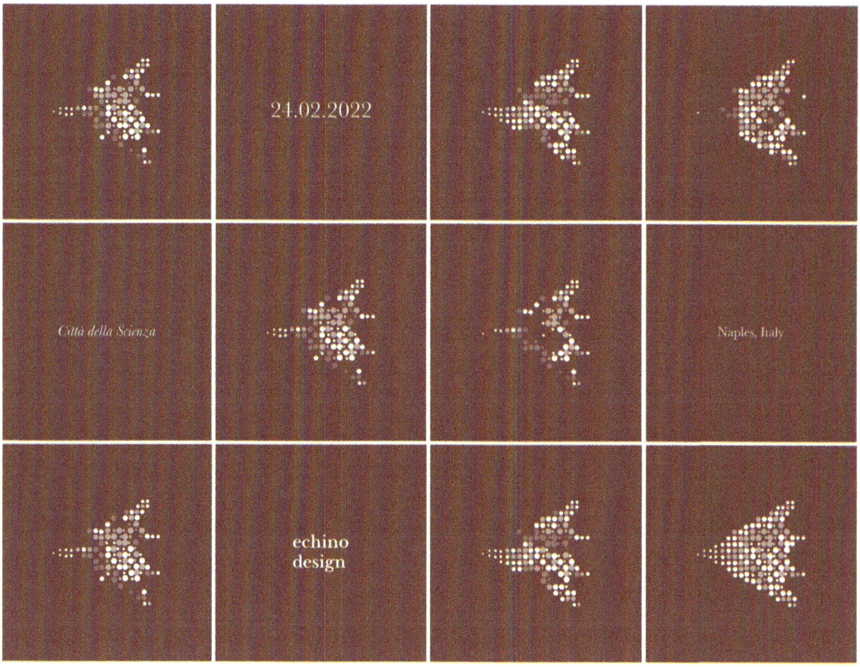

Fig. 32 Image of the logo with the information about the exhibition opening

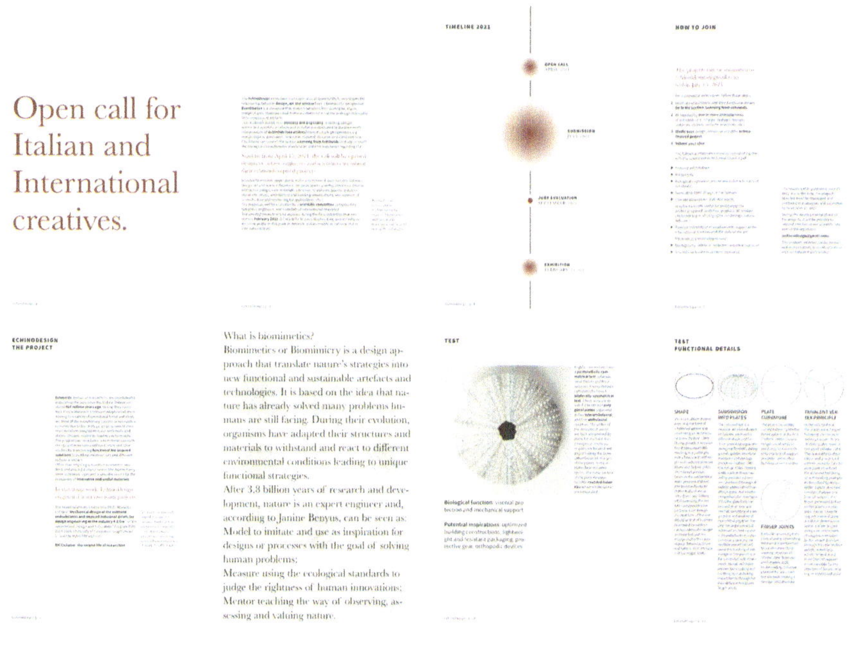

Fig. 33 Some of the pages of the call

the research group involved in the project and the scientific committee (Fig. 33). At the same time, data on the timing and modalities of participation have been provided. The call is completed by the presence of contents and cards from which designers could draw inspiration for the transfer of aesthetic and functional aspects of echinoids to their concept and project. Some SEM images were also included, to highlight the complexity of the structures analysed.

In conclusion, the reported examples that compose the VIS are useful to evidence as every artefact is a unique project, but through shrewdness and declinations, they emphasise the identity characters identified and synthesised in the project of the logo.

The convergence between these artefacts, as well as the declinations, is not simply a sum or a combination of elements, but each one acts triggering different types of access to knowledge for different types of users. In this regard, the familiarity [68] among the artefacts helps users in perceiving different levels of the message, gradually discovering what are the biological, conceptual, and design contents, which accompanied the identity project, and allowing to integrate different disciplines and fields in the exhibition project into a coherent image in the colours, typography, and signs used.

7 Conclusions

The described experience shows a new approach to carrying out biomimetic research and design experimentation projects, in which dissemination can overcome the limited academic channels and scientific community. The exhibition led to the translation of the proper scientific results into knowledge and products accessible to society and its need, creating synergic collaboration with local industries. In this way, public funds allocated to research can support technology transfer actions to professionals and companies promoting new opportunities for cultural and economic growth for the design sector.

The described approach aims to surmount the observed trend in the international design landscape, where biomimicry is predominantly confined to the realms of research and education. The major part of the developed biomimetic products and processes remains in fact in a conceptual stage or laboratory level. Consequently, the biomimetic products that make their way into production and find a place in the market are rare. However, the market has never been as ready and receptive to products inspired by nature, given the increasing ecological crisis and biophilic attention. Hence, academic design research must strive to engage manufacturers and industrial sectors in order to translate ideas and concepts into products that can become part of people's daily lives and actions, ultimately enhancing their quality of life.

Acknowledgements Although the approach and contents of the essay are shared by the authors, it should be noted that V. Perricone wrote and edited the introduction, paragraphs 2, 3, and 4 and conclusion; the introduction, paragraphs 2.1, 3, and 4 and conclusion are edited by C. Langella; G. Pontillo edited paragraph 5; and R. Angari edited paragraph 6 and provided graphics and pictures.

References

1. Lefebvre, B., et al. (2013). Palaeobiogeography of Ordovician echinoderms. *Geological Society, London, Memoirs, 38*, 173–198.
2. Boivin, S., Saucède, T., Laffont, R., Steimetz, E., & Neige, P. (2018). Diversification rates indicate an early role of adaptive radiations at the origin of modern echinoid fauna. *PLoS One, 13*.
3. Nebelsick, J. H., Dynowski, J. F., Grossmann, J. N., & Tötzke, C. (2015). Echinoderms: Hierarchically organized light weight skeletons. In C. Hamm (Ed.), *Evolution of lightweight structures: Analyses and technical applications* (pp. 141–156). Springer.
4. Perricone, V., Grun, T. B., Marmo, F., Langella, C., & Carnevali, M. D. C. (2020). Constructional design of echinoid endoskeleton: Main structural components and their potential for biomimetic applications. *Bioinspiration & Biomimetics, 16*(1), 011001.
5. Benyus, J. M. (1997). *Biomimicry: Innovation inspired by nature*. Quill-William Morrow.
6. Speck, O., Speck, D., Horn, R., Gantner, J., & Sedlbauer, K. P. (2017). Biomimetic bio-inspired biomorph sustainable? An attempt to classify and clarify biology-derived technical developments. *Bioinspiration & Biomimetics, 12*.
7. Fayemi, P. E., Wanieck, K., Zollfrank, C., Maranzana, N., & Aoussat, A. (2017). Biomimetics: Process, tools and practice. *Bioinspiration & Biomimetics, 12*(1), 011002.
8. Seilacher, A., & Gishlick, A. D. (2014). *Morphodynamics*. CRC Press.
9. Gould, S. J. (2010). *The panda's thumb: More reflections in natural history*. W.W. Norton.
10. Perricone, V., Santulli, C., Rendina, F., & Langella, C. (2021). Organismal design and biomimetics: A problem of scale. *Biomimetics, 6*(4), 56.
11. Helms, M. E., Vattam, S. S., Goel, A. K., Yen, J., & Weissburg, M. (2008). *Problem-driven and solution-based design: Twin processes of biologically inspired design*.
12. Badarnah, L., & Kadri, U. (2015). A methodology for the generation of biomimetic design concepts. *Architectural Science Review, 58*(2), 120–133.
13. Langella, C. (2007). *Hybrid design. Progettare tra tecnologia e natura*. FrancoAngeli.
14. Reich, M., Smith, A. B. (2009). Origins and biomechanical evolution of teeth in echinoids and their relatives. *Palaeontology, 52*, 1149–1168.
15. Hendricks, J. R., Stigall, A. L., & Lieberman, B. S. (2015). The Digital Atlas of Ancient Life: Delivering Information on Paleontology and Biogeography. *Palaeontologia Electronica*, 1.
16. Allasinaz, A. (1995). *Paleontologia generale e sistematica degli invertebrati*. ECIG.
17. Ebert, T. A. (1996). Adaptive aspects of phenotypic plasticity in echinoderms. *Oceanologica Acta, 19*(3–4), 347–355.
18. Fernandez, C., & Boudouresque, C. F. (1997). Phenotypic plasticity of *Paracentrotus lividus* (Echinodermata: Echinoidea) in a lagoonal environment. *Marine Ecology Progress Series, 152*, 145–154.
19. Telford, M. (1985). Domes, arches and urchins: The skeletal architecture of echinoids (Echinodermata). *Zoomorphology, 105*, 114–124.
20. Ellers, O., & Telford, M. (1992). Causes and consequences of fluctuating coelomic pressure in sea urchins. *The Biological Bulletin, 182*(3), 424–434.
21. Philippi, U., & Nachtigall, W. (1996). Functional morphology of regular echinoid tests (Echinodermata, Echinoida): A finite element study. *Zoomorphology, 116*, 35–50.
22. Ellers, O., Johnson, A. S., & Moberg, P. E. (1998). Structural strengthening of urchin skeletons by collagenous sutural ligaments. *The Biological Bulletin, 195*(2), 136–144.
23. Vogel, S. (2013). *Comparative biomechanics: Life's physical world*. Princeton University Press.

24. Smith, A. B. (1980). The structure and arrangement of echinoid tubercles. *Philosophical Transactions of the Royal Society B: Biological Sciences, 289*, 1–54.
25. Oldfield, S. C. (1976). Surface ornamentation of the echinoid test and its ecologic significance. *Paleobiology, 2*, 122–130.
26. Wester, T. (1984). *Structural order in space: The plate-lattice dualism*. Plate Laboratory, Royal Academy of Arts, School of Architecture.
27. Wester, T. (2002). Nature teaching structures. *International Journal of Space Structures, 17*, 135–147.
28. Mancosu, A., & Nebelsick, J. H. (2020). Tracking biases in the regular echinoid fossil record: The case of Paracentrotus lividus in recent and fossil shallow-water, high-energy environments. *Palaeontologia Electronica, 23*(2).
29. Marmo, F., Perricone, V., Cutolo, A., Daniela Candia Carnevali, M., Langella, C., & Rosati, L. (2022). Flexible sutures reduce bending moments in shells: From the echinoid test to tessellated shell structures. *Royal Society Open Science, 9*(5), 211972.
30. Grun, T. B., & Nebelsick, J. H. (2018). Structural design of the minute clypeasteroid echinoid Echinocyamus pusillus. *Royal Society Open Science, 5*(5), 171323.
31. Perricone, V., Grun, T. B., Rendina, F., Marmo, F., Candia Carnevali, M. D., Kowalewski, M., et al. (2022). Hexagonal Voronoi pattern detected in the microstructural design of the echinoid skeleton. *Journal of the Royal Society Interface, 19*(193), 20220226.
32. Smith, A. B. (1980). *Stereom microstructure of the echinoid test.*
33. Candia Carnevali, M. D., Bonasoro, F., & Melone, G. (1991). Microstructure and mechanical design in the lantern ossicles of the regular sea-urchin *Paracentrotus lividus*. A scanning electron microscope study. *The Italian Journal of Zoology, 58*, 1–42.
34. Andrietti, F., Candia Carnevali, D. M., Wilkie, I. C., Lanzavecchia, G., Melone, G., & Celentano, F. C. (1990). Mechanical analysis of the sea-urchin lantern: The overall system in *Paracentrotus lividus*. *Journal of Zoology, 220*, 345–366.
35. Stock, S. R., Ignatiev, K. I., Veis, A., De Carlo, F., & Almer, J. D. (2004, October). Micro-CT of sea urchin ossicles supplemented with microbeam diffraction. In *Developments in X-ray tomography IV* (Vol. 5535, pp. 11–20). SPIE.
36. Markel, K., Gorny, P., & Abraham, K. (1977). Microarchitecture of sea urchin teeth. *Fortschritte der Zoologie, 24*, 103–114.
37. Espinosa, H. D., Zaheri, A., Nguyen, H., Restrepo, D., Daly, M., Frank, M., McKittrick, J. (2019). In situ Wear Study Reveals Role of Microstructure on Self-Sharpening Mechanism in Sea Urchin Teeth. *Matter, 1*, 1246–1261.
38. Ma, Y., Aichmayer, B., Paris, O., Fratzl, P., Meibom, A., Metzler, R. A., et al. (2009). The grinding tip of the sea urchin tooth exhibits exquisite control over calcite crystal orientation and Mg distribution. *Proceedings of the National Academy of Sciences, 106*(15), 6048–6053.
39. Smith, D. S., Wainwright, S. A., Baker, J., & Cayer, M. L. (1981). Structural features associated with movement and 'catch'of sea-urchin spines. *Tissue and Cell, 13*(2), 299–320.
40. Wainwright, S. A., Biggs, W. D., Currey, J. D., & Gosline, J. M. (1976). *Mechanical design in organism*. Edward Arnold Publishers.
41. Tsafnat, N., Fitz Gerald, J. D., Le, H. N., & Stachurski, Z. H. (2012). Micromechanics of sea urchin spines. *PLoS One, 7*.
42. Presser, V., Schultheiß, S., Berthold, C., & Nickel, K. G. (2009). Sea urchin spines as a model-system for permeable, light-weight ceramics with graceful failure behavior. Part I. Mechanical behavior of sea urchin spines under compression. *Journal of Bionic Engineering, 6*(3), 203–213.
43. Grossmann, J. N., & Nebelsick, J. H. (2013). Stereom differentiation in spines of *Plococidaris verticillata*, *Heterocentrotus mammillatus* and other regular sea urchins. In C. Johnson (Ed.), *Echinoderms in a changing world* (pp. 97–104). Taylor & Francis.
44. Stock, S. R., Ebert, T. A., Ignatiev, K., & De Carlo, F. (2006). Structures, structural hierarchy, and function in sea urchin spines. In *Developments in X-Ray tomography* (Vol. 6318, p. 63180A). International Society for Optics and Photonics.

45. Toader, N., Sobek, W., & Nickel, K. G. (2017). Energy absorption in functionally graded concrete bioinspired by sea urchin spines. *Journal of Bionic Engineering, 14*, 369–378.
46. Seto, J., Ma, Y., Davis, S. A., Meldrum, F., Gourrier, A., Kim, Y. Y., et al. (2012). Structure-property relationships of a biological mesocrystal in the adult sea urchin spine. *Proceedings of the National Academy of Sciences, 109*(10), 3699–3704.
47. Coppard, S. E., Kroh, A., & Smith, A. B. (2012). The evolution of pedicellariae in echinoids: An arms race against pests and parasites. *Acta Zoologica, 93*, 125–148.
48. Cavey, M. J., & Märkel, K. (1994). Echinodermata. In F. W. Harrison (Ed.), *Microscopic anatomy of invertebrates* (pp. 345–400). Wiley.
49. Campbell, A. C. (1973). Observations on the activity of echinoid pedicellariae: I. Stem responses and their significance. *Marine Behaviour and Physiology, 2*, 33–61.
50. Santos, R., & Flammang, P. (2006). Morphology and tenacity of the tube foot disc of three common European sea urchins species: A comparative study. *Biofouling, 22*, 187–200.
51. Smith, A. B. (1979). Peristomial tube feet and plates of regular echinoids. *Zoomorphologie, 94*, 67–80.
52. Jensen, M. (1985). *Functional morphology of test, lantern, and tube feet ampullae system in flexible and rigid sea urchins (Echinoidea).* In Proceedings of the International Echinoderms Conferences, Galway, Eire, 1984 (pp. 281–288). A. A. Balkema.
53. Perricone, V., Grun, T., Raia, P., & Langella, C. (2022). Paleomimetics: A conceptual framework for a biomimetic design inspired by fossils and evolutionary processes. *Biomimetics, 7*(3), 89.
54. Prince, J. D. (2014). 3D printing: An industrial revolution. *Journal of Electronic Resources in Medical Libraries, 11*(1), 39–45. https://doi.org/10.1080/15424065.2014.877247
55. The Economist. (2012). *The third industrial revolution.* https://www.economist.com/leaders/2012/04/21/the-third-industrial-revolution. Accessed 18 Mar 2022.
56. Sondhi, P., & Stine, K. J. (2021). Methods to generate structurally hierarchical architectures in nanoporous coinage metals. *Coatings, 11*(12), 1440.
57. Koerner, J. (2017). *Digitally crafted couture* (pp. 40–47). Wiley.
58. Atzeni, E., & Salmi, A. (2012). Economics of additive manufacturing for end-usable metal parts. *The International Journal of Advanced Manufacturing Technology, 62*(9), 1147–1155.
59. Lélis, E. C., Paulino, M. C., dos Santos Jesus, M. B., & Bueno, M. J. C. (2020). Um estudo de acessibilidade em uma instituição de ensino de São Paulo. *South American Development Society Journal, 6*(17), 408.
60. Van Nes, I. (2012). How to create a living brand. In D. Raposo (Ed.), *Design, visual communication and branding* (pp. 72–81). IS Publishers.
61. Wheeler, A. (2017). *Designing brand identity: An essential guide for the whole branding team*. Wiley.
62. Kapferer, J. N. (2012). *The new strategic brand management: Advanced insights and strategic thinking*. Kogan Page Publishers.
63. Keller, K. L., Parameswaran, M. G., & Jacob, I. (2011). *Strategic brand management: Building, measuring, and managing brand equity*. Pearson Education India.
64. Bartholmé, R. H., & Melewar, T. C. (2011). Remodelling the corporate visual identity construct: A reference to the sensory and auditory dimension. *Corporate Communications: An International Journal, 16*(1), 53–64.
65. Machado, J. C., de Carvalho, L. V., Torres, A., & Costa, P. (2015). Brand logo design: Examining consumer response to naturalness. *Journal of Product & Brand Management, 24*(1).
66. MacInnis, D. J., Shapiro, S., & Mani, G. (1999). Enhancing brand awareness through brand symbols. *ACR North American Advances.*
67. Zoppè, M. (2014). Comunicare l'invisibile. La rappresentazione visiva di concetti biofisici. *Progetto Grafico, 25*, 50–55.
68. Jackson, R. B., Bejarano, A., Winkle, K., & Williams, T. (2021). Design, performance, and perception of robot identity. In *Workshop on robo-identity: Artificial identity and multi-embodiment at HRI* (Vol. 2021).

Valentina Perricone, A marine biologist, she achieved a bachelor's degree (cum laude) in Natural Science at the Federico II, University of Naples; master's degree (cum laude) in Marine Biology at the University of Bologna; and PhD in Environment, Design, and Innovation at University of Campania Luigi Vanvitelli. Currently, she is a postdoc at Biomimetics and Nanostructured Materials Lab, University of California, Irvine. Her research mainly regards the echinoid biomechanics and inspired biomaterials.

Carla Langella, Architect, associate professor of Industrial Design at the Department of Architecture of University of Naples "Federico II", and currently teaching User Experience Design and Industrial Design. She has founded and coordinated since 2006 the Hybrid Design Lab, a project and research laboratory dedicated to mutual relations between design and biosciences with a particular interest in biomimetics, design-driven materials, design for health, and design of cultural experiences mediated by digital technologies.

Gabriele Pontillo, Product designer and researcher at the Department of Architecture of the University of Florence on the issues of sustainability and Made in Italy. He obtained a PhD in Environment, Design, and Innovation at the University of Campania "Luigi Vanvitelli". The main focus of his line of research is parametric design, medical design, and advanced manufacturing.

Roberta Angari, Graphic designer and contract professor at the Department of Architecture and Industrial Design of the University of Campania "Luigi Vanvitelli". She obtained a PhD in Architecture, City and Design, Design Science Curriculum, at the IUAV University of Venice. The main focus of her line of research is data visualisation, digital design, and digital archives.

Biodigital Product Design: Experiences in the Field of Applied Research

David A. Torreblanca-Díaz

1 Introduction

Evolution in nature has created living creatures with extraordinary patterns and features to protect them from predators, temperature changes and aggressive agents in the environment, thus giving them evolutionary benefits to survive. These unique properties have garnered the attention of researchers, who seek to reveal the secrets behind these structures and systems to transfer these evolutionary features into science and technology projects [1]. *Learning from nature* has a long history with a well-known potential for innovation and many success stories; the current discoveries from nature are progressively increasing the possibilities of application. Consequently, robust, repeatable and scalable methodologies have originated in this emerging field of research [2].

Different *nature-based design approaches* have been emerging such as *biomimicry, biodesign, biofabrication, biomimetics and bioinspiration*, among others; these concepts propose new and innovative paradigms for design and manufacturing. However, these new concepts located around biology and its various subdisciplines are sometimes ambiguous and difficult to understand, especially by researchers from nonscientific sectors [3]. In response to this problem, the idea of *bioinformed disciplines* (BID) was proposed as an umbrella term to consolidate the range of bio-related activities [4].

On the other hand, digital design (DD) is growing progressively, challenging the paradigms of conventional design, a unique phenomenon that goes beyond to replace drawing techniques with digital technologies; it is rethinking the entire

D. A. Torreblanca-Díaz (✉)
Universidad Pontificia Bolivariana/Universidad Internacional de Cataluña,
Medellín, Colombia
e-mail: david.torreblanca@upb.edu.co

A. J. V. Arruda, F. L. Palombini (eds.), *Biomimetics, Biodesign and Bionics*,
Environmental Footprints and Eco-design of Products and Processes,
https://doi.org/10.1007/978-3-031-51311-4_5

design process and reconfiguring its basic components such as generation, presentation, performance and evaluation, as a consequence establishing new and complex relationships between designers, design media and digital design toolmakers [5]. With scripting, computer programming becomes integral to the digital design process and opening new opportunities for innovation; it allows designers to customize the software according to their interests and requirements, freeing them from routine activities to dedicate more time to *design-thinking*. By integrating digital design with emerging digital fabrication technologies, time is saved at the *front end*, and new *file-to-factory* protocols can be taken advantage of [6]. Parametric design (PD) can be defined as a formation process of parametric structures of associative morphologies that generates the geometry of a design object. In this digital environment, the designer interacts with the parameterized modeling system through visual codes; these associative systems can generate or modify the design simply by changing the parameters without rewriting the entire algorithm [7]. Among the wide range of emerging digital manufacturing technologies, additive manufacturing (AM) stands out. The American Society for Testing and Materials (ASTM) defines additive manufacturing as "the process of joining materials to create objects from 3D model data, usually layer by layer, as opposed to fabrication methods by subtraction" [8]. AM is an innovative technology; it opens new opportunities and lends itself to many possibilities for companies looking to improve manufacturing efficiency. This novel technology simplifies significantly manufacturing processes compared to conventional techniques [9]. AM is considered one of the main achievements in the recent 30 years of manufacturing, and this technology has fewer restrictions and more freedom. Some representative applications include aerospace, automotive, bio-constructs and jewelry, among others; this technology has used different materials such as polymers, ceramics, and metals. Design for additive manufacturing (DfAM) is a new concept referring to the research field that investigates the design tools and methods to be used in AM processes to obtain *easy-to-print* components, supported on computer-aided design (CAD) files [10].

In recent decades, at the height of digital technologies and new discoveries in the fields of biology and genetics, *biodigital architecture and design* was born, an emerging twenty-first-century *avant-garde* originated as a consequence of *digital organicism*. This proposal is a fusion between the biological and the digital domains, *genetics* and *cybernetics, natural intelligence* and *artificial intelligence, biolearning* and *machine learning, and biomanufacturing* and *digital manufacturing*; these are the keywords that shape the *biodigital* cloud. *Biodigital* approach must give our planet better future through architecture and design – perhaps the only future to live with dignity [11].

It is possible to use synergistically nature and digital technologies for design processes, to work with DNA as if it was a *natural software*, and to work with software as if it was *digital DNA*, since software and DNA end up being the same: information chains that determine the process order, the emergence of forms, *self-organization* and *autonomous growth* [IDEM].

This text shows a selection of applied research projects in the field of product design, under the biodigital bases, with emphasis on the synergy of *parametric-generative design* and AM, embracing the DfAM concept. These projects are diverse and are situated at different contexts, scales and complexities; the research projects selected are (i) bioinspired parametric textures, (ii) biodigital stool series and (iii) biodigital columns. In methodologies, different approaches were considered, always from the biodigital point of view. First, two perspectives were considered: *design looking to biology* (*top-down approach*) and *biology-influencing design* (*bottom-up approach*). In *design looking to biology* approach, the designer must recognize exactly the design problem and match this problem with organisms and creatures that have solved similar problems in nature; in *biology-influencing design,* the designer and his team study natural systems, characteristics and their advantages, and then they look for possible applications in design problems [12]. In addition, three levels of mimicry were considered: the organism level, the behavior level and the ecosystem level. The organism level is the mimicking of an organism or the mimicry of part from the whole organism. The second level is the mimicry of behavior of how the organism, or a group of organisms, behaves; this level includes the emergent behavior. The third level is the mimicking of the ecosystem; this level is considered the most complex because it included the two previous levels and the dynamic relationships between organisms and the environment [13]. Finally, six project strategies were considered: *morphogenetic, genetic, morphodynamic, biomimetic, biomorphic and parametric*. *Morphogenetic* strategy is based on endogenous processes and internal formation processes. *Genetic approach* is part of *morphogenetic strategy* and is based on genetic techniques and applications exclusively; thus, DNA design will define the development of the future creature or species. *Morphodynamic* strategy is based on exogenous forces and dynamic systems that generate and model forms as a result of stress or other external factors. *Biomimetic* strategy is the mimicking of natural systems according to project goals, imitating functions, mechanisms, processes, materials, etc. *Biomorphic* strategy is the mimicking of biological forms without knowing why they exist, their functions, processes and consequences. Finally, *parametric strategy* is focused on using *parametric-associative* and *generative software*, based on algorithms [14].

2 Research Projects

2.1 Bioinspired Parametric Textures Project

Species on earth have faced over millions of years hostile environments, with changing conditions in a dynamic ecosystem; living beings have developed mechanisms, systems and different adaptive responses to survive. Organisms interact with their surrounding environment through their skin; this surface (and *sub-surface*) is an important interface between the organisms and their environment, with a critical role in the process of adaptation and survival. Skin surfaces interact with aggressive environmental factors such as temperature, moisture, abrasive agents such as sand and ice, pH, bacteria and viruses, light, mating abrasive or super smooth surfaces,

Fig. 1 Image of an insect of the species *Umbonia spinosa*. (Source: photography by David A. Torreblanca-Díaz. Reprinted with permission from Ref. [16], Copyright 2023, Springer Nature Singapore Pte Ltd)

etc. Skins have developed unique designs with multifunctional properties including (but not limited to) adhesion, hydrophobicity, hydrophilicity, thermal management, anti-reflection and structural colors, among others [15] (Fig. 1).

Experimental Morphology Research Line at Universidad Pontificia Bolivariana in Medellín, Colombia, proposed an applied research project aimed to transfer morphological characteristics of textures from nature to artificial elements with the goal to solve human problems, in the context of product design projects. The research highlighted the advantages of parametric design and digital fabrication technologies as integrated creative tools. The textures of species from the plant and *fungi kingdom* in the Aburrá Valley in Colombia were studied; they were selected because thanks to evolutionary strategies they have adapted to complex environmental conditions. In addition, they are abundant and biodiverse, and it is easy to get samples. All these conditions and characteristics make them an almost inexhaustible set of study specimens. Furthermore, with the results obtained, it was proposed to create a *digital repertoire of bioinspired parametric surfaces* to be used by designers, architects and other creative professionals using the *open-source* modality [17] (Fig. 2).

2.1.1 Methodology

An applied research project was proposed with a multidisciplinary approach, with an *empirical-experimental* method, qualitative analysis of the information, synergistic use of parametric design and digital fabrication technologies. The collaboration between the university, companies and external professionals was relevant, as

Fig. 2 Texture of pitahaya, fruit of *Selenicereus megalanthus* species. (Source: photography by David A. Torreblanca-Díaz. Reprinted with permission from Ref. [16], Copyright 2023, Springer Nature Singapore Pte Ltd)

well as between researchers and industrial design students. In the *design based on nature* methods, *biology-influencing design approach* [12], *organism level* [13] *biomorphic* and *parametric* project strategies [14] were considered to develop the research project. The research process began with a photographic record of the species in the Aburrá Valley in Colombia, and then the digital abstraction was done in the Rhinoceros© computer-aided design (CAD) software, after that the digital abstraction and parameterization of the textures were carried out using Grasshopper©, a parametric tool that works within the Rhinoceros© CAD software environment. Subsequently, various samples were made with digital fabrication technologies, including additive manufacturing technologies (3D printing) and subtractive manufacturing technologies (milling through computer numerical control – CNC); a methodological proposal (for applications of bioinspired textures in design projects) was formulated, and the application of bioinspired textures in degree projects was carried out. Finally, an analysis of the information and a proposal for improvement were made [17] (Fig. 3).

2.1.2 Results

In the research, samples of bioinspired textures, a methodology for the application of bioinspired textures in design and architecture and final degree projects were proposed. Samples were fabricated with digital fabrication technologies, additive manufacturing technologies (3D printing) and subtractive manufacturing technologies (milling through computer numerical control – CNC) were used. Additionally, concrete samples were made, using conventional methods. In additive

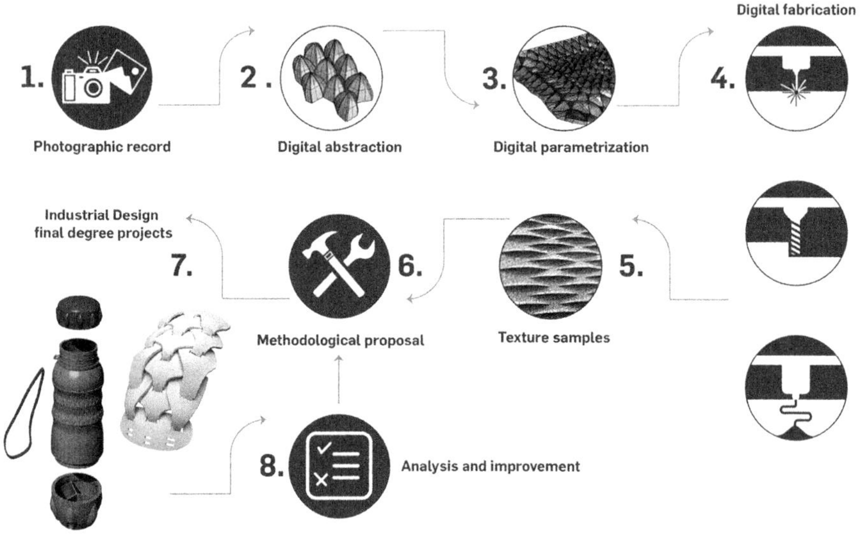

Fig. 3 Methodological sequence. (Source: the author. Reprinted with permission from Ref. [17], Copyright 2023, Springer Nature Singapore Pte Ltd)

manufacturing, samples of 50 × 50 millimeters were made through digital light processing (DLP) technology and samples of 100 × 100 millimeters through fused deposition modeling (FDM) technology (Figs. 4, 5, 6, and 7).

Samples of 200 × 200 millimeters were manufactured through milling (CNC), using wood (Fig. 8). The concrete samples were made in different stages, first milled wooden models (with CNC technology) were made, then molds through vacuum thermoforming were done with a high-impact polystyrene (HIPS), and finally the concrete was poured into the molds; the dimensions of these samples were 200 × 200 millimeters [16] (Figs. 9, 10, and 11).

A methodology for the application of bioinspired textures in design and architecture projects was proposed; the methodology proposes six sequential steps: (i) problem and opportunities identification, (ii) selection of textures, (iii) digital experimentation and transformations, (iv) digital fabrication, (v) evaluation and (vi) feedback and improvement [17] (Fig. 12).

The first application of bioinspired textures through the methodology described above was carried out in the academic field, in the degree projects of industrial design students. Two projects stood out, and both made an appropriate selection of a problem in the local context, an adequate selection of bioinspired textures with future product development possibilities. In the first project, the students worked with the different problems that originated in the use of food bottles used by elderly people, with emphasis on the ergonomic aspects. Elderly people have a progressive decrease in strength in arms and hands, loss of fine motor skills and visual abilities limitations; all these problems make them difficult to manipulate and use plastic bottles. The students proposed a line of food supplement plastic bottles. In this

Fig. 4 Texture samples made using digital light processing (DLP) and epoxy thermosensitive resin. From top to bottom, based on *Pinus pinea*, *Artocarpus heterophyllus and Eugenia uniflora* species. Size: 50 × 50 mm. (Source: photography by David A. Torreblanca-Díaz. Reprinted with permission from Ref. [16], Copyright 2023, Springer Nature Singapore Pte Ltd)

Fig. 5 Texture sample made through digital light processing (DLP) and epoxy thermosensitive resin. Based on *Artocarpus heterophyllus* species. Size: 50 × 50 mm. (Source: photography by David A. Torreblanca-Díaz. Reprinted with permission from Ref. [16], Copyright 2023, Springer Nature Singapore Pte Ltd)

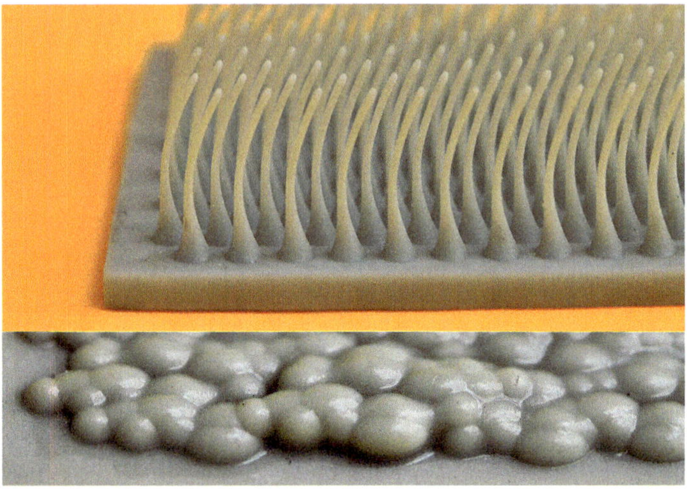

Fig. 6 Texture sample made through digital light processing (DLP) and epoxy thermosensitive resin. From top to bottom, based on *Nephelium lappaceum* and *Theobroma cacao* species. Size: 50 × 50 mm. (Source: photography by David A. Torreblanca-Díaz. Reprinted with permission from Ref. [16], Copyright 2023, Springer Nature Singapore Pte Ltd)

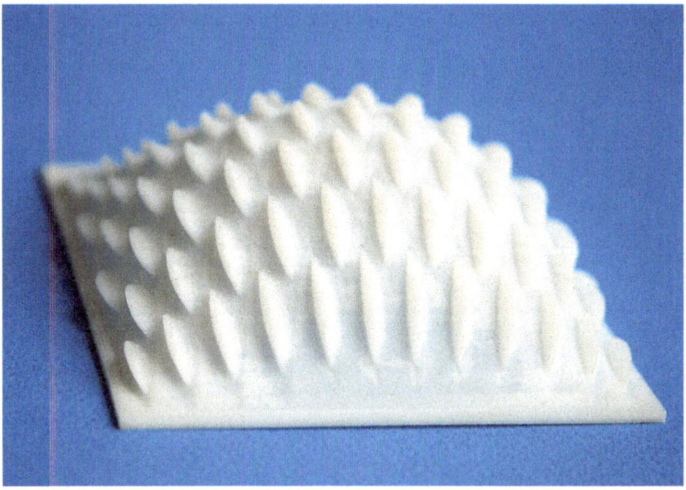

Fig. 7 Texture sample made through fused deposition modeling (FDM) and acrylonitrile butadiene stirene (ABS) material. Based on *Phoenix canariensis* species, complex morphologic transformation. Size: 100 × 100 × 40 mm. (Source: photography by David A. Torreblanca-Díaz. Reprinted with permission from Ref. [16], Copyright 2023, Springer Nature Singapore Pte Ltd)

Fig. 8 Texture sample made through control numeric milling and wood. Based on *Pinus pinea* species, complex morphologic transformation. Size: 200 × 200 × 33 mm. (Source: photography by David A. Torreblanca-Díaz. Reprinted with permission from Ref. [16], Copyright 2023, Springer Nature Singapore Pte Ltd)

Fig. 9 Concrete texture sample and its mold. Based on *Artocarpus heterophyllus* species. Size: 200 × 200 × 18 mm. (Source: photography by David A. Torreblanca-Díaz. Reprinted with permission from Ref. [16], Copyright 2023, Springer Nature Singapore Pte Ltd)

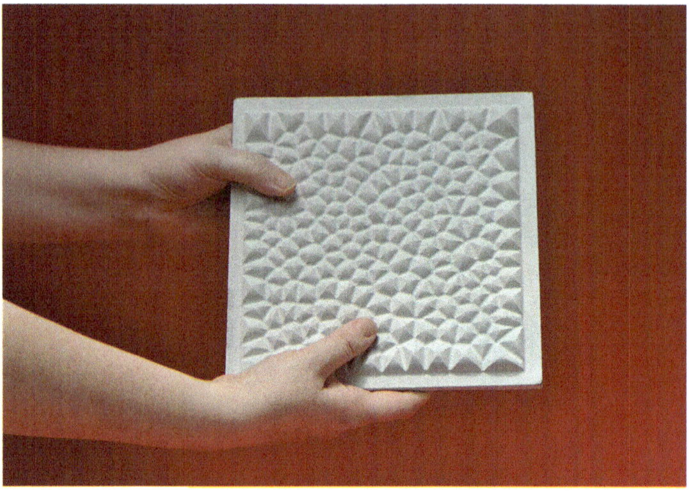

Fig. 10 Concrete texture sample. Based on *Artocarpus heterophyllus* species. Size: 200 × 200 × 18 mm. (Source: photography by David A. Torreblanca-Díaz. Reprinted with permission from Ref. [16], Copyright 2023, Springer Nature Singapore Pte Ltd)

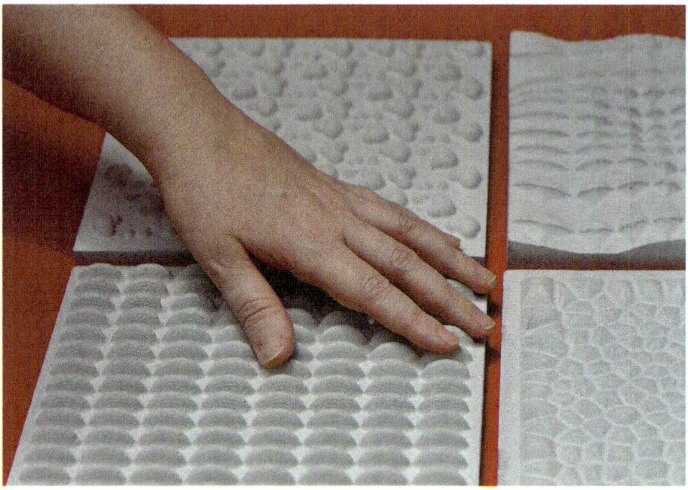

Fig. 11 Concrete texture samples. Size: 200 × 200 mm. (Source: photography by David A. Torreblanca-Díaz. Reprinted with permission from Ref. [16], Copyright 2023, Springer Nature Singapore Pte Ltd)

proposal, the integration of bioinspired textures helped to reduce the effort to open the cap, changing the cap size and using large textures, improve grips to prevent slipping and differentiate the parts for easier visual reading. The prototype was manufactured through FDM technology and using Acrilonitrilo Butadieno Estireno (ABS) (Figs. 13, 14, and 15). The second project aimed to design a vest to protect

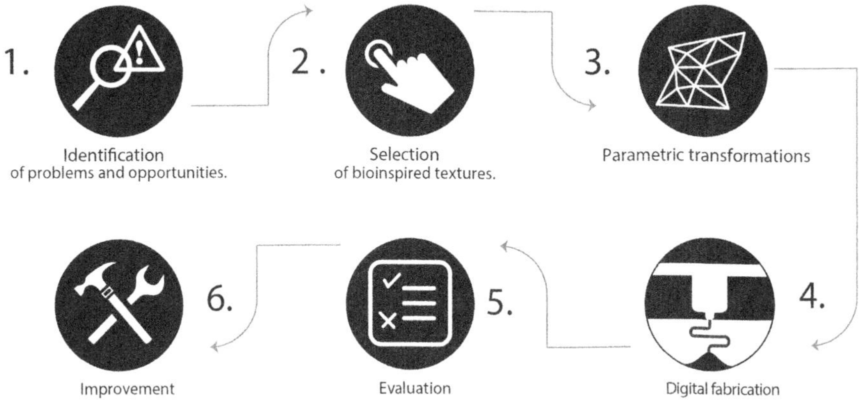

Fig. 12 Methodology for the application of bioinspired textures. (Reprinted with permission from Ref. [16], Copyright 2023, Springer Nature Singapore Pte Ltd)

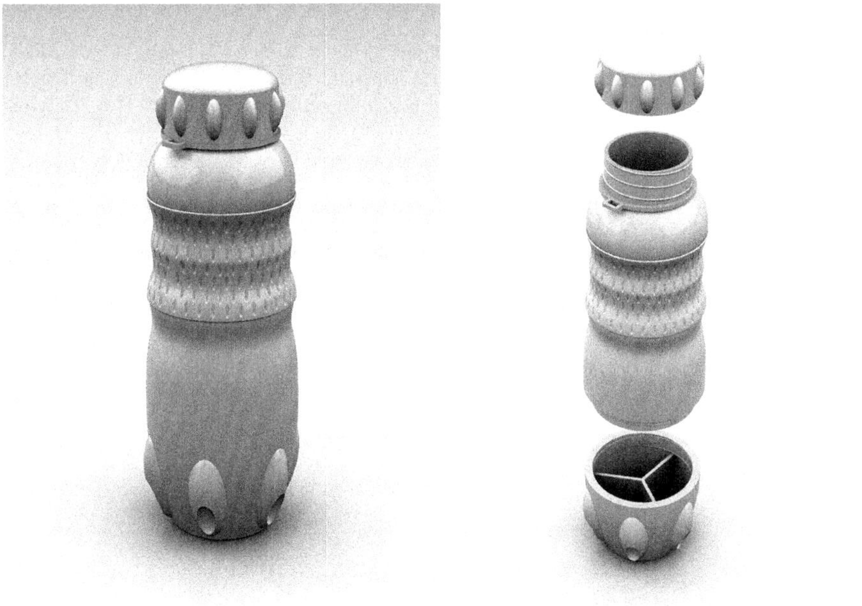

Fig. 13 Digital design of bottle for nutritional supplements oriented to elderly people, Rhinoceros CAD software. (Source: students Mariana Betancur Sánchez and Valentina Cardona Echeverri. Reprinted with permission from Ref. [16], Copyright 2023, Springer Nature Singapore Pte Ltd)

users in motorcycle accidents situations in the city of Medellín in Colombia. In Medellín, there is a high rate of motorcycle accidents; this is the result of different factors such as a temperate, rainy and changeable climate, progressive increase in the use of motorcycles, excessive speed and other recklessness of people. The students proposed a multifunctional system using the morphologies of modular natural

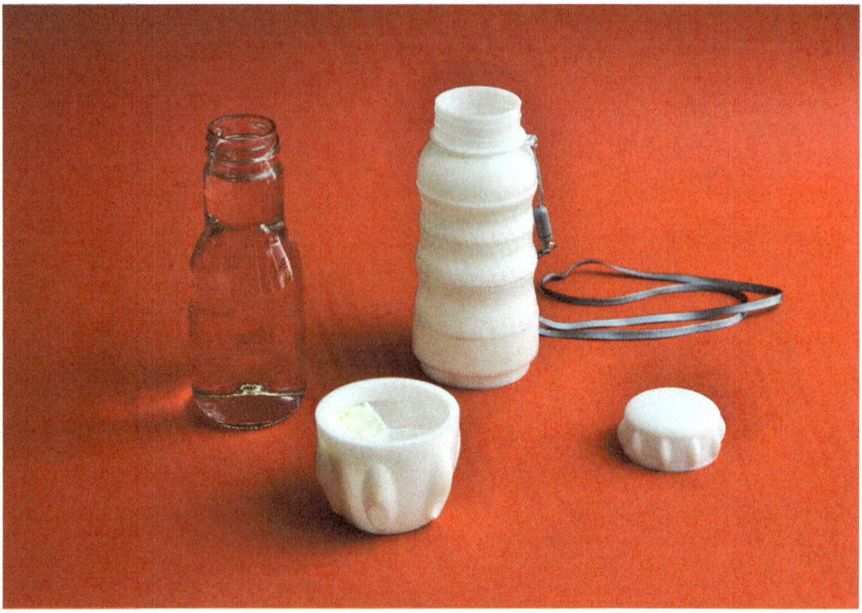

Fig. 14 Prototype of bottle for nutritional supplements made through fused deposition modeling (FDM) and acrylonitrile butadiene stirene (ABS) material. (Source: students Mariana Betancur Sánchez and Valentina Cardona Echeverri. Reprinted with permission from Ref. [16], Copyright 2023, Springer Nature Singapore Pte Ltd)

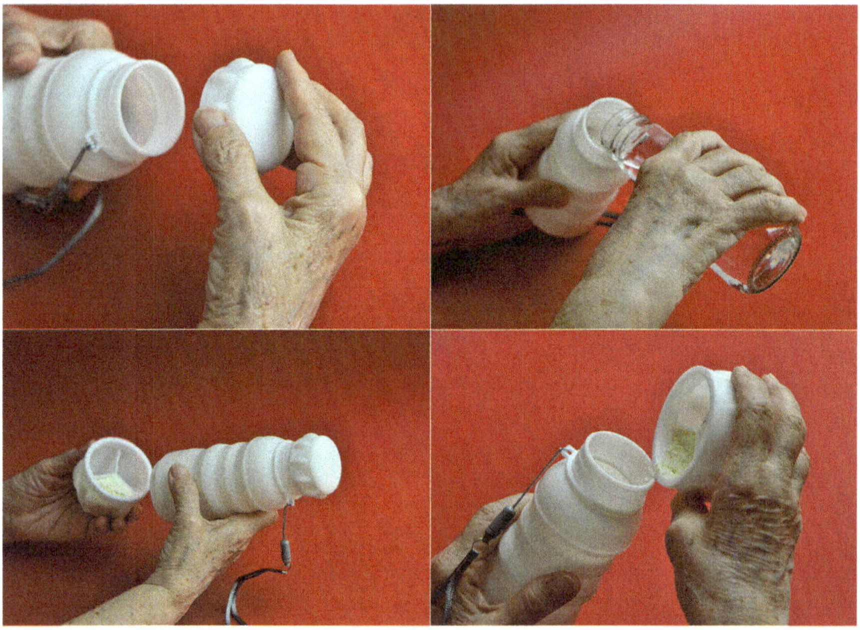

Fig. 15 User test: prototype of bottle for nutritional supplements oriented to elderly people. (Source: students Mariana Betancur Sánchez and Valentina Cardona Echeverri. Reprinted with permission from Ref. [16], Copyright 2023, Springer Nature Singapore Pte Ltd)

Fig. 16 Digital design of a vest to protect users in motorcycle accidents, Rhinoceros CAD software. (Source: student Juan Diego Buriticá. Reprinted with permission from Ref. [16], Copyright 2023, Springer Nature Singapore Pte Ltd)

elements with the possibility of customizing. At the beginning, the body of the user is scanned to obtain the precise dimensions; the vest is resistant to falls and impacts, is flexible to allow movement and adjusts to the biomechanics of the person and blocks movements that could cause injuries. The modular parts allow disassembly for replacement; finally, the aesthetics of the product communicate safety and protection. The prototype was manufactured through FDM technology and ABS [IDEM] (Figs. 16, 17, 18, and 19).

2.1.3 Conclusions of Bioinspired Parametric Textures Project

After the experience in the academic field, it is possible to say that bioinspired textures allowed us to solve and enhance aspects of usability and sensoriality, functionality and *aesthetic-communicative*, among others. The integration of parametric design with digital fabrication technologies had broad technical possibilities, such as the generation of complex morphologies with dynamic transformations, high level of geometric accuracy, customization possibilities, among other advantages [17].

For next stages, it would be valuable to study other textures of plant species, even from other kingdoms from nature. With this, it will be possible to expand the patterns of the repertoire created. In addition, it would be a contribution to fabricate

Fig. 17 Digital design of a vest to protect users in motorcycle accidents, Rhinoceros CAD software. (Source: student Juan Diego Buriticá. Reprinted with permission from Ref. [16], Copyright 2023, Springer Nature Singapore Pte Ltd)

Fig. 18 Digital design of a vest to protect users in motorcycle accidents, Rhinoceros CAD software. (Source: student Juan Diego Buriticá. Reprinted with permission from Ref. [16], Copyright 2023, Springer Nature Singapore Pte Ltd)

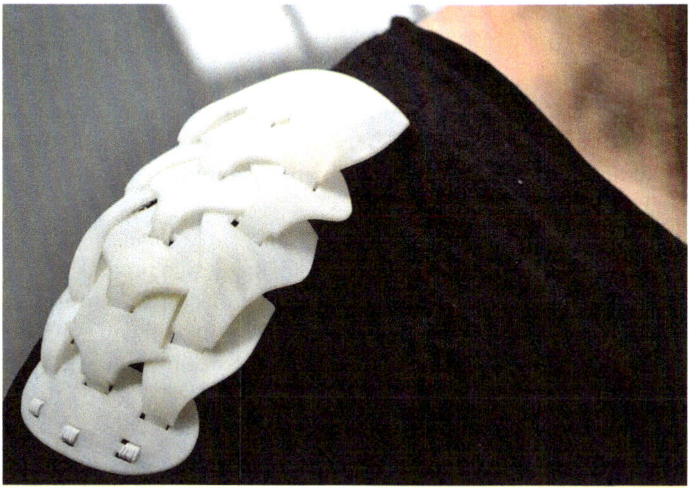

Fig. 19 Focused prototype of a vest to protect users in motorcycle accidents, made using fused deposition modeling (FDM) and acrylonitrile butadiene stirene (ABS) material. (Source: student Juan Diego Buriticá. Reprinted with permission from Ref. [16], Copyright 2023, Springer Nature Singapore Pte Ltd)

skins with biopolymers, responsive skins at a more advanced level, they react to environmental stimuli, such as pressure, temperature, humidity, and others, just as it happens in nature; in that order of ideas, 4D printing could be an interesting tool to use. Finally, the transfer of morphological characteristics from the surfaces of plant species to artificial elements has a wide range of applications at different scales and complexities; therefore, it is expected in the future to use these bioinspired textures in product design, architecture and engineering, in the context of companies, industrial or entrepreneurial fields, thereby improving the methodology proposed and the quality of the results.

2.2 *Biodigital Stool Series Project*

In the different morphologies in nature, we can accept *the evolutionary theory* according to which everything develops or evolves until it reaches the forms more adapted to its environment. In the same way, the elements in nature achieve a configuration that uses the least possible energy for adaptation; we can also admit that the morphologies that exist in nature are those that within the possibilities and restrictions are those that have a greater probability to exist. Furthermore, every phenomenon in nature, apparently elementary or isolated, is part of a larger system, which in turn operates within other systems of increasing complexity. We can see an overwhelming variety of morphologies and geometric patterns in nature, but this diversity is the result of an elaboration and continuous reworking of a restricted number of basic geometries. These morphologies have originated in response to different contexts, limited by the scarcity of resources, as well as the restrictions imposed by three-dimensional space and the relationships between the different sizes of the elements. The most important restriction for geometries and patterns in nature is that defined by space – understanding the space as an active entity whose structure exerts a decisive influence on the morphology of the elements to adapt appropriately to their environment [18].

A research project is proposed, aimed to study *three-dimensional growth morphologic patterns in nature* to be integrated into the design process of a stool series and developed under a biodigital approach. This project highlights the advantages of the integration between parametric design and additive manufacturing technologies; in addition, a structural performance analysis of different *three-dimensional growth morphologic patterns* was made.

2.2.1 Methodology

An applied research project was proposed, using an *empirical-experimental method* with a multidisciplinary approach, a synergy of parametric design and additive manufacturing and a quantitative analysis of the information. In the *design based on nature methods*, biology-influencing design approach [12], organism level [13], biomorphic and parametric project strategies [14] were considered to develop the research project.

The methodological sequence followed the following steps: (1) pattern from nature selection, (2) abstraction and parametric design, (3) compression test/1:5 scale models, (4) 1:1 scale prototype fabrication, (5) user test and (6) conclusions and improvement proposals in an iterative and cyclical process [19] (Fig. 20).

2.2.2 Results

Different *natural growth patterns* in nature were studied, of which three were chosen; these chosen patterns belonged to diverse environments with different morphological characteristics. The chosen patterns were *differential growth, branching* and *the Voronoi diagram.* Differential growth occurs in the plant kingdom and corals (animals called polyps or zooids); its non-homogeneous growth is due to the non-uniform structure of materials or intercellular transport and physical limitations in cell walls force adjacent cells to deform in a coordinated manner generating complex and irregular morphologies (Fig. 21). Branching is a pattern that occurs in different living beings in nature, for example, in the vegetable kingdom to study species such as plants and trees. In this case, they give structural support to the organism, allow maximum exposure to the sun, and distribute nutrients; they start from a trunk, from which different branches fork at angles that generally vary between 75° and 90°; in this project, plants and trees were chosen as branching morphological models (Fig. 21). The Voronoi diagram is a morphological pattern

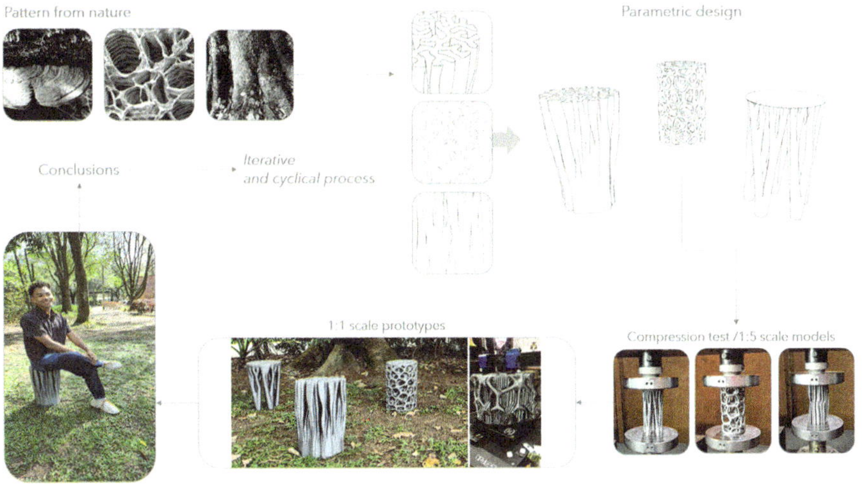

Fig. 20 Methodological sequence, biodigital stool series project. (Source: Juan Pablo Velásquez, David A. Torreblanca-Díaz and Alberto T. Estévez (Voronoi pattern). Reprinted with permission from Ref. [19], Copyright 2023, Springer Nature Singapore Pte Ltd)

Fig. 21 Pattern from nature used in biodigital stool series project. From the left to the right: differential growth and branching pattern. (Source: Juan Pablo Velásquez and David A. Torreblanca-Díaz. Reprinted with permission from Ref. [19], Copyright 2023, Springer Nature Singapore Pte Ltd)

that can be found in different organisms, on a micro and macro scale. The pattern selected is a kind of Voronoi three-dimensional growth and is found at the microscopic level in bones, cactus and others. In bones, this three-dimensional pattern is a light and structurally efficient structure, because it has more volume in areas that require greater mechanical effort and in areas that require less effort, and it has a reduced volume or is empty [IDEM]. In Fig. 22, a microscopic cactus section can be seen.

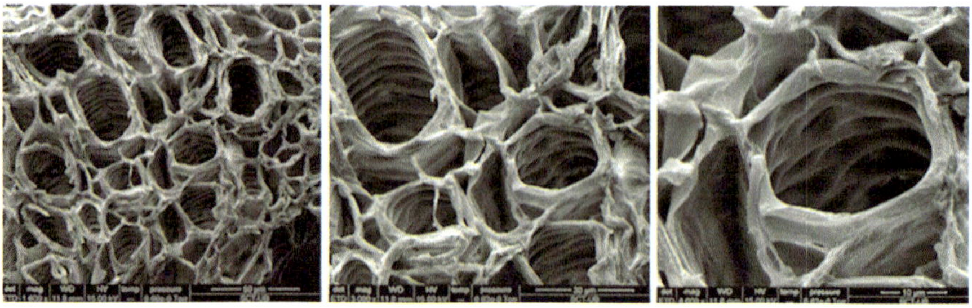

Fig. 22 Cactus section (1600×, 3000×, 6000×), 2008: photo through a scanning electron microscope (SEM), it is possible to see the three-dimensional Voronoi diagram because it also appears on the walls perpendicular to the section. (Source: © Alberto T. Estévez, 2023)

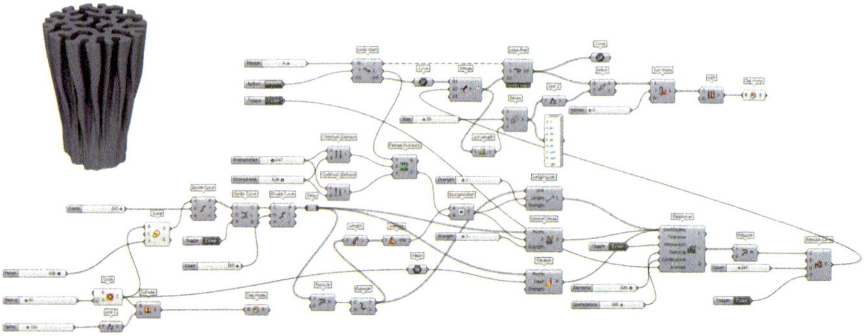

Fig. 23 Differential growth pattern algorithm. (Source: the author. Reprinted with permission from Ref. [19], Copyright 2023, Springer Nature Singapore Pte Ltd)

Fig. 24 Branching pattern algorithm. (Source: the author. Reprinted with permission from Ref. [19], Copyright 2023, Springer Nature Singapore Pte Ltd)

Once the morphological patterns were chosen, the abstraction and parameterization were developed, and an algorithmic process was carried out using Grasshopper©. These morphologies were integrated into a cylindrical volume of 420 mm height and 300 mm diameter, which correspond to dimensions of target object (stools). In Figs. 23, 24, 25, and 26, algorithms and 3D models can be seen [IBID].

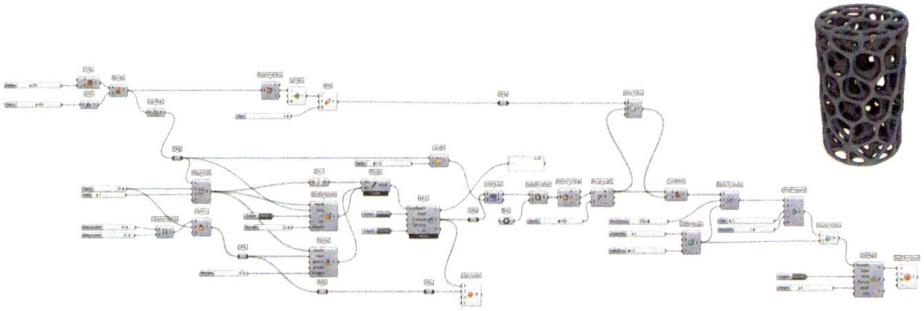

Fig. 25 Voronoi diagram algorithm. (Source: the author. Reprinted with permission from Ref. [19], Copyright 2023, Springer Nature Singapore Pte Ltd)

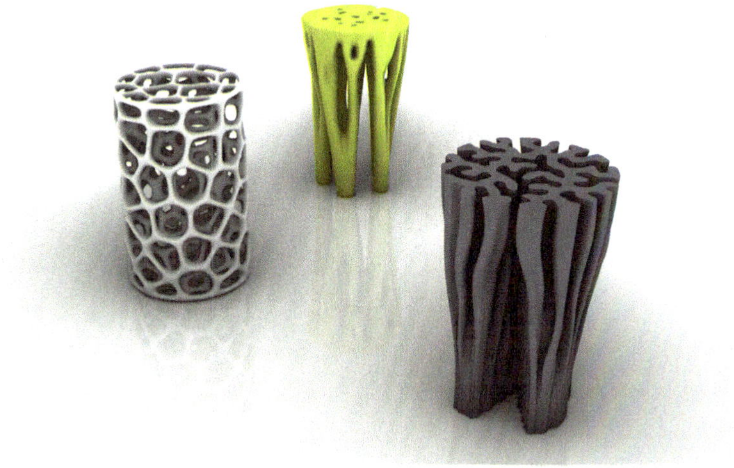

Fig. 26 3D models of three patterns based on nature. (Source: the author. Reprinted with permission from Ref. [19], Copyright 2023, Springer Nature Singapore Pte Ltd)

To do the compression test, 1:5 scale models (samples) were printed, four samples were made through fused deposition modeling (FDM) technology and polylactic acid (PLA) material, with 120mm high and 60mm diameter dimensions, one cylindrical and three with morphologies extracted from nature; the cylindrical piece allowed to characterize this type of test with material and process chosen [IBID]. In Table 1 and Figs. 27, 28, 29, 30, and 31, sample information and compression test process can be seen.

In Table 2, it can be seen that the branching pattern had the highest specific resistance (2.60 MPa/g), followed by Voronoi pattern (0.24 MPa/g) and differential growth pattern (0.21 MPa/g), and the cylindrical sample had the lowest specific resistance (0.13 MPa/g). Next to that, the branching pattern had the highest specific resistance (2.60 MPa/g), a more rigid behavior and was lighter (56 g); on the contrary, the cylindrical sample had the greatest tendency to deformation and was the heaviest (171 g) [IBID].

Table 1 Sample information

	Name	Equivalent diameter (mm)	Area (mm^2)	Length (mm)	Mass (g)
1	Cylinder	60.00	2827.43	120.00	171.00
2	Voronoi	18.14	258.56	125.54	79.00
3	Branching	6.24	30.58	120.82	56.00
4	Differential growth	41.95	1381.85	120.23	128.00
Average		31.58	1124.61	121.65	108.50
Standard deviation		24.07	1279.68	2.62	51.36

Source: author. Reprinted with permission from Ref. [19], Copyright 2023, Springer Nature Singapore Pte Ltd

Fig. 27 1: 5 scale models for compression test, fabricated through fused deposition modeling (FDM) technology with polylactic acid (PLA) material. (Source: the author. Reprinted with permission from Ref. [19], Copyright 2023, Springer Nature Singapore Pte Ltd)

Fig. 28 Compression test, cylindric sample. (Source: the author. Reprinted with permission from Ref. [19], Copyright 2023, Springer Nature Singapore Pte Ltd)

Fig. 29 Compression test, differential growth pattern sample. (Source: the author. Reprinted with permission from Ref. [19], Copyright 2023, Springer Nature Singapore Pte Ltd)

Fig. 30 Compression test, Voronoi pattern sample. (Source: the author. Reprinted with permission from Ref. [19], Copyright 2023, Springer Nature Singapore Pte Ltd)

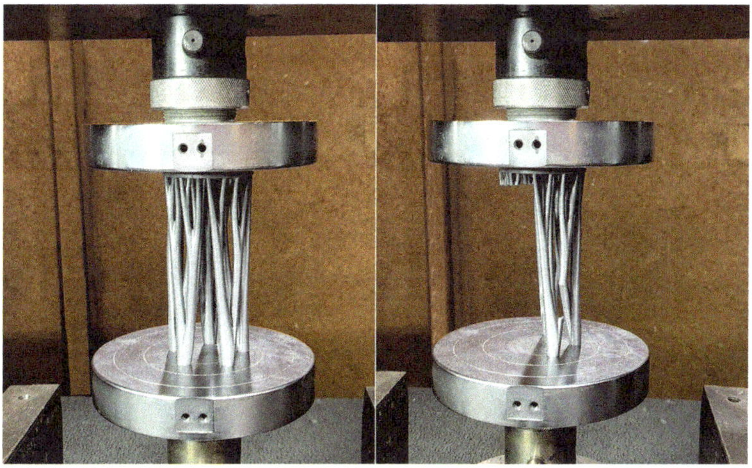

Fig. 31 Compression test, branching pattern sample. (Source: the author. Reprinted with permission from Ref. [19], Copyright 2023, Springer Nature Singapore Pte Ltd)

Table 2 Loads and stress

	Maximum load (kgf)	Maximum stress (MPa)	Module (automatic) (MPa)	Specific maximum effort (MPa/g)
1	<−6424.67	>22.28	908.88	0.13
2	<−493.21	>18.71	811.30	0.24
3	<−454.25	>145.67	15218.70	2.60
4	<−3811.56	>27.05	1289.01	0.21
Average	−2795.90	53.43	4556.97	0.79
Standard deviation	2885.90	61.59	7110.81	1.21

Source: the author. Reprinted with permission from Ref. [19], Copyright 2023, Springer Nature Singapore Pte Ltd

Table 3 3D printing parameters/differential growth pattern

3D printing machine	Artillery Sidewinder X2
Technology	FDM
Nozzle diameter	0.8 mm
Layer height	0.4 mm
Wall thickness	2.4 mm
Wall line count	2,4 mm (3 lines)
Fill density	10%
Speed	45 mm/seg
Support structure	No supports used
3D printing time	96h (4 days)

Source: the author. Reprinted with permission from Ref. [19], Copyright 2023, Springer Nature Singapore Pte Ltd

After the compression test, the 1:1 scale prototypes were fabricated, through AM technology. The Creality 3D Ender-3 (FDM technology) 3D printer was used, and the material used was polylactic acid (PLA). It was not possible to print the stools in one piece, so it was necessary to print them in parts and glue them together. In Tables 3, 4, 5, and 6, fabrication parameters and comparative fabrication information can be seen. The branching pattern took less time and used less material to print. On the other hand, the Voronoi pattern took the longest time and used the most material to print and the differential growth pattern was in the middle [IBID].

In Figs. 32, 33, 34, 35, 36, 37, and 38, a 1:1 scale prototype fabrication process (3D printing) and user test can be seen. After the user test, it was possible to verify basic aspects; the size, use and functionality of the furniture were adequate, and the stools were resistant and had an appropriate weight to be handled [IBID].

2.2.3 Conclusions of Biodigital Stool Series Project

The synergistic use of PD and AM was an advantage for the project; it was possible to materialize complex morphologies based on nature (impossible to make with other techniques) with the possibility of being edited, without molds, a controlled

Table 4 3D printing parameters/branching pattern

3D printing machine	Artillery Sidewinder X2
Technology	FDM
Nozzle diameter	0.8 mm
Layer height	0.32 mm
Wall thickness	2.4 mm
Wall line count	2,4 mm (3 lines)
Fill density	10%
Speed	45 mm/seg
Support structure	Standard
3D printing time	60 h (2.5 days)

Source: the author. Reprinted with permission from Ref. [19], Copyright 2023, Springer Nature Singapore Pte Ltd

Table 5 3D printing parameters/Voronoi pattern

3D printing machine	Artillery Sidewinder X2
Technology	FDM
Nozzle diameter	0.8 mm
Layer height	0.4 mm
Wall thickness	2.4 mm
Wall line count	2,4 mm (3 lines)
Fill density	15%
Speed	45 mm/seg
Support structure	Tree support
3D printing time	264 h (10.7 days)

Source: the author. Reprinted with permission from Ref. [19], Copyright 2023, Springer

Table 6 Comparative information 3D printing process

Pattern	3D printing time (days)	Material quantity (kg)
Voronoi	10.7	12
Branching	2.5	6
Differential growth	4	8

Source: the author. Reprinted with permission from Ref. [19], Copyright 2023, Springer Nature Singapore Pte Ltd

accuracy to obtain functional objects. In the case of a compression test with samples, branching was the lightest pattern and the one with the highest specific resistance; furthermore, this pattern took less time and used less material to print; this information may be useful for future projects. For future work, an optimization of the fabrication process will be considered to obtain a balance between structural performance, manufacturing times and amount of material used. In addition, it would be ideal to integrate topological optimization tools to improve the performance of the proposals, as well as evaluate new and diverse geometric patterns extracted from nature. On the other hand, there are design issues that have not been studied, such as the perception of people and emotional response to each

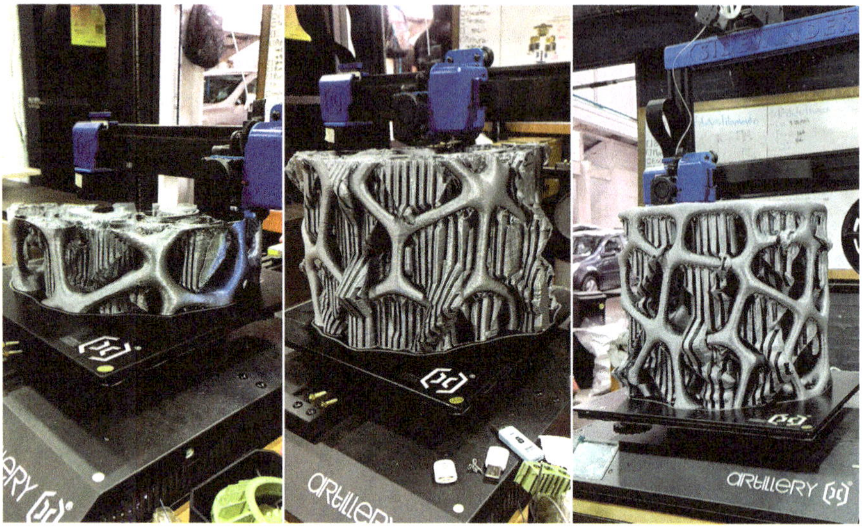

Fig. 32 3D printing process. (Source: the author. Reprinted with permission from Ref. [19], Copyright 2023, Springer Nature Singapore Pte Ltd)

Fig. 33 1:1 scale prototype, Voronoi pattern. (Source: the author. Reprinted with permission from Ref. [19], Copyright 2023, Springer Nature Singapore Pte Ltd)

morphological pattern and comfort and ergonomic aspects, all of which start new research to complement what has already been achieved. Finally, this experience and methodology could be applied to different design projects; with this, it would be valuable to improve the methodological proposal in a constant evolution based on experimentation and feedback [IBID].

Fig. 34 1:1 scale prototype, differential growth pattern. (Source: the author. Reprinted with permission from Ref. [19], Copyright 2023, Springer Nature Singapore Pte Ltd)

Fig. 35 1:1 scale prototype, branching pattern. (Source: the author. Reprinted with permission from Ref. [19], Copyright 2023, Springer Nature Singapore Pte Ltd)

2.3 Biodigital Columns Project

In recent decades, we have witnessed a growing expansion of digital applications in the field of architecture, which has increased the agency of architects to design and materialize proposals; they are willing to embrace computing to connect the virtual and the physical domain, linking architectural design with matter. It could be stated that with this paradigm shift, the boundaries that separate the conception of the idea

Fig. 36 Biodigital stool series, 1:1 scale prototype. (Source: the author. Reprinted with permission from Ref. [19], Copyright 2023, Springer Nature Singapore Pte Ltd)

Fig. 37 Biodigital stool series, 1:1 scale prototype. (Source: the author. Reprinted with permission from Ref. [19], Copyright 2023, Springer Nature Singapore Pte Ltd)

and its construction can be ignored; these architects are designing and fabricating more and more at the same time [20]. In this context of a progressive assimilation of digital technologies in the field of architecture, great transformations are expected, automation in construction will drastically change the standard processing systems, giving rise to disruptive technologies such as 3DConcrete printing (3DCP). This construction method has experienced great progress in the last decade due to its potential to break with some practices rooted in the conventional construction industry; this technique supported by robotic systems makes the construction process less dependent on labor. With the support of PD, it is possible to materialize complex geometries without using formwork (which are not feasible using conventional methods). These advantages reduce costs and accident risks and improve

Fig. 38 User test of differential growth pattern stool, 1:1 scale prototype. (Source: the author)

overall productivity [21], [22] and [23]. Pioneering companies such as Contour Crafting, Apis-Cor, Concrete Printing and D-Shape have developed different machine configurations and variations of this technique. Although the concrete 3D printing method is considered a promising technique, there are several factors that affect the performance of the process that are not fully studied. For example, printing performance depends largely on the properties of the fresh concrete, as well as its rheology and strength. Printing parameters, extrusion speed, nozzle type and time interval between layers also influence printing performance. The variables affect the mechanical properties of stamped concrete in the hardened state [24].

Research proposal was based on a biodigital architecture approach; the aim was oriented to design architectural components based on nature through a synergy between PD and 3DCP to solve local problems in Medellín, Colombia. This text shows the results of the first stage of research aimed at column design [25].

2.3.1 Methodology

An applied research project was proposed, using an *empirical-experimental method* with a multidisciplinary approach, a synergy of PD and 3DCP and a quantitative analysis of the information. In the *design based on nature* methods, *design looking to biology* [12], *organism level* [13], *biomimetic* and *parametric* project strategies [14] were considered. In the first stage related to the design of columns, the following methodological sequence is used. (1) Organism from nature, (2) parameterized morphological synthesis, (3) parameterized morphological transformations, (4) finite element analysis, (5) selection of alternatives, (6) detailed design, (7) 3D concrete printing of 1:1 scale prototype and (8) conclusions and proposals for improvement, an empirical, cyclical and iterative process [25] (Fig. 39).

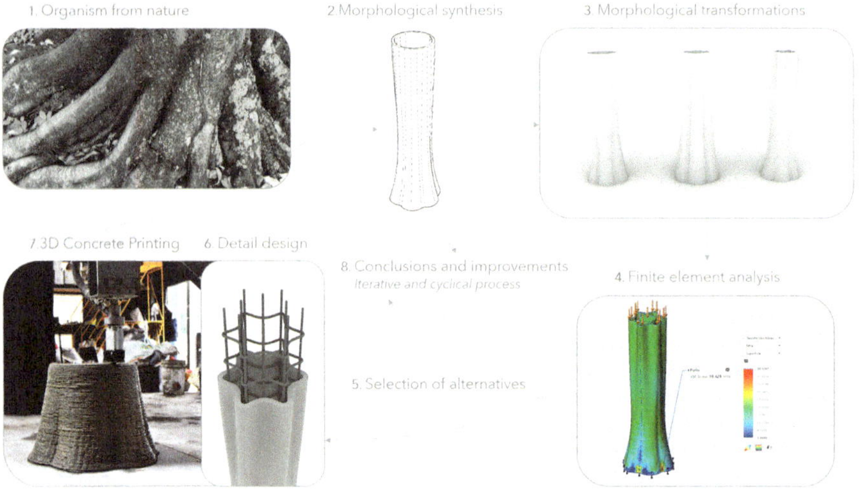

Fig. 39 Methodological sequence. (Source: the author. Reprinted with permission from Ref. [25], Copyright 2023, Springer Nature Singapore Pte Ltd)

Fig. 40 Trees of Medellín analyzed in the biodigital column project. (Source: the author. Reprinted with permission from Ref. [25], Copyright 2023, Springer Nature Singapore Pte Ltd)

A decision was taken to analyze tree morphologies for the column design; for this purpose, different individuals were observed in Medellín Colombia, Fig. 40. Basic morphological variables were studied, such as tree height, trunk diameter, root height and root length, as well as the relationships between height with trunk diameter and height with root length [26], Fig. 41.

The parametric digital tool Grasshopper© in the Rhinoceros© software was used for the morphological synthesis process; the geometric transformations aimed at the design of alternatives, and finally a detailed design of the columns was made. This work environment offers wide geometric freedoms that can then be modified in an unlimited way, making it possible to create families of products with common

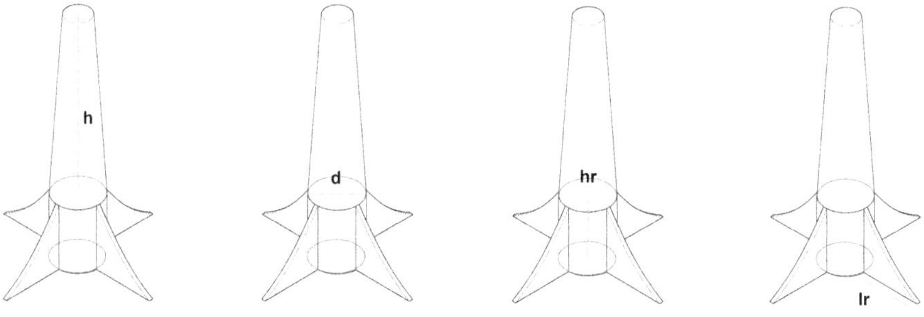

Fig. 41 Basic morphological variables and proportions used in the biodigital column project. (Source: drawing by the author, based on Ref. [26]. Reprinted with permission from Ref. [25], Copyright 2023, Springer Nature Singapore Pte Ltd)

Table 7 Finite element analysis parameters

Wall thickness	20 mm
Material	Concreto
Force	1×10^6 N
Mass	Kg
Effort	MPa

Source: the author. Reprinted with permission from Ref. [25], Copyright 2023, Springer Nature Singapore Pte Ltd

geometric features. Base dimensions were defined for digital modeling: 220 cm high and external diameter of 50 cm, in accordance with the Colombian construction standard. To evaluate the mechanical efficiency of the columns, a digital simulation of a compression test was carried out, using a finite element analysis with a linear elastic model in the Creo Parametric© software, which allows parameterized simulations. A load of 1,000,000 Newtons was applied to each column on the upper surface, and the base was completely restricted in the Y axis. Simulations were carried out on three models, model 1 a cylinder (as a reference) and models 2 and 3, based on tree trunk morphologies; these two models were compared considering the maximum stress in relation to their weight. In Table 7, parameters used for the simulation can be seen, and in Fig. 42, the models are used [25].

In concrete 3D printing, the large format printer from Constructora Conconcreto in Medellín, Colombia, was used, designed in 2016 by the engineers of that company, a unique machine in Latin America and a milestone for the construction sector in Colombia. The printer works by extrusion technique; it has an available printing volume of 6 m in the X axis, 3 m in the Y axis and 3 m in the Z axis; it has speeds between 3000 mm and 12,000 mm per minute, three nozzles of 12.7 mm, 15 mm and 25 mm in diameter. The concrete used is a mixture developed by this company and protected by industrial secret. In Table 8, it is possible to see the technical characteristics of this large format printer [25].

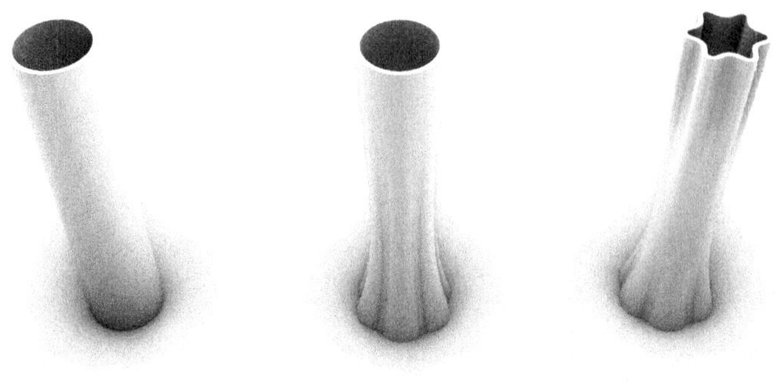

Fig. 42 Three models of biodigital column. (Source: the author. Reprinted with permission from Ref. [25], Copyright 2023, Springer Nature Singapore Pte Ltd)

Table 8 Concrete printer features

Technology	3D concrete printing
Model	Developed by the company in 2016
Printing area	6 m (X axis), 3 m (Y axis), and 3 m (Z axis)
Nozzle (diameter)	12.7 mm/15 mm/25.4 mm
Technology	Concrete extrusion

Source: Conconcreto company. Reprinted with permission from Ref. [25], Copyright 2023, Springer Nature Singapore Pte Ltd

2.3.2 Results

A morphological synthesis of different tree trunks was done; in Fig. 43, a geometric abstraction can be seen of one of the individuals observed in the city of Medellín. Then parameterized geometric transformations were made, changing the height and shape of the roots, the horizontal cross sections at the base and the upper area [25]; some of these transformations can be seen in Fig. 44.

Analyses of finite elements were carried out using the Creo Parametric© software. When comparing the results (Figs. 45, 46 and 47), it was observed that models 2 and 3, having a cross section that is not constant, tend to distribute in a non-homogeneous manner the forces along the column, finding the maximum forces at the base of the column. On the other hand, it is observed that the relationship between the maximum stress and the mass of the column is greater compared to model 1 (reference cylindrical geometry), Table 9. This information demonstrates that *nature-based morphologies* are more efficient structures under compression loads for this case [25].

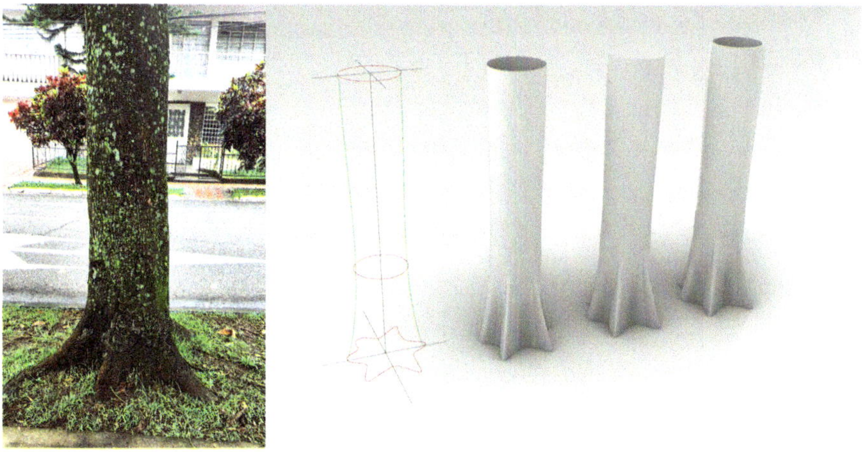

Fig. 43 Geometric abstraction of trees. (Source: the author. Reprinted with permission from Ref. [25], Copyright 2023, Springer Nature Singapore Pte Ltd)

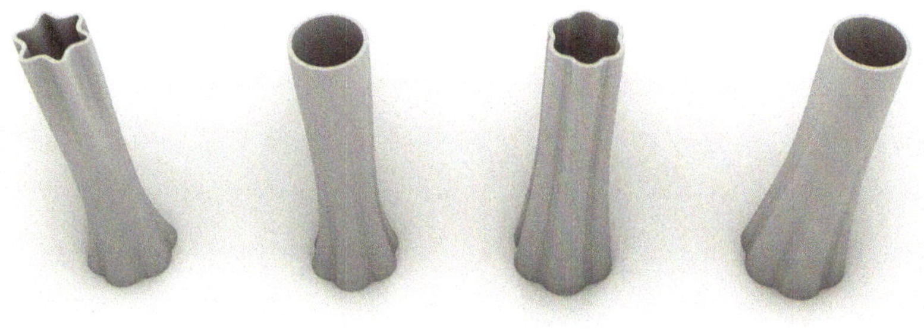

Fig. 44 Morphologic transformations. (Source: the author. Reprinted with permission from Ref. [25], Copyright 2023, Springer Nature Singapore Pte Ltd)

The detailed design was made in compliance with Standard NSR-10, the earthquake-resistant regulation of Colombia. Two construction alternatives are considered: the first is a prefabricated pillar that will be installed on-site and the second option is a hollow 3D printed pillar; it will be finished on-site. For both options, it is considered to pour a conventional concrete mixture inside, which includes sand, Portland cement, coarse aggregate (gravel) and water, reinforced with corrugated iron rods of ½″ diameter (vertical pieces) and ¾″ diameter (horizontal sections); the concrete mix for a column must have a compressive strength between 21 and 49 Mpa [25] (Figs. 48 and 49).

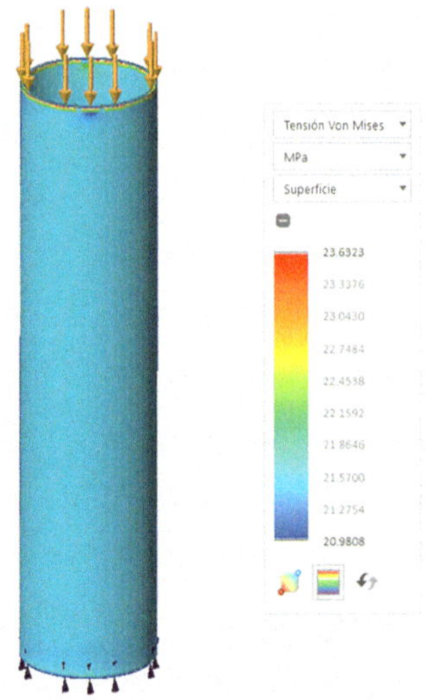

Fig. 45 Finite element analysis of cylindrical column (model 1). (Source: the author. Reprinted with permission from Ref. [25], Copyright 2023, Springer Nature Singapore Pte Ltd)

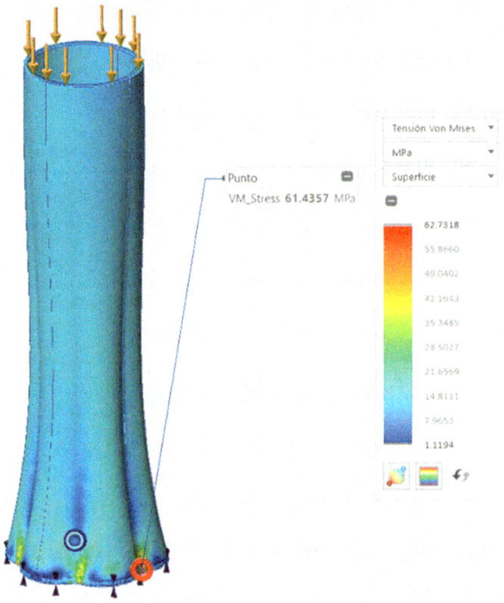

Fig. 46 Finite element analysis of cylindrical column (model 2). (Source: the author. Reprinted with permission from Ref. [25], Copyright 2023, Springer Nature Singapore Pte Ltd)

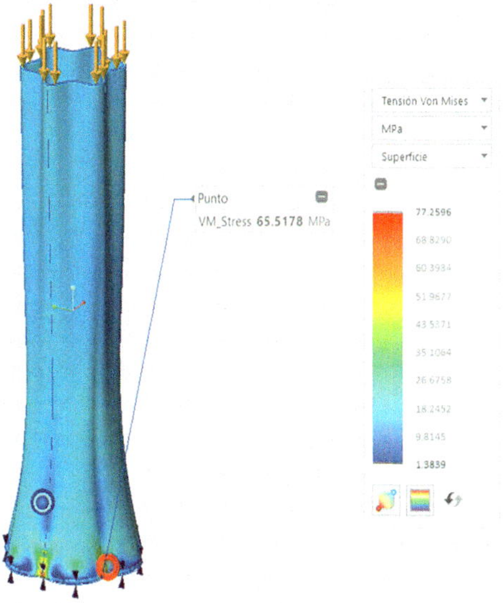

Fig. 47 Finite element analysis of cylindrical column (model 3). (Source: the author. Reprinted with permission from Ref. [25], Copyright 2023, Springer Nature Singapore Pte Ltd)

Table 9 Information of finite elements analysis

Model	Mass [kg]	Effort [MPa]	Effort/mass [MPa/Kg]
Model 1	328,86	23,6323	0,071861278
Model 2	299,15	74,3196	0,248435902
Model 3	310,49	61,5178	0,198131341

Source: the author. Reprinted with permission from Ref. [25], Copyright 2023, Springer Nature Singapore Pte Ltd

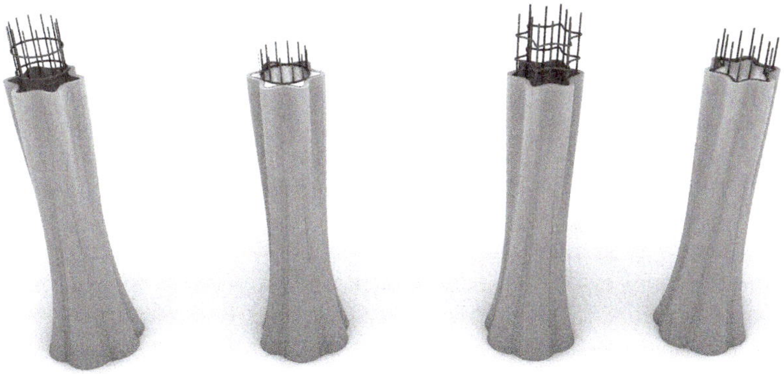

Fig. 48 Detailed design of biodigital columns. (Source: the author. Reprinted with permission from Ref. [25], Copyright 2023, Springer Nature Singapore Pte Ltd)

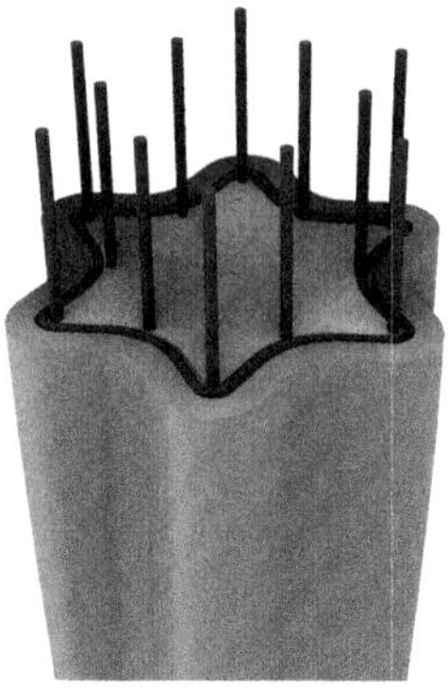

Fig. 49 Detailed design of biodigital column. (Source: the author. Reprinted with permission from Ref. [25], Copyright 2023, Springer Nature Singapore Pte Ltd)

Table 10 Parameters of 3D printing process

Nozzle	25.4 mm diameter
Speed	6500 mm/minutes
Material	Developed by Conconcreto company
Layer height	10 mm
Wall thickness	35 mm
Printing time	45 min

Source: Conconcreto company. Reprinted with permission from Ref. [25], Copyright 2023, Springer Nature Singapore Pte Ltd

The column prototype was fabricated through a large format printer at Constructora Conconcreto, in Medellín, Colombia. It was planned to print models 2 and 3, but due to time and resource limitations, only model 3 materialized. Table 10 shows the printing parameters used; Figs. 50, 51, 52, and 53 show the fabrication process, tests, errors and prototype. As can be seen in Fig. 51, the failures caused the prototype to collapse; this has been mainly attributed to the mixture of the material that must be sufficiently fluid and consolidate effectively, hardening layer by layer, in synergy with the printing speed and nozzle diameter, key aspects in this construction process [25].

2.3.3 Conclusions of Biodigital Columns Project

The first stage was oriented to the design of columns; the advantages of a *biodigital approach* were evident, and tree-based proposals were designed and fabricated with superior structural performance (compression stress) compared to the geometries commonly used in column design. According to the results obtained with the finite element analysis, this information was also useful to select the alternatives to fabricate. In the construction process, it was possible to fabricate complex geometries based on nature. It was concluded that a key factor is the

Fig. 50 3D concrete printing process. (Source: the author. Reprinted with permission from Ref. [25], Copyright 2023, Springer Nature Singapore Pte Ltd)

Fig. 51 3D concrete printing process, failure in material mixing. (Source: the author. Reprinted with permission from Ref. [25], Copyright 2023, Springer Nature Singapore Pte Ltd)

concrete mix, in balance with other variables such as printing speed, nozzle diameter and ambient temperature, among others. In this context, constant experimentation; trial and error; and doing and redoing have been essential. In the validations, due to cost and time limitations, it was not possible to finish the prototype of the printed pillar (with the pouring of concrete poured inside and reinforced with metal rods). Likewise, if it is finished, it is a very complex element to validate due to its dimensions and weight; this would make it difficult to do a compression test. On the other hand, since it is not included in a real architectural project, it would not be easy to verify its operation in relation to other structural

Fig. 52 3D concrete printing process. (Source: the author. Reprinted with permission from Ref. [25], Copyright 2023, Springer Nature Singapore Pte Ltd)

Fig. 53 Biodigital column prototype. (Source: the author. Reprinted with permission from [25], Copyright 2023, Springer Nature Singapore Pte Ltd)

elements, such as walls. Therefore, the only validations to test the structural performance of these proposals have been made with digital simulations with the finite element techniques. In the next stage of column design, it is necessary to continue doing finite element analysis, testing a greater number of morphological variations, making improved proposals, ideally with topological optimization tools and analyzing how the columns behave linked to other architectural elements, to make reduction scale prototypes, and testing them in compression tests will be useful. On the other hand, other architectural elements such as walls or envelopes will be designed based on elements from nature, and characteristics such as structural performance, lightness and multifunctionality, among others, will continue to be emphasized. Finally, it is expected that this research will be a milestone in the construction sector in Colombia in the future, showing and expanding new ways of approaching architectural design, through the confluence between nature, PD and 3DCP, erasing the limits between digital and physical environments, with a multidisciplinary approach, and consequently generating *dynamic-parameterized*, efficient design proposals for the local context [25].

3 Final Conclusions

The three projects shown in this text have different scales and complexities, designed for different contexts and users; however, all have a common goal: to materialize morphologies based on nature through the integration of parametric design and digital fabrication technologies in the context of product design projects. It can be said that, in the three cases presented, the synergy between PD and digital fabrication (emphasis on AM) allowed to achieve the goals of each project and highlighted the advantages of these technologies. In the parametric design, it was possible to generate automatically geometries based on nature and their transformations through algorithms; this is a dynamic system in which some variables are defined and manipulated to generate a family of editable geometries with infinite morphological possibilities, furthermore with the possibility of customizing. This type of algorithmic tools can respond more efficiently to requirements than the conventional computer-aided design (CAD) methods. AM allowed to fabricate complex geometries with an adjustable definition (including high definition) and without molds; technologies such as FDM, DLA, 3DCP and various materials were used. It is possible to conclude that, as a result of the combination of these technologies, complex morphologies based on nature can be generated effectively, with wide geometric and manufacturing freedoms to obtain functional objects; the results can have a performance appropriate for their function and context. Furthermore, the importance of practical work and experimentation was evident, continuous and dynamic learning in a cyclical process of trial, error and reconfiguration in multidisciplinary work teams. On the other hand, all the projects presented were developed in Medellín in Colombia; this shows that it is possible to implement, use and assimilate *cutting-edge technologies* in a Latin American context, in other words, using global digital technologies for a local context, a place with fewer resources and access to these technologies than the United States of America, Europe, Asia or other developed countries. Along with this, the importance of the link between universities, companies and government entities is evident; these alliances have made possible to obtain resources, obtain access to specialized technical advice and carry out feasibility evaluations and validations, among others.

3.1 Limitations

In the PD, the generation of the modules (bioinspired texture project) was not a difficult task, and many times the algorithms were simple to do, but in the case of morphological transformations, in several cases, the algorithms were complex; it was the same in the *growing morphologies pattern* (stool project). In these cases, the files were very large, and the changes in the parameters were not continuous to originate the transformations, so the need to have a user with an advanced level of Grasshopper® and a high-performance computer became evident. In regard to the fabrication of prototypes through digital technologies, in some cases, the processes were very extensive and a lot of material was used. This is the case of the stool based on the Voronoi pattern where the prototype took 10.7 days to be fabricated, and 12 kg of PLA material were used, and it was also necessary to use custom supports (branching type). In the case of bioinspired parametric textures, a problem was difficult access and high cost of high-definition 3D printing technologies, such as digital light processing (DLP) technology. In the case of textures manufactured by CNC milling, there were geometric limitations according to the axes and ranges of freedom of the machine, the roughing of material results in a considerable expenditure of energy and tools. These situations had an impact and consequences on the use of resources, and finally a high cost of the prototypes, in response to the problem alliances with companies were important to be able to solve it. In other projects, it was difficult to make validations, for example, in the case of the biodigital column, due to cost and time limitations, it was not possible to finish the prototype of the printed part (this should have concrete poured inside and reinforced with metal rods). Therefore, the only validations to test the structural performance of these proposals have been carried out with digital simulations with finite element techniques.

Finally, a limitation was identified in aspects of communication in multidisciplinary teams. This situation could be seen at different levels, for example, with respect to the use of certain words or concepts that are different between one professional and the other, work methods, sometimes the type of activities for each stage were not clear, etc.. All of this was solved through a constant dialogue, working to build a common language and reaching agreements.

3.2 Future Work and Prospects

The projects presented show the advantages of the biodigital proposal; this can open new *methodological-conceptual* paths to develop design projects. Regarding methodological approaches these investigations used the perspectives of *design looking to biology* (top-down approach) and *biology-influencing design* (bottom-up approach) [12], the *organism level* according [13], finally *biomimetic, biomorphic and parametric strategies* were used [14]. In this sense, it would be valuable to explore new approaches, for example, to propose projects based on the *behavioral* and *ecosystem level* [13], to continue using the perspectives of *design looking to biology* and *biology-influencing design* [12] and to explore *morphogenetic*, *genetic* and *morphodynamic* project strategies [14]. With these new approaches, it would be possible to think prospectively in *responsive products or systems*; they could react to external stimuli (temperature, pressure, wind, etc.) and have collective behavior and work collaboratively, as occurs with flocks of birds or schools of fish in nature (emergent behavior). Furthermore, these systems could improve their morphological, functional and behavioral characteristics to evolve over time autonomously; with this, it is possible that *biodigital evolutionary systems* may be born to respond efficiently to the functional requirements and stimuli of a dynamic environment.

These ideas could be possible as a result of an intersection of different technologies, materials and processes, such as biomaterials (including living and inert organic elements) and nanomaterials (responsive materials), advanced robotic systems with multiple sensors, 4D additive manufacturing (the parts or systems react to external stimuli once manufactured) and digital algorithms enhanced with artificial intelligence, among others. It would be ideal to include a systemic and sustainable vision in new projects to contribute to achieving ecological balance on the planet.

Finally, local problem solving is expected to continue in different places in the world, through the use of *cutting-edge technologies* as a result of a collaborative synergy between universities, companies, governmental and nongovernmental institutions, at a national and international level.

References

1. Tee, Y. L., Maconachie, T., Pille, P., Leary, M., Do, T., & Tran, P. (2021). From nature to additive manufacturing: Biomimicry of porcupine quill. *Materials and Design, 210*, 110041. https://doi.org/10.1016/j.matdes.2021.110041
2. Chirazi, J., Wanieck, K., Fayemi, P. E., Zollfrank, C., & Jacobs, S. (2019). What do we learn from good practices of biologically inspired design in innovation? *Applied Sciences (Switzerland), 9*(4), 1–16. https://doi.org/10.3390/app9040650
3. Kapsali, V. (2022). All things bio: A conceptual domain-based approach to mapping practice within the landscape of biologically informed disciplines. *The Design Journal, 25*(4), 516–536. https://doi.org/10.1080/14606925.2022.2058449
4. Iouguina, A., Dawson, J. W., Hallgrimsson, B., & Smart, G. (2014). Biologically informed disciplines: A comparative analysis of bionics, biomimetics, biomimicry, and bio-inspiration among others. *International Journal of Design & Nature and Ecodynamics, 9*(3), 197–205. https://doi.org/10.2495/DNE-V9-N3-197-205
5. Oxman, R. (2006). Theory and design in the first digital age. *Design Studies, 27*(3), 229–265. https://doi.org/10.1016/j.destud.2005.11.002
6. Burry, M. (2011). *Scripting cultures: Architectural design and programming*. Wiley.
7. Oxman, R. (2017). Thinking difference: Theories and models of parametric design thinking. *Design Studies, 52*, 4–39. https://doi.org/10.1016/j.destud.2017.06.001
8. ASTM International. (2013). *F2792-12a – Standard terminology for additive manufacturing technologies* (pp. 10–12). Rapid Manufacturing Association. https://doi.org/10.1520/F2792-12A.2
9. Attaran, M. (2017). The rise of 3-D printing: The advantages of additive manufacturing over traditional manufacturing. *Business Horizons, 60*(5), 677–688. https://doi.org/10.1016/j.bushor.2017.05.011
10. Trovato, M., & Cicconi, P. (2023). Design tools for metal additive manufacturing: A critical and perspective overview. *Procedia CIRP, 119*, 1084–1090. https://doi.org/10.1016/j.procir.2023.03.151
11. Estévez, A. T. (2020). *4th international conference for biodigital architecture & genetics* (A. T. Estévez, Ed., 1st ed.). Bubok Publishing S.L. https://www.bubok.es/autores/biodigital?fbclid=IwAR2n2ey4uYtxStxWHd93GBKGGqZgX2k-oe8pH6VrnieyrmkVv-3dKbzI0w4
12. Aziz, M. S., & El Sherif, A. Y. (2016). Biomimicry as an approach for bio-inspired structure with the aid of computation. *Alexandria Engineering Journal, 55*(1), 707–714. https://doi.org/10.1016/j.aej.2015.10.015
13. Zari, M. P. (2006, April). *Biomimetic approaches to architectural design*.

14. Estévez, A. T. (2015). *Arquitectura biodigital y genética: escritos/Biodigital architecture genetics: writings*. ESARQ (Escola Técnica Superior d'Arquitectura, Universitat Internacional de Catalunya).
15. Malshe, A. P., Bapat, S., Rajurkar, K. P., & Haitjema, H. (2018). Bio-inspired textures for functional applications. *CIRP Annals, 67*(2), 627–650. https://doi.org/10.1016/j.cirp.2018.05.001
16. Torreblanca-Díaz, D. A., & Patiño Mazo, E. (2021). *Texturas bioinspiradas y fabricación digital*. Editorial Universidad Pontificia Bolivariana. https://doi.org/10.18566/978-958-764-957-4
17. Torreblanca Díaz, D. A. (2018). *Aplicaciones de texturas paramétricas bioinspiradas en proyectos de diseño académico* (pp. 997–1003). Blucher. https://doi.org/10.5151/sigradi2018-1370. ISSN 2318-6968.
18. Stevens, P. (1987). *Patrones y pautas en la naturaleza*. Salvat Editores.
19. Torreblanca-Díaz, D. A., & Velásquez Peña, J. P. (2023). Diseño de productos biodigitales mediante tecnologías de fabricación aditiva: Proyecto Taburete. In *XXVI Congreso Internacional de la Sociedad Iberoamericana de Gráfica Digital* (pp. 1189–1200). Blucher. https://doi.org/10.5151/sigradi2022-sigradi2022_116. ISSN 2318-6968.
20. Carpo, M. (2011). *The alphabet and the algorithm*. MIT Press.
21. Ahmed, G. H. (2023). A review of "3D concrete printing": Materials and process characterization, economic considerations and environmental sustainability. *Journal of Building Engineering, 66*, 105863. https://doi.org/10.1016/j.jobe.2023.105863
22. Nguyen-Van, V., Li, S., Liu, J., Nguyen, K., & Tran, P. (2023). Modelling of 3D concrete printing process: A perspective on material and structural simulations. In *Additive manufacturing* (Vol. 61). Elsevier B.V. https://doi.org/10.1016/j.addma.2022.103333
23. Rehman, A. U., Birru, B. M., & Kim, J.-H. (2023). Set-on-demand 3D Concrete Printing (3DCP) construction and potential outcome of shotcrete accelerators on its hardened properties. *Case Studies in Construction Materials, 18*, e01955. https://doi.org/10.1016/j.cscm.2023.e01955
24. Joh, C., Lee, J., Bui, T. Q., Park, J., & Yang, I. H. (2020). Buildability and mechanical properties of 3d printed concrete. *Materials, 13*(21), 1–24. https://doi.org/10.3390/ma13214919
25. Torreblanca-Díaz, D. A., Montoya Tobón, A. F., Carmona Duque, A. M., Saldarriaga Londoño, J. D., Ossa Henao, E. A., Camargo Piedrahita, R. A., & Gil Durán, S. (2013). Arquitectura Biodigital a través de Impresión 3D de Concreto: Caso en la Ciudad de Medellín, Colombia. In A. T. Estévez (Ed.), *5th international conference for biodigital architecture & genetics* (pp. 274–285). iBAG-UIC Barcelona. ISBN: 978-84-127604-7-7
26. Nölke, N., Fehrmann, L., Jaya, I. N. S., Tiryana, T., Seidel, D., & Kleinn, C. (2015). On the geometry and allometry of big-buttressed trees – A challenge for forest monitoring: new insights from 3D-modeling with terrestrial laser scanning. *iForest, 8*(October 2015), 574–581. https://doi.org/10.3832/ifor1449-007

David A. Torreblanca-Díaz, Chilean product designer, master's degree in design technologies. His research projects are focused on *biodigital design* through a transdisciplinary approach. He has experience in product design, teaching and research. He is an international lecturer, workshop instructor and has been part of international exhibitions. He is currently an associate professor at the Universidad Pontificia Bolivariana (Medellín, Colombia) and a PhD student in the *Institute for Biodigital Architecture and Genetics*, at the Universitat Internacional de Catalunya (Barcelona, Spain).

Method and Processes for Abstraction of Natural Microstructures for New Product Development

Antônio Roberto Miranda de Oliveira and Amilton José Vieira Arruda

1 Introduction

Biomimicry has emerged as an innovative approach to addressing technical challenges, combining biology with technology and design to create nature-inspired solutions [1]. Biological structures, such as those found in fruits, have evolved over millions of years to achieve high structural efficiency, providing protection against mechanical damage and negative environmental influences. The reductive application of biomimicry, aimed at finding solutions without necessarily creating an ecologically correct product, has been seen as a complement to the existing toolkit of the designer [2].

Over the past few decades, biomimicry has been recognized as a rich and still unexplored source of inspiration for innovation in various areas, including product development. Nature, through evolutionary processes, has developed efficient strategies and mechanisms to solve complex challenges related to the survival and reproduction of species. These mechanisms, when studied and understood, can be abstracted and applied to solve human challenges across various disciplines, including engineering, medicine, and, of course, product design [3].

The application of biomimicry in product development is not limited to merely imitating natural forms but extends to abstracting and applying biological principles and strategies to create innovative and sustainable solutions. For instance, nature has been a source of inspiration for the development of advanced materials and

A. R. M. de Oliveira (✉)
Federal University of Pernambuco, Recife, PE, Brazil
e-mail: antonio.rmoliveira@ufpe.br

A. J. V. Arruda
BioDesign Lab, Federal University of Pernambuco, Recife, PE, Brazil
e-mail: amilton.arruda@ufpe.br

A. J. V. Arruda, F. L. Palombini (eds.), *Biomimetics, Biodesign and Bionics*,
Environmental Footprints and Eco-design of Products and Processes,
https://doi.org/10.1007/978-3-031-51311-4_6

optimized structures, which are lightweight, robust, and functional. The ability to observe, understand, and abstract these characteristics from biological entities can open new horizons in product design and manufacturing, allowing for the creation of solutions that are not only efficient but also ecologically sustainable.

The relevance of biomimicry in product development can also be viewed through the lens of sustainability. In a world where resources are limited and the pursuit of ecologically viable solutions is imperative, biomimicry provides a pathway for the development of products and technologies that not only meet human needs but are also in harmony with the environment. Biological strategies often incorporate energy efficiency, optimized use of materials, and adaptability, which are crucial principles for the development of sustainable products [4].

The exploration of microscopic structures in tropical fruits, as proposed in this study, represents a fascinating foray into the universe of biomimicry, where natural cellular structures are studied with the aim of abstracting their fundamental principles and applying them in the development of new products. Through the observation and analysis of these structures, we can not only understand the mechanisms by which nature solves structural and functional challenges but also apply these findings in a practical and innovative way in product design. In nature, we often find simple or multifunctional systems that are associated with other systems which, due to their development of various functions, are assigned different tasks. Fruits serve as an example, developing various functionalities and becoming true natural packaging developed by nature that presents various strategies. Thus, this research will focus on one of its main functions, protection. In fruits, we find in their peels and seeds strategies for protection against mechanical damage, ultraviolet radiation, and strategies that can also attract animals that are vectors for seed dispersal [5–8].

Exploring the nature of natural packaging means observing how life envelops, protects, contains, preserves, transports, and communicates its strategies [9]. Thus, applying these solutions to industrial problems leads to the following research questions: Which structures present in natural packaging allow for better use of space and material for improved storage and transport? Which fruits have a hierarchical structure developed for protection against impacts and collisions, providing greater protection to what is being transported/stored?

The need to develop new products and technologies that are sustainable and efficient has directed researchers and designers to explore innovative strategies in product development. Biomimicry, by interceding between design issues, offers a unique perspective, allowing designers to explore solutions that nature has already optimized over millions of years [1]. Furthermore, the integration of noninvasive observation technologies and 3D prototyping techniques provides a new dimension in the practical application of biomimicry, enabling the materialization of bioinspired structures that can be applied in the development of new products.

2 State of Art

In recent years, biomimicry has been increasingly utilized in the field of applied research for the development of new products and materials. It has high innovation potential and a great possibility of offering new bioinspired products and sustainable production chains [10]. Technological advancement has brought enormous possibilities, especially in the field of design, for the development of new techniques and devices that are linked to the potential of new materials, interaction, human factors, perception, cognition, and user experience. This progress has brought profound consequences in various fields of science, both intellectual and practical [4].

Over millions of years, nature has been developing high-performance materials and structures, providing valuable sources of inspiration for the design of structural materials with the aim of enhancing or developing new products, given the variety of excellent properties found in nature. Aiming to improve traditional materials and structures for energy absorption, scientists, engineers, and designers seek to learn from biological organisms that have resilient structures that have evolved optimally over generations [11]. Thus, natural biology offers excellent structures with remarkable absorption capacity, energy dissipation, and high impact resistance.

The term biomimicry (from bios, meaning life, and mimesis, meaning to imitate), coined by Janine Benyus in her 1997 book, *Biomimicry: Innovation Inspired by Nature*, has generally been defined as "the examination of nature, its models, systems, processes, and elements to emulate or take inspiration from in order to solve human problems." Solving human problems has been a significant driving force in this research field. Biomimicry as a scientific concept has been capturing the imagination of researchers far beyond the objectives of producing better drugs, artificial organs, or new bioinspired products, but rather as a foundation and inspiration for new innovative ways to design and understand matter on a deeper level [12].

Never before has biology been such a significant inspiration for innovations. Today, the vast majority of engineers and materials scientists are aware of this source of ideas for technology development. However, since the functionality of biological systems is often based on their extreme complexity, there are not many examples of ideas successfully implemented in modern technology. For this reason, bioinspiration, biomimicry, and bioreplication require not only intense experimental research in the field of biology but also an abstraction strategy to extract essential features responsible for a specific function of a biological system. Therefore, before transferring biological principles to technology, we must recognize and distinguish these principles amid enormous biological complexity. The first approach is classical biology, in which a specific functional system is studied comparatively in various organisms. In this case, we can potentially recognize the same functional solutions that have evolved several times, independently, in the evolution of different groups of organisms. Moreover, we obtain information about the diversity of solutions. This comparative approach is time-consuming but can be very effective for bioinspiration, leading to biomimicry and bioreplication [13].

The fundamental principle of biomimicry is to use the natural biological world as a source of inspiration and as a guide in the development of new materials. A detailed study of the systems and organisms present in natural systems, which can be used as models in the science and engineering of new materials, is necessary. However, it must be recognized that such a study should focus on elucidating the relationships between structure and function in natural systems, in order to apply the results of this study in engineering [14].

2.1 Natural Fruit Structures

Fruits, in their diversity, exhibit a variety of microscopic structures that have been the subject of study in the development of innovative materials, especially in the context of impact absorption and damping. The hierarchical structure of the pummelo peel, for example, is known for its impact resistance, attributed to the organization of the fruit peel, termed pericarp. This organization can be transferred to engineering materials to develop enhanced damping and impact properties [15].

Another notable example is the fibrous structure of the coconut, where the orientation of the fibers can enhance energy absorption capacity, influencing the stress wave propagation to protect the coconut's endocarp [16]. The hierarchical structure and compressive behavior of *Fomes fomentarius* fruit bodies also suggest an adaptive role to natural dry/wet conditions, with drying leading to plateau stresses about seven times higher and extensive damage, but rehydration restoring the properties of the wet state [17].

The exploration of natural fruit structures for the development of innovative materials is not limited to impact absorption but also extends to other applications, such as the encapsulation of phenolic extracts from food waste, where natural fruit fibers have shown promising results as a support material for encapsulation [18].

These studies indicate that biomimicry, inspired by the natural structures of fruits, can offer innovative solutions for the development of materials with enhanced mechanical properties, such as impact absorption and damping, which can be applied in various industries, including automotive, aerospace, and packaging. The continuous and in-depth exploration of these natural structures, along with the application of advanced technologies such as scanning electron microscopy and X-ray computed tomography, may unveil more secrets about the strategies nature uses to create efficient and sustainable materials.

Material engineering has extensively explored hierarchical and gradient structures, drawing inspiration from natural systems to develop materials with superior mechanical, thermal, and acoustic properties. Gradient structures in engineering materials produce intrinsic synergistic strengthening, which can provide an innovative strategy for designing material structures with superior properties, exploring principles from both mechanics and material science [19].

Hierarchical architecture is a common feature in many natural materials and has been used to access unique mechanical properties that are unattainable in traditional materials [20]. For instance, the structure of the Menger sponge, a mathematical fractal form, was explored to understand how different combinations of individual architectures can be stacked to construct hierarchical structures, allowing the engineering of the directional dependence of the mechanical response [21].

Hierarchical structure in materials has also been explored in contexts beyond mechanics, such as in environmental chemistry and technology, where hierarchical nanostructures open many possibilities for material design and device fabrication [22]. Furthermore, the analysis of materials with a hierarchical structure, like proteins, can serve as prototypes for future lightweight structural materials, exposing the systematic benefits of hierarchical structuring in nanoscale network structures [23].

The ongoing exploration of hierarchical and gradient structures, inspired by natural systems and supported by robust theoretical models, promises to pave new pathways for the development of materials with exceptional properties, which can be customized for a wide range of applications across various engineering and science disciplines.

Modeling and simulation of bioinspired structures have been a prominent research area, exploring methods and techniques to replicate and better understand biological systems and their applications in engineering and design. The simulation of complex structures, for example, can be conducted by placing the modeling process in the context of overall research planning, operationalizing the synthetic description into smaller models specified at three scalar hierarchical levels, and using model evaluation to establish the veracity of the links between the implemented model, the model design specification, and the synthesis [24].

Modeling of bioinspired structures has also been applied to create constructions that serve as important research tools in experimental biology, allowing the creation of model organisms that can be easily manipulated to fill a desired parameter space of structural and functional repertoires [25]. Modeling and simulation have also been utilized to explore and understand the effective response of cellular materials with cubic anisotropy and structures made from them, enabling the simulation of structures, parts, and components that consist of or contain such cellular materials [26]. The application of simulations in bioinspired structures has also been explored in creating materials with programmable morphologies and movements, demonstrating 3D structures with various morphologies, including bioinspired structures with programmed sequential movements [27]. Modeling has also been applied to understand the difference between different polymer models that simulate the behavior of biological structures, based on their reaction to a given mechanical factor [28].

The simulation of bioinspired structures has also been explored in materials engineering, where biomimetic modeling can simulate the fidelity of analogous interior osteostructural microarchitecture [29]. Modeling and simulation of bioinspired structures, therefore, provide a robust platform to explore, understand, and apply biological principles in design and engineering, opening new possibilities for the development of materials and structures with superior properties and functionalities. The continuous integration of these techniques with practical and theoretical experiments promises significant advances in the application of bioinspired principles across various engineering and science disciplines.

The search for materials that combine robustness, self-recovery, and self-healing has been an area of intense research, especially when it comes to elastomers that exhibit superior shape recovery, drawing inspiration from bioinspired multiphase design [30]. Bioinspiration is not limited only to the conception of materials but also extends to the creation of manufacturing techniques that are both effective and ecologically sound, enabling large-scale production and application in various material systems, paving an advanced path to build structural materials with high mechanical performance in an efficient manner for practical applications [31].

The exploration of bioinspired materials has proven to be an effective strategy for developing new materials with self-healing and reprocessable properties, as well as enhancing both stiffness and toughness [32]. The structural complexity of biogenic materials also offers opportunities for practical applications, where structural diversity can be explored to develop new materials with properties comparable to biological ones, providing enhanced mechanical performance [33].

The integration of science and art through the categorization of structure-property relationships presents a new paradigm for creating new bioinspired materials, translating structures and mechanisms from distinct hierarchical systems in the context of the limited number of building blocks that universally govern these systems [34]. Moreover, the exploration of bioinspired materials for energy-relevant conversions has been a significant research area, where the benefits of bioinspired systems are combined with practical technological requirements to achieve the maturity of these materials [35]. The application of bioinspired strategies is not limited only to the creation of new materials but also to the conception of manufacturing and design strategies that can be applied across a variety of material systems, providing an advanced reinforcement route for constructing structural materials with high mechanical performance in an efficient manner for practical applications. The continuous exploration of bioinspired strategies and mechanisms, along with the integration of technological advancements, can pave new pathways for the development of advanced materials with unprecedented properties and functionalities, providing innovative solutions to practical and industrial challenges.

Nature, with its extraordinary diversity, offers a seemingly "infinite" array of possibilities for the development of innovative products and materials. This statement encompasses biomimetic materials that range from micro- and nanoscale dimensions to studies at the molecular level. In nature, we find principles that need to be understood and interpreted with a systematic character, in order to establish analogies for the development of bioinspired designs. In this way, these principles can be applied in solving design problems [36]. The exploration of natural concepts and materials can take various forms, from using actual biomolecules for human applications, like "prospecting" the rainforest for substances with pharmaceutical potential, to manufacturing new materials whose structural characteristics and hierarchical organization are inspired by nature. Bioinspiration encompasses and discusses examples of systems, emphasizing new functionalities that incorporate the integration of biological concepts on one side and the world of material science on the other. As such, this multidisciplinary field, primarily based on biology, nanotechnology, material science, and related disciplines, continues to produce new materials, processes, and concepts used in a variety of applications [12]. For this, it is essential to understand the biological phenomenon, natural biological materials, and the processes involved in their natural production and thus develop biofabrication or bioinspired manufacturing approaches. Understanding the functions provided by species and processes found in living nature can guide us in producing nanomaterials, nanodevices, and processes with desirable functionalities. The materials found in nature possess numerous intriguing characteristics, such as complexity, miniaturization, hierarchical structure, hybridization, resilience, and adaptability. A multidisciplinary approach is needed to elucidate the underlying components and construction principles determined by evolution, in order to provide more reliable, efficient, and ecologically correct materials [37].

Parametric design involves a computational method for designing a system of specified functions that will guide the creation of the object, allowing the generation of a wide range of geometric variations. Parametric design starts with a base model and uses parameters to control the geometry, enabling the creation of various design possibilities through a planned set of rules and algorithms. Parametric modeling is a modeling process that can be simulated on a computer, allowing a modeling approach similar to the genetic development studied in biology.

According to [38], interest in natural design also draws attention to biological issues related to the life of organisms and their gradual and continuous remodeling, which has parallels with current additive formation processes. In these processes, the structure starts with a straight line and then conforms and adapts, while in nature, evolution occurs from a spiral, allowing the development of creatures in which linearity exists only as a conceptual abstraction. However, by focusing on design issues, designers are successfully able to suggest similarities with the evolutionary strategies used by biological organisms.

The shapes and principles used in parametric design are derived, in some cases, from the transposition of growth logics studied and translated by biologists, mathematicians, and engineers into ideal models and algorithms recurrent in nature. This transfer is based on strategic analogies but does not always follow an effective biological-mathematical scientific basis. It is fundamental that designers interested in the dimension of parametric design understand that the use of biology-inspired algorithms allows for achieving not only aesthetically pleasing and complex morphologies but also structural optimization, hierarchization, energy conservation, resilience, synergistic collaboration, adaptability, and flexibility, which provide the basis for the formation of these morphologies in nature and can be highly relevant in the project [38].

Sustainability is a growing concern in various areas of science and technology, and biomimicry offers an innovative approach to developing sustainable solutions in various practical applications. Bioinspired structures not only mimic nature in terms of functionality but can also incorporate principles of sustainability intrinsic to natural systems [39]. indicate that bioinspired design principles derived from ecosystems can lead to both environmentally and economically sustainable industrial resource networks, serving as guidelines for designing sustainable industrial networks. This suggests that the analysis and abstraction of natural structures can provide valuable insights into the development of industrial networks and systems that align with sustainability principles. A study by [40] highlights that bioinspired structures offer an innovative pathway for cooling buildings, presenting great application prospects in zero-energy buildings. This suggests that incorporating bioinspired structures into architectural design can contribute to creating more energy-efficient and sustainable buildings [41]. highlight that the use of bioinspired structural complexity can avoid the need for unnecessary chemical diversity, facilitating recycling, and thus promoting a more sustainable materials economy. This principle can be fundamental for the development of new materials and products that not only mimic the functionality of natural structures but also incorporate sustainable production and disposal practices.

Reference [42] emphasize that, despite current challenges, bioinspired materials have become an important part of promoting innovations and advancements in the modern materials industry. This suggests that the research and development of bioinspired structures can play a crucial role in promoting innovation in sustainable materials and technologies [43] present an approach to assess the sustainability of structures in autonomous single-family dwellings, allowing solutions to be prioritized to ensure their overall sustainability. This indicates that bioinspired structures can be integrated into urban development practices to promote the construction of more sustainable housing and infrastructure. The integration of bioinspired structures with sustainability practices and principles offers a promising path for the development of technologies, materials, and practices that are not only efficient and effective but also environmentally responsible and viable in the long term. Through the analysis and application of structures found in nature, researchers and developers can create solutions that meet current needs without compromising the ability of future generations to meet their own needs.

3 Methodological Steps/Activity/Method and Tools

In this research, four phases were developed with their respective methodological steps, addressing the main activities, theoretical references, as well as methods and tools necessary for the empirical development of the research. The phases of this research are integrated into the steps of the diagrams that make up the Biomimicry DesignLens. This methodological approach was developed by the Biomimicry Institute 3.8 and includes Biomimicry Design Thinking. The details of this approach are framed in the stages and phases of the Biology to Design diagram, which is a flexible methodology for creative processes with the application of models from nature in creating innovative bio-inspired solutions [44] presents four phases of the biomimetic process that describe the position of teaching it to developers of products and services, processes, structures, and systems designed for sustainable futures and potential solutions to human problems. The four phases of the biomimetic design thinking process to address challenges using biology are described by the author.

3.1 Stage 1. Scope Definition: Input

The process begins with the contextualization of the design problem, describing the chosen design challenge/problem and the context in which the issue arises. A potential design solution is presented, listing the specific function that this solution must fulfill, based on an expected outcome and the most relevant life principles for the specific type of project challenge.

- **Problem-Driven: Natural Structures for Energy and Impact Absorption (Research Question):** Can the structural analysis and transposition of biological principles, which allow certain fruits to resist impacts and collisions during falls in various contexts, be utilized as a reference in the development of bio-inspired structures?
- **Continuous Dialogue:** The dialogue transpires continuously throughout all stages of this chapter's development. The Biology to Design approach was selected due to its didactic method that aligns with the bottom-up type of project approach, facilitating the attainment of objectives given the presented problem. Biomimicry thinking comprises four developmental areas: scope definition, discovery, creation, and evaluation. Adhering to the steps of the Biology to Design method and its specific stages, this research introduces the necessary requirements and successfully integrates with life's strategies and principles in the field of bio-inspired projects.

3.2 Stage 2. Discovery: Biological Identification and Understanding Strategy Principles

This begins with an investigation based on the function of the design, starting from a biology-based research question. A very useful tool for addressing questions in the field of biological research is the biomimetic taxonomy, which especially assists non-biologists in finding the correct terminology to research specific functions in scientific literature.

- **A Systematic Literature Review Was Conducted:** A Systematic Bibliographic Review (SBR) was utilized, conducted according to the simplified model proposed by [45] for theoretical foundation. Through this review, it was possible to map and process data, synthesizing them for the formulation of the theoretical framework of the thesis. In this way, a set of scientific articles was evaluated with the intention of identifying the state of the art in the respective theoretical fields in recent years for the development of the scientific base and listing the main authors cited in these collected studies to substantiate the theoretical construction.

Plant anatomy studies cells and plant tissues, observing internal structures in plant organisms. The precise observation of biological materials under an optical microscope involves a series of preliminary procedures, described below. These procedures are often referred to as histology techniques, which study cells and plant tissues. Plant anatomy analyses aim to prepare samples for study under light microscopy. Microscopy is generally performed by transmitted light, meaning that light must pass through the object being examined. Therefore, it is necessary to obtain plant tissue fragments that will be collected on very thin and transparent slides. In this context, samples from the median (equatorial) region of the fruits were collected for anatomical analyses. The collected material was fixed in FAA50 [46] for 48 h and subsequently preserved in 70% ethanol. The samples preserved in 70% ethanol were subjected to post-dehydration in a 50–100% ethanol/butanol series, infiltrated and embedded in histological paraffin, and then sectioned transversely and longitudinally with the aid of the rotary microtome (Leica RM2125RT), with a thickness of 10 μm. The obtained sections were deparaffinized, stained with 1% Alcian blue and safranin, both in 50% ethanol, and permanently mounted in Canada balsam [47].

A plant anatomy study was conducted using optical microscopy, which reveals the structures of cellular tissues and their distributions in the fruit pericarp, through the use of a light beam. Plant anatomy is a discipline that studies the cells, tissues, and internal structures of plant organisms. To perform this analysis, it is necessary to use an optical microscope and follow a series of preliminary procedures. These procedures, known as histology techniques, allow for the detailed study of plant cells and tissues. The aim of plant anatomy analysis is to prepare samples for study under a light microscope. In microscopy, observation is conducted through transmitted light, meaning that light must pass through the object being analyzed. Therefore, it is necessary to obtain plant tissue fragments that are collected on very thin, transparent slides to enable the observation of microscopic structures [48].

After exploratory research, it was possible to identify inclusion and exclusion criteria for fruits based on the analysis of optical microscopy images. This analysis allowed for the selection of a new group of fruits for more in-depth investigation through scanning electron microscopy (SEM). In this study, the biological elements of the selected fruits were characterized, identifying those that presented the best characteristics for the development of bioinspired structures, aligned with design principles. This phase of abstracting biological strategies seeks to find an application area for the energy dissipation structures.

- The scanning electron microscopy (SEM) study was conducted with orange (*Citrus*) following the anatomical analyses through conventional optical microscopy. The JSM-IT300 was utilized, which is a scanning electron microscope that can be operated with high efficiency using touch panel operation and high-speed phase. The low vacuum pressure range of 10–650 Pa extends the range of materials that can be readily observed. In Touch Interface A, a graphical interface software was used to produce the images and was adjusted so that the workspace could be optimized during the analysis or data processing.

3.3 Biological Entity Selection: Morphology and Anatomy

The exocarp, or the outermost layer, encompasses a structural complexity that includes the cuticle (Cu), epidermis, and hypodermis, the latter containing oil glands (GO). These glands exhibit variations in number and size, influenced by the specific species and variety of *Citrus* under investigation. SEM provides a detailed visualization of these glands, offering a unique perspective on the distribution and morphology of the cells and the oil glands themselves.

Conversely, the external mesocarp (or albedo) is an internal layer that, under SEM, reveals a distinct and complex cellular structure. A notable feature of the mesocarp is the presence of parenchymal cells (Cp) and aerenchyma (Ae). The aerenchyma, in particular, is a specialized tissue that develops large intercellular spaces filled with gases, forming a continuous network that permeates the entire tissue, facilitating gaseous exchange and buoyancy in aquatic contexts.

SEM analysis also highlights the vascular bundles (FV) of *Citrus*, which are pivotal for the transport of nutrients and water throughout the fruit. Detailed visualization of the vascular bundles, their connections, and orientation within the tissue can offer insights into the mechanisms of nutrient transport and distribution within the fruit. The application of SEM in studying these structures in *Citrus* not only provides a deeper understanding of the fruit's anatomy and functionality but can also inspire practical applications and innovations in various fields, such as agriculture, to enhance fruit resistance and quality, and material engineering, to develop synthetic materials that mimic the unique properties observed in the natural structures of *Citrus*.

Through SEM analysis, researchers can explore and quantify microscopic characteristics, such as cell wall thickness, cell size and shape, and the distribution and morphology of oil glands and vascular bundles. This information is vital for understanding how microscopic structures contribute to the macroscopic properties and functionalities of the fruit, such as its ability to resist environmental stresses, its texture, and its nutritional and aromatic content.

- **Microscopic Exploration of *Citrus* Exocarp and External Mesocarp**

The anatomical complexity of *Citrus* fruits is revealed through a meticulous examination of their structural layers, notably the exocarp and mesocarp. Employing scanning electron microscopy (SEM), this study delves into the intricate cellular architecture and its implications for the fruit's physiological functionalities. The *Citrus* is anatomically divided into three distinct layers. However, the endocarp or pulp, not represented in Fig. 1, does not correspond to the area of interest in this study. The flavedo or exocarp (Ex), evidenced in the outermost part, includes the

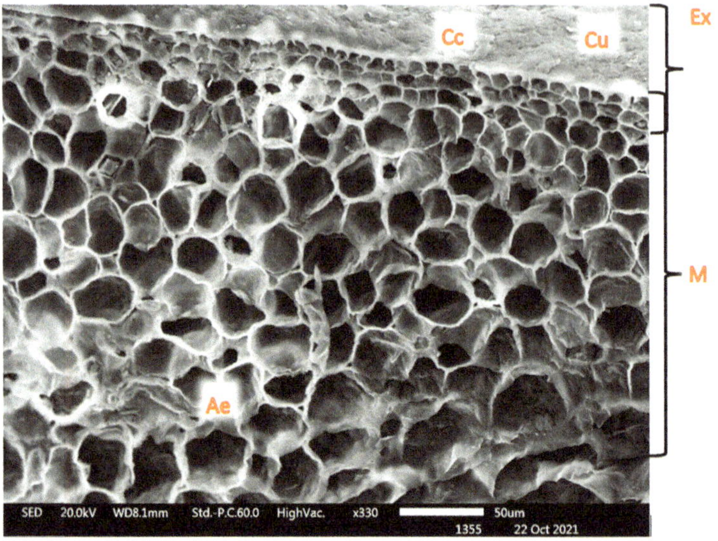

Fig. 1 SEM of *Citrus*

cuticle (Cu), epidermis, and hypodermis, which contain oil glands (GO) that present a variable number and size according to the species and citrus variety (Fig. 1). The innermost part of the peel is called albedo or mesocarp (M), compacted cells (Cc), parenchymal cells (Cp), and vascular bundles (FV), which correspond to one of the elements that constitute the transport system of vascular plants. Aerenchyma (Ae) is a specialization of the parenchymal tissue in which large intercellular spaces filled with gases develop, usually interconnected, forming a continuous gaseous phase that branches throughout the tissue [49].

3.4 Abstraction, Detachment from the Biological Model

- **Micro-CT (X-ray Microtomography):** Scanning procedures were carried out on four fruits: cacao, orange, passion fruit, and pomegranate. Through the scanning analyses, the research cutout was made, and the orange (*Citrus sinensis*) was selected for analysis through specialized software for industrial computed tomography (CT) scanning, and specific modules were selected for the study. This resulted in the creation phase with 3D reconstruction, surface determination, and material determination. Subsequently, in the evaluation phase, the structure was investigated with the 3D porosity analysis module.

(a) **CT Reconstruction:** Through algorithms, the scanned images can be digitized three dimensionally. Techniques for artifact reduction were used for CT reconstruction to obtain the maximum from the utilized X-ray data.

(b) **Analyze Designed Material Properties**: The result reveals what is beneath the surface of your scanned parts. In this way, you can comprehensively investigate and quantify the material properties of your parts using nondestructive testing (NDT). Analyze the designed material properties, as well as discontinuities, and visualize them – from research and development to production stages.
(c) **Foam Analysis:** Determines the cellular structures in porous foams. In this manner, connected cells can be subdivided to individually quantify the sizes and shapes of the cells. The thicknesses of the stems can be visualized locally and aggregated to global statistics, such as through histograms.
(d) **Porosity/Inclusion Analysis**: Provides various porosity/inclusion analysis algorithms to ensure optimal porosity detection for each part, suitable for multimaterial datasets with customized porosity evaluation and porosity selection.

The XT H 225 ST offers a microfocus X-ray source, an inspection volume to accommodate small to medium-sized parts and high image resolution. The generation of X-rays occurs between a tungsten filament (cathode) and a tungsten target (anode) with voltage applied to the X-ray tube (Fig. 2) of up to 225 kV and a maximum electrical current of 2 mA.

Computed tomography, also used to gain insight into the internal view of complex industrial parts, observes internal structure. In this study, it was used for noninvasive observation in plant samples. Thus, the use of the micro-CT resource was qualified to quantify any internal or external dimension in a smooth, fast, and nondestructive (noninvasive) process in plant samples (**configuration:** 60 kV, 200 μA, and 10 μm). Despite this procedure of obtaining images in a noninvasive manner, it was necessary to scan smaller pieces to obtain a better resolution of the internal

Fig. 2 Preparation for scanning in the X-ray

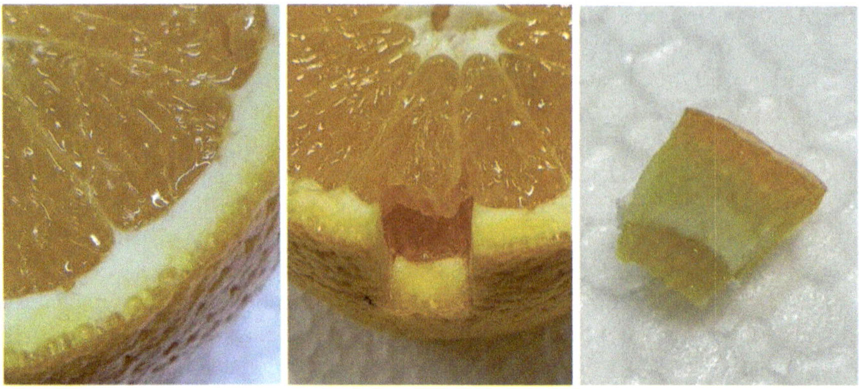

Fig. 3 Fresh orange samples for analysis with 5 × 5 mm

structures of the fruits. X-ray images were obtained from pieces of the samples and whole fruits, as shown in Fig. 2.

The smallest sample piece that can be enlarged in the scan is 5.5 × 5.5 mm. For this measurement, a digital caliper was used, and cuts were made at the time of scanning with blades and carefully to avoid damaging the structures of the cell walls of the samples (Fig. 3).

The three-dimensional reconstruction involves assembling three-dimensional images from various flat images resulting from the scanning of the measured element at the selected angular position. The piece is affixed to a rotary table that allows movement and linear and angular positioning of the component to be scanned. While the measured object is illuminated by an X-ray beam generated in the tube, image detection is performed at specific angular positions of the rotary table on the detector array. The acquired images are processed by specialized software, producing a three-dimensional model of the measured element [50].

3.5 Biological Abstraction and Inspiration

Biological abstraction and inspiration involve deriving insights and strategies from nature to solve human challenges and innovate in various fields, such as design, engineering, and medicine. This approach referred entails observing, understanding, and emulating natural systems and organisms to develop sustainable solutions.

- **Selection of the Area of Interest (RoI):** All metrology based on computed tomography relies on an accurate determination of the surface in the voxel model. The better the surface determination, the lower the measurement uncertainty. With surface determination, a multimaterial mode is obtained (Fig. 4) that allows the simultaneous determination of the surfaces of various materials within a volume, resulting in one component per material.

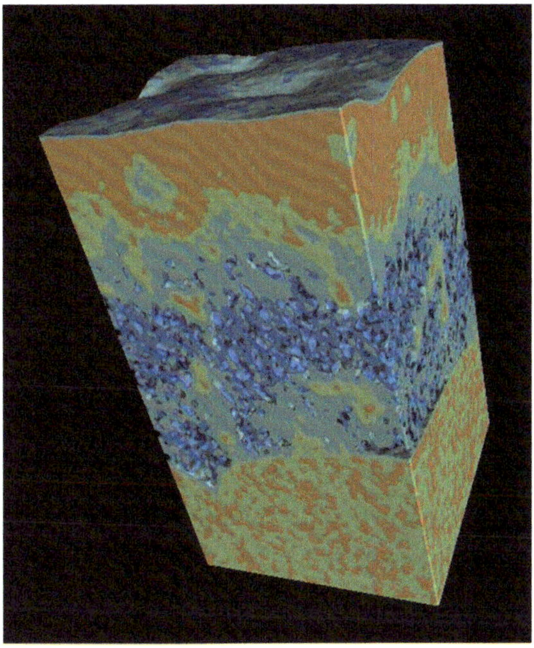

Fig. 4 Selection of RoI

In the context of computed tomography (CT) scanning, the region of interest (RoI) selection is a crucial step to ensure that the analysis is focused on the pertinent area of the sample or object being studied. This selection aims to optimize the scanning process, focusing on the area that contains the relevant information or features for the analysis, thereby reducing data complexity and processing time. Below are some key aspects related to RoI selection in CT-based metrology:

(a) **Precision and Accuracy**: Ensuring that the RoI encompasses all necessary details while excluding irrelevant areas to maintain the precision and accuracy of the measurements.
(b) **Data Management**: Managing the volume of data generated during CT scanning by focusing only on the relevant region, which can also reduce computational requirements and enhance processing speed.
(c) **Material Differentiation**: In multimaterial components, accurately defining the RoI helps in differentiating between various materials and their respective surfaces within the scanned volume.
(d) **Analysis Enhancement**: Focusing on a specific RoI allows for a more detailed and enhanced analysis of the particular area, ensuring that the resolution and quality of the data are optimized for the features of interest.
(e) **Reduction of Artifacts**: By concentrating on a specific RoI, potential artifacts from surrounding structures or materials can be minimized, improving the quality of the reconstructed images.
(f) **Optimization of Resources**: Efficiently utilizing computational and analytical resources by focusing on the most relevant section of the sample.

In practical applications, the selection of RoI might involve considering the structural complexity, material properties, and the specific research question or inspection criteria. Advanced software tools are often employed to define and isolate the RoI, ensuring that subsequent analyses, such as surface determination and 3D reconstruction, are focused and relevant.

Tetrahedral meshes can be exported for FEM (finite element method) simulation in another software to create integration meshes. In this way, it was possible to convert a monomaterial volume dataset (2D images) of plant samples into a tetrahedral mesh (Fig. 5). The technique was used to create, optimize, and export mesh data with custom entity sets for FEM use and three-dimensional abstraction of the cell walls observed in scanning electron microscopy (SEM) scans.

In the context of utilizing volume mesh in computational simulations and analyses, especially in the realm of bio-inspired design and research, several key aspects are considered:

- **Mesh Generation**: The conversion of 2D images into a 3D tetrahedral mesh involves utilizing algorithms that can accurately represent the internal and external geometries of the sample, ensuring that the mesh accurately represents the structural and material properties observed in SEM scans.

In the development of bio-inspired designs, especially those derived from plant samples, the accurate representation and analysis of the biological structures through volume mesh and FEM simulations provide a robust methodology to understand, abstract, and apply the inherent biological strategies and principles in innovative solutions. This approach not only enhances the understanding of biological

Fig. 5 Volume mesh

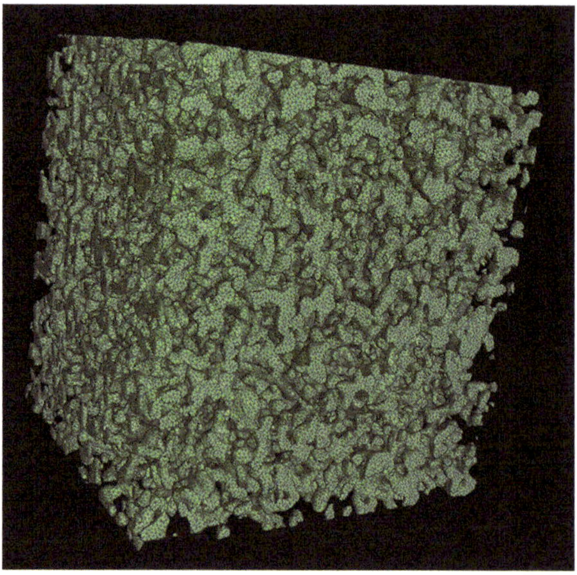

systems but also facilitates the translation of these insights into practical, sustainable, and efficient design solutions in engineering and technology.

- **Porosity Analysis of the Structure:** Computed tomography enables the conduction of nondestructive tests on specimens. The examination of simple cross sections allows for the detection of existing pores and inclusions of other materials. With the aid of algorithms, the volume of pores can be calculated. To verify the accurate functioning of the algorithms, appropriate gauges must be constructed. The gauges should be built in a way that enables the estimation of the actual volume of the pores [50].

In Fig. 6, a porosity analysis of the structure from which the mesh was extracted is presented, corresponding to approximately 44% porosity through the 3D porosity analysis that recognizes and characterizes porosity in three dimensions. In the context of bio-inspired design and material science, understanding the porosity of biological structures is crucial for several reasons:

(a) **Structural Integrity**: Understanding how porosity affects the structural integrity and mechanical properties of the material, which can inform the design and development of bio-inspired materials and structures.
(b) **Material Efficiency**: Analyzing the porosity can provide insights into how biological structures achieve strength and functionality while utilizing minimal material, contributing to lightweight and material-efficient designs.
(c) **Functional Properties:** Porosity can influence various functional properties of the material, such as thermal and acoustic insulation, permeability, and more,

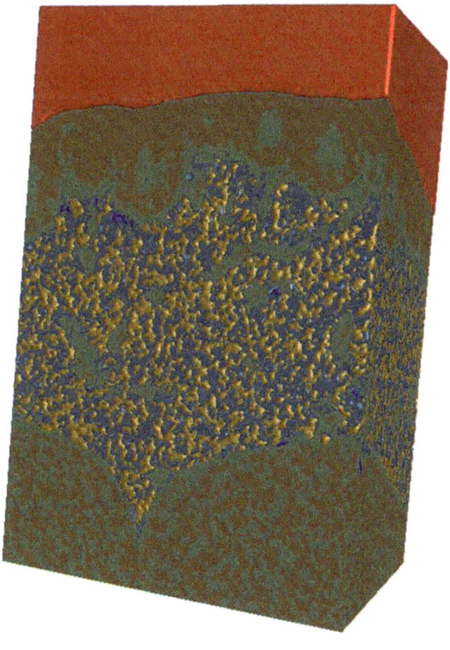

Fig. 6 Porosity analysis of the structure

which can be abstracted and applied in the design of bio-inspired materials and structures.

(d) **Simulation and Modeling**: Accurate porosity analysis allows for the development of more accurate computational models and simulations, which can be used to predict the behavior of bio-inspired designs under various conditions.

(e) **Innovation in Design**: Understanding the role and distribution of porosity in biological structures can inspire innovative design strategies that mimic nature's ability to optimize material distribution for specific functional and structural purposes.

(f) **Sustainability:** By understanding and mimicking how nature utilizes porosity to achieve specific properties and functionalities, designers and engineers can develop solutions that are not only efficient but also sustainable, minimizing material usage and potentially enhancing recyclability.

In the application of porosity analysis in the development of bio-inspired designs, it is crucial to consider how the inherent properties and strategies of biological structures can be abstracted and applied in a practical, innovative, and sustainable manner across various applications and industries. The insights derived from such analyses can inform and inspire the development of new materials, structures, and technologies that mimic the efficient and optimized properties of biological systems.

In this phase, the researcher will formulate alternative structures derived from the fruits under study, generating designs based on the examined biological entity to optimize material usage (volumetrics) versus the energy absorption of the devised structure. This will subsequently be evaluated using physical simulation software, through mathematical calculations in the ensuing research phase. Biomimetics entails understanding biological structures and processes, their applications, and analogous technological methods or procedures. This activity is anchored in the principle that every living organism on Earth is the culmination of millions of years of evolution. The fundamental principle of biomimetics is to utilize the natural biological world as a source of inspiration and a guide in the development of new materials [5].

Utilizing flexible resin printing with the Creality Ld-002 h printer, rapid prototyping of the bioinspired structure was achieved (Fig. 7), abstracted from the orange albedo, aligning with phase 5 (technical implementation). The Chitubox slicer was employed, with a printing duration of approximately 2 h. The flexible resin was selected for its commendable flow, processing ease, low shrinkage, superior dielectric properties with insulation, and elevated chemical and moisture resistance. A detailed exploration of systems and organisms within natural systems, which can serve as models in the science and engineering of new materials, is requisite [6].

The attained results are encouraging and furnish a foundation for future research wherein various facets of this design strategy and bioinspired products for mechanical shock absorption, corresponding to phase 6 (design solution), can be explored. Through biomimetics, the development of new solutions and strategies, unveiling new perspectives and opportunities in the realms of design, engineering, and biology, is achievable. This study pursues a crucial process of integration for the new

Fig. 7 Prototype in flexible resin of bioinspired structure abstracted from orange albedo

industrial era, namely, the fourth industrial revolution, projecting for the technical world from natural inspiration. Observing nature and employing specific methods enables the abstraction of functional structures for application in specific problems, culminating in new products.

4 Conclusion

The examination of organisms in nature, from the simplistic to the highly complex, provides a plethora of materials, structural architectures, systems, and functions that can establish analogies and be integrated into the design process, applicable across various domains. Consequently, this study explored strategies inherent in fruits that perform a protective function, which can be applied in diverse activity fields as a strategic tool for designing bioinspired products targeting damping systems, mechanical shock protection, and energy absorption.

Exploratory research steered the focus toward certain structures found in both fleshy and dry fruits, analogous in achieving efficiency in structures with significant energy dissipation, resistance, and damping capabilities. The results, obtained through systematic review, unveiled 22 key studies that guided this research, seeking to generate concepts for new structures inspired by fruits.

The studies produced various arrangements, primarily structural, revealing the methods, research objectives, and mechanical tests with fruits, from which various methodological stages of this research were abstracted, utilizing software and methods essential for attaining the anticipated results. In the various research presented, similar methodology, methods, and research techniques were found, applied to biological models, resulting in bioinspired artifacts or cell conformation studies that spurred inspiration for developing bioinspired products.

Observing cellular microstructures fosters the creation of innovative materials and forms in the development of new products. Characterizing these structures, at a cellular level, along with the mechanical properties found in these natural materials, can assist in understanding characteristics related to morphology and guide structural optimization in industrial design projects. Engaging in a research endeavor that seeks to intertwine the realms of biology and design, particularly through the lens of biomimicry and structural analysis of fruits for bioinspired structural development, presents its own set of challenges and limitations.

- **Biological Complexity**: The inherent complexity and variability found within biological systems, such as fruits, can pose a significant challenge. The structural and mechanical properties of fruits can be influenced by various factors, including species, growing conditions, and maturity, which may introduce variability in the research data.
- **Translational Applicability**: Translating biological structures and mechanisms into applicable design principles and materials in engineering and product development is not always straightforward. The complexity of biological systems and their multifunctionality can be difficult to abstract and implement in a synthetic context.
- **Technological Limitations**: Utilizing technologies like micro-CT and scanning electron microscopy (SEM) for detailed analysis of biological structures demands a meticulous and precise methodology. The resolution, accuracy, and interpretation of obtained data can be influenced by the technology's limitations and the sample preparation methods.
- **Material Challenges**: Developing materials that mimic the properties of biological structures, especially at a commercial or industrial scale, can be challenging. The availability, cost, and manufacturability of such materials can pose limitations to the practical application of the research findings.
- **Multidisciplinary Collaboration**: The integration of knowledge across biology, materials science, engineering, and design is crucial for the success of bioinspired development. Ensuring effective communication and collaboration across these diverse fields can be challenging but is paramount to navigating the complexity of biomimetic design.
- **Ethical and Sustainability Considerations**: Ensuring that the research and resultant applications are developed with a keen eye on sustainability and ethical considerations is pivotal. The sourcing of biological materials, the environmental impact of material development, and the broader ecological implications of implementing bioinspired designs need thorough consideration and alignment with sustainability principles.
- **Validation and Testing**: Validating the performance and reliability of bioinspired designs, especially in various practical and real-world scenarios, requires comprehensive testing. Establishing testing protocols that accurately reflect the intended application and ensuring that the bioinspired designs meet the necessary safety and performance standards can be complex and resource-intensive.

Addressing these challenges and limitations necessitates a holistic, multidisciplinary approach that not only seeks to understand and abstract from biological systems but also considers the practical, ethical, and sustainability implications of applying this knowledge in design and engineering contexts. Future research and development in this field will undoubtedly benefit from collaborative efforts that bring together expertise from biology, materials science, engineering, and design to navigate the complexities and realize the potential of bioinspired design and development.

References

1. Rossin, K. (2010). Biomimicry: Nature's design process versus the designer's process, *138*, 559–570. https://doi.org/10.2495/DN100501
2. Volstad, N., & Boks, C. (2012). On the use of biomimicry as a useful tool for the industrial designer. *Sustainable Development, 20*, 189–199. https://doi.org/10.1002/SD.1535
3. Kennedy, E., Miller, D., & Niewiarowski, P. (2018). Industrial and biological analogies used creatively by business professionals. *Creativity Research Journal, 30*, 54–66. https://doi.org/10.1080/10400419.2017.1411631
4. Oliveira, P., Silva, D., Silva, L., Lopes, M., & Helleno, A. (2016). Factors that influence product life cycle management to develop greener products in the mechanical industry. *International Journal of Production Research, 54*, 4547–4567. https://doi.org/10.1080/00207543.2015.1071893
5. Antreich, S. J., et al. (2019). The puzzle of the walnut shell: A novel cell type with interlocked packing. *Advanced Science, 6*(16), 1900644.
6. Bührig-Polaczek, A., et al. (2016). Biomimetic cellular metals—Using hierarchical structuring for energy absorption. *Bioinspiration & Biomimetics, 11*(4), 045002.
7. Kaupp, G., & Naimi-Jamal, M. R. (2011). Nutshells' mechanical response: From nanoindentation and structure to bionics models. *Journal of Materials Chemistry, 21*(23), 8389–8400.
8. Sonego, M., Fleck, C., & Pessan, L. A. (2020). Hierarchical levels of organization of the Brazil nut mesocarp. *Scientific Reports, 10*(1), 1–13.
9. Grijalva, S. F. (2018). *La naturaleza del embalaje: la naturaleza como fuente de innovación para empaques*. Caligrama.
10. Speck, T., et al. (2018). Biomechanics and functional morphology of plants—Inspiration for biomimetic materials and structures. In *Plant biomechanics* (pp. 399–433). Springer.
11. San Ha, N., et al. (2020). Mechanical properties and energy absorption characteristics of tropical fruit durian (*Durio zibethinus*). *Journal of the Mechanical Behavior of Biomedical Materials, 104*, 103603.
12. Jelinek. (2013). *Biomimetics: A molecular perspective*. ISBN: 978-3-11-028117-0.
13. Lakhtakia, A., & Martín-Palma, R. J. (Eds.). (2013). *Engineered biomimicry*. Newnes.
14. Ellison, M. S. (2013). Chapter 10: Biomimetic textiles. In *Engineered biomimicry*. Elsevier.
15. Fischer, S., Thielen, M., Loprang, R., Seidel, R., Fleck, C., Speck, T., & Bührig-Polaczek, A. (2010). Pummelos as concept generators for biomimetically inspired low weight structures with excellent damping properties. *Advanced Engineering Materials, 12*. https://doi.org/10.1002/adem.201080065
16. Lu, C., Hou, S., Zhang, Z., Chen, J., Li, Q., & Han, X. (2020). The mystery of coconut overturns the crashworthiness design of composite materials. *International Journal of Mechanical Sciences, 168*, 105244. https://doi.org/10.1016/j.ijmecsci.2019.105244

17. Müller, C., Klemm, S., & Fleck, C. (2021). Bracket fungi, natural lightweight construction materials: Hierarchical microstructure and compressive behavior of Fomes fomentarius fruit bodies. *Applied Physics A, 127*. https://doi.org/10.1007/s00339-020-04270-2
18. Kaderides, K., & Goula, A. (2019). Encapsulation of pomegranate peel extract with a new carrier material from orange juice by-products. *Journal of Food Engineering*. https://doi.org/10.1016/J.JFOODENG.2019.02.019
19. Wu, X., Jiang, P., Chen, L., Zhang, J., Yuan, F., & Zhu, Y. (2014). Synergetic strengthening by gradient structure. *Materials Research Letters, 2*, 185–191. https://doi.org/10.1080/21663831.2014.935821
20. Sha, Y., Jiani, L., Haoyu, C., Ritchie, R., & Jun, X. (2018). Design and strengthening mechanisms in hierarchical architected materials processed using additive manufacturing. *International Journal of Mechanical Sciences*. https://doi.org/10.1016/J.IJMECSCI.2018.09.038
21. Ambekar, R., Mohanty, I., Kishore, S., Das, R., Pal, V., Kushwaha, B., Roy, A., Kar, S., & Tiwary, C. (2021). Atomic scale structure inspired 3D-printed porous structures with tunable mechanical response. *Advanced Engineering Materials, 23*. https://doi.org/10.1002/adem.202001428
22. Ren, Z., Guo, Y., Liu, C., & Gao, P. (2013). Hierarchically nanostructured materials for sustainable environmental applications. *Frontiers in Chemistry, 1*. https://doi.org/10.3389/fchem.2013.00018
23. Shi, S., Li, Y., Ngo-Dinh, B., Markmann, J., & Weissmüller, J. (2021). Scaling behavior of stiffness and strength of hierarchical network nanomaterials. *Science, 371*, 1026–1033. https://doi.org/10.1126/science.abd9391
24. Zuo, S., & Wei, Y. (2007). Effective elastic modulus of bone-like hierarchical materials. *Acta Mechanica Solida Sinica, 20*, 198–205. https://doi.org/10.1007/s10338-007-0723-z
25. Flammang, B., & Porter, M. (2011). Bioinspiration: Applying mechanical design to experimental biology. *Integrative and Comparative Biology, 51*(1), 128–132. https://doi.org/10.1093/icb/icr014
26. Pettermann, H., & Hüsing, J. (2012). Modeling and simulation of relaxation in viscoelastic open cell materials and structures. *International Journal of Solids and Structures, 49*, 2848–2853. https://doi.org/10.1016/J.IJSOLSTR.2012.04.027
27. Nojoomi, A., Arslan, H., Lee, K., & Yum, K. (2018). Bioinspired 3D structures with programmable morphologies and motions. *Nature Communications, 9*, 3705. https://doi.org/10.1038/s41467-018-05569-8
28. Kmiecik, B., Łabowska, M., & Detyna, J. (2019). Determination of the difference between two complex polymer models simulating the behaviour of biological structures. *Biocybernetics and Biomedical Engineering, 39*, 503. https://doi.org/10.1016/J.BBE.2019.04.001
29. Chen, Z., Su, Z., Ma, S., Wu, X., & Luo, Z. (2007). Biomimetic modeling and three-dimension reconstruction of the artificial bone. *Computer Methods and Programs in Biomedicine, 88*(2), 123–130. https://doi.org/10.1016/j.cmpb.2007.08.001
30. Yoshida, S., Ejima, H., & Yoshie, N. (2017). Tough elastomers with superior self-recoverability induced by bioinspired multiphase design. *Advanced Functional Materials, 27*. https://doi.org/10.1002/adfm.201701670
31. Chen, K., Zhang, S., Li, A., Tang, X., Li, L., & Guo, L. (2018). Bioinspired interfacial chelating-like reinforcement strategy toward mechanically enhanced lamellar materials. *ACS Nano, 12*(5), 4269–4279. https://doi.org/10.1021/acsnano.7b08671
32. Wang, F., Yang, Z., Li, J., Zhang, C., & Sun, P. (2021). Bioinspired polyurethane using multifunctional block modules with synergistic dynamic bonds. *ACS Macro Letters, 10*(5), 510–517. https://doi.org/10.1021/ACSMACROLETT.1C00054
33. Petkov, V., Ren, Y., Saratovsky, I., Pastén, P., Gurr, S., Hayward, M., Poeppelmeier, K., & Gaillard, J. (2009). Atomic-scale structure of biogenic materials by total X-ray diffraction: A study of bacterial and fungal MnOx. *ACS Nano, 3*(2), 441–445. https://doi.org/10.1021/nn800653a
34. Wong, J., McDonald, J., Taylor-Pinney, M., Spivak, D., Kaplan, D., & Buehler, M. (2012). Materials by design: Merging proteins and music. *Nano Today, 7*(6), 488–495. https://doi.org/10.1016/J.NANTOD.2012.09.001

35. Artero, V. (2017). Bioinspired catalytic materials for energy-relevant conversions. *Nature Energy, 2*. https://doi.org/10.1038/nenergy.2017.131
36. Arruda, A. J. V. (2002). *Bionic Basic: verso un nuovo modello di ricerca progettuale* [Tese de Doutorado, Politecnico di Milano–Doutorado em Desenho Industrial e Comunicação Multimídia].
37. Sanchez, C., Arribart, H., & Giraud Guille, M. M. (2005). Biomimetism and bioinspiration as tools for the design of innovative materials and systems. *Nature Materials, 4*(4), 277–288. https://doi.org/10.1038/nmat1339
38. Langella, C., & Santulli, C. (2017). Processi di crescita biologica e Design parametrico. *MD Journal, 3*(1), 14–27.
39. Layton, A., Bras, B., & Weissburg, M. (2016). Designing industrial networks using ecological food web metrics. *Environmental Science & Technology*. https://doi.org/10.1021/acs.est.6b03066
40. Yang, Z., & Zhang, J. (2021). Bioinspired radiative cooling structure with randomly stacked fibers for efficient all-day passive cooling. *ACS Applied Materials & Interfaces*. https://doi.org/10.1021/acsami.1c12267
41. Eder, M., Shahrouz Amini, P., & Fratzl. (2018). Biological composites—Complex structures for functional diversity. *Science*. https://doi.org/10.1126/science.aat8297
42. Yayun Wang, S., & Naleway, B. W. (2020). Biological and bioinspired materials: Structure leading to functional and mechanical performance. *Bioactive Materials*. https://doi.org/10.1016/j.bioactmat.2020.06.003
43. Sánchez-Garrido, A. J., & Yepes, V. (2020). Multi-criteria assessment of alternative sustainable structures for a self-promoted, single-family home. *Journal of Cleaner Production*. https://doi.org/10.1016/j.jclepro.2020.120556
44. Rowland, R. (2017). Biomimicry step-by-step. *Bioinspired, Biomimetic and Nanobiomaterials, 6*(2), 102–112.
45. Conforto, E. C., Amaral, D. C., & da Silva, S. L. (2011). Roteiro para revisão bibliográfica sistemática: aplicação no desenvolvimento de produtos e gerenciamento de projetos. *Trabalho apresentado, 8*.
46. Johansen, D. (1940). *Plant microtechnic*. Mc Graw Hill Book Company.
47. Kraus, J., & Arduin, M. (1997). *Manual básico de métodos em morfologia vegetal*. EDUR.
48. Ruzin, S. E., et al. (1999). *Plant microtechnique and microscopy*. Oxford University Press.
49. Mauseth, J. D., & Ross, R. G. (1988). Systematic anatomy of the primitive cereoid cactus Leptocereus quadricostatus. *Bradleya, 1988*(6), 49–64.
50. Kowaluk, T., & Woźniak, A. (2012). Study of porosity measurement using the computer tomograph. *Pomiary Automatyka Robotyka, 16*, 82–86.

Antônio Roberto Mirandade Oliveira, Currently, a substitute professor at UFPE, in the Design department, and Adjunct Professor at UNIAESO, where he teaches in the Graphic Design and Animation Design courses. He holds a Ph.D. in Design, focusing on Design and Technology (UFPE). He has a degree in Design and a master's degree in Design also by the Federal University of Pernambuco (UFPE). Researcher at BIoDesign Lab (UFPE) and VirtuHab - Sustainable Technologies (UFSC).

Amilton José Vieira Arruda, Graduated in Industrial Design Product Project by UFPE, Master in Design and Bionics by IED in Milan, Ph.D. in Ricerca in Disegno Industriale by University of Milan Politecnico. Postdoctoral in Design and Bionics at IADE European University UNIDCOM Lisbon and postdoctoral in Bionics at University Luigi Vanvitelli Naples (IT). He is currently associate professor IV at the UFPE Design Course and coordinates the BioDesign Lab.

Biomimetics and 4D Printing: A Synergy for the Development of Innovative Materials

Santina Di Salvo

1 Introduction

1.1 Biomimetics and Relevance in Material Innovation

Historically, innovation in materials has been a driving force behind technological advancements and industrial applications. The conventional approach to discovering new materials involves a combination of practical needs and a scientific understanding of material properties. However, in recent decades, a nature-inspired approach known as biomimicry has gained prominence [1].

While the concept of bio-inspired materials, or biomimetic materials, has ancient roots in human attempts to imitate nature's designs, the formalization of biomimicry as a scientific discipline began in the latter half of the twentieth century. This recognition stems from the acknowledgment of nature's efficient and adaptive solutions that have evolved over millions of years.

In the 1950s and 1960s, Janine Benyus coined the term "biomimicry" and advocated the study of nature's designs to address modern human challenges. This movement adopts a multidisciplinary approach and has led to the emergence of biomimetic materials as a field dedicated to creating synthetic materials inspired by biology. Essentially, biomimicry, derived from the Greek words "bios" (life) and "mimesis" (imitation), represents an effective design paradigm that leverages nature's evolutionary engineering [2].

By analyzing evolution's optimized solutions, biomimicry applies biological design principles to human problems through the creation of materials, systems, and technologies that dynamically adapt and respond to the environment. It involves

S. Di Salvo (✉)
Polis University, Tirana, Albania
e-mail: santina_disalvo@universitetipolis.edu.al

A. J. V. Arruda, F. L. Palombini (eds.), *Biomimetics, Biodesign and Bionics*, Environmental Footprints and Eco-design of Products and Processes,
https://doi.org/10.1007/978-3-031-51311-4_7

emulating biological structures, functions, processes, and systems found in nature and applying them to human-made technologies.

The principles of biomimetics have gained significant traction across various fields, including materials science, due to their potential to transform the way of designing and engineering innovative materials. Therefore, in materials science, biomimetics provides an innovative perspective on creating materials with high-performing properties, functionalities, and benefits, including mechanical strength, durability, adaptability, and responsiveness to environmental stimuli (Fig. 1) [3].

By closely studying the intricate designs and mechanisms employed by living organisms, both researchers and engineers engage in experiments aimed at developing novel technologies and solutions that have demonstrated enduring efficacy.

The relevance of biomimetics in materials science lies in its ability to overcome traditional limitations in material design. Traditional materials often face trade-offs between conflicting properties, such as strength and flexibility or stiffness and resilience. Biomimetic approaches allow us to break free from these constraints by mimicking the hierarchical structures and self-assembly mechanisms found in biological materials [4].

This approach opens doors to the creation of materials with tailored properties, such as lightweight composites with exceptional strength or self-healing materials capable of repairing damage.

Furthermore, as the demand for sustainable and environmentally friendly solutions continues to rise, biomimetics offers a path to developing materials that mirror nature's efficiency and resourcefulness. Nature rarely wastes energy or resources, and by emulating these efficient strategies, it becomes feasible to craft materials that excel not only in performance but also in environmental consciousness.

Fig. 1 Leaf hygroscopicity. (Reproduced from Flickr, License CC0)

1.2 4D Printing: Fundamentals and Materials and Technologies

Concurrently, another revolution in the materials and manufacturing sector emerged, thanks to 4D printing. This advancement, an evolution beyond traditional 3D printing, has ushered in a new dimension in object production. In this paradigm, "4D" signifies the integration of time as the fourth dimension, complementing the three spatial dimensions typically associated with 3D printing.

In 2013, Skylar Tibbits, a researcher at the Massachusetts Institute of Technology (MIT) and the founder of the Self-Assembly Lab, introduced the concept of the fourth dimension in printing during a TEDx conference [5]. Put simply, 4D printing enables the creation of objects with intricate shapes and structures capable of adapting over time in response to specific stimuli or environmental factors, including humidity and light.

The pioneering research on 4D printing at the MIT Self-Assembly Lab started with the utilization of a Stratasys Connex machine. This approach is rooted in the principle of material ejection, constituting a multimaterial process that has paved the way for groundbreaking advancements in this disruptive field.

Incorporating programmable structures into the manufacturing process enables the post-printing modification of object functionalities. Bastien E. Rapp, President of the NeptunLab Process Technology Lab, describes 4D printing as an extension of 3D printing, wherein a code is embedded into the material (Figs. 2, and 3; [6]). Upon activation, the material carries out its pre-programmed function, providing directives for the 3D-printed component. However, it's essential to highlight that the available range of materials for 4D printing is relatively limited compared to those utilized in additive manufacturing, mainly due to the early stage of development of this technology.

Nevertheless, the scientific community and industry leaders have undertaken numerous studies and made substantial advancements, collectively providing a comprehensive panorama for understanding the fundamental applications, advantages, and challenges in the field of 4D printing.

A key aspect of 4D printing technology is that it requires the creation of structures with remarkable capacity for a controlled transition from one shape to another [7].

A category of 4D printing materials is shape-memory polymers (SMPs). These smart materials have the ability to memorize a macroscopic shape, retain it for a specified duration, and subsequently return to their original shape, stiffness, or other predetermined properties when exposed to specific triggers. Remarkably, this process occurs without any residual deformation. Various indirect stimuli can also induce transformation in 4D printing materials, such as magnetic fields, electric fields, or immersion in water, providing versatile mechanisms for shape alteration (Fig. 4).

Other materials suitable for 4D printing are liquid crystal elastomers (LCEs). As the name implies, LCEs contain heat-sensitive liquid crystals, allowing for precise

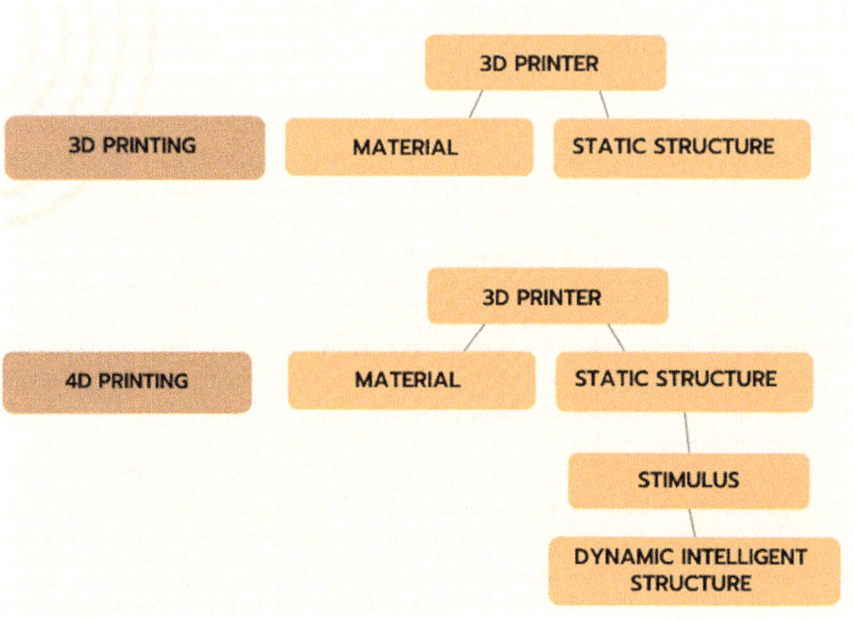

Fig. 2 Scheme with the features of 3D and 4D. (Reproduced by Author)

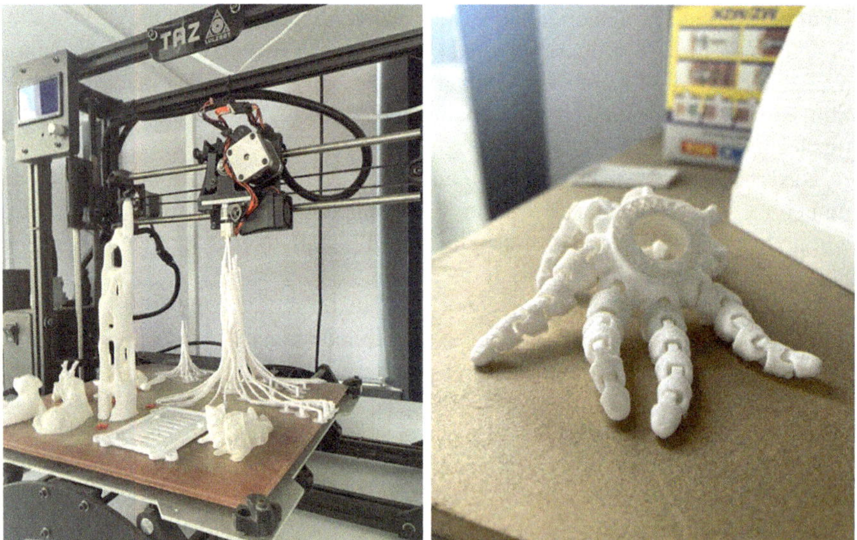

Fig. 3 3D Printer and 3D printed objects. (Reproduced by Author)

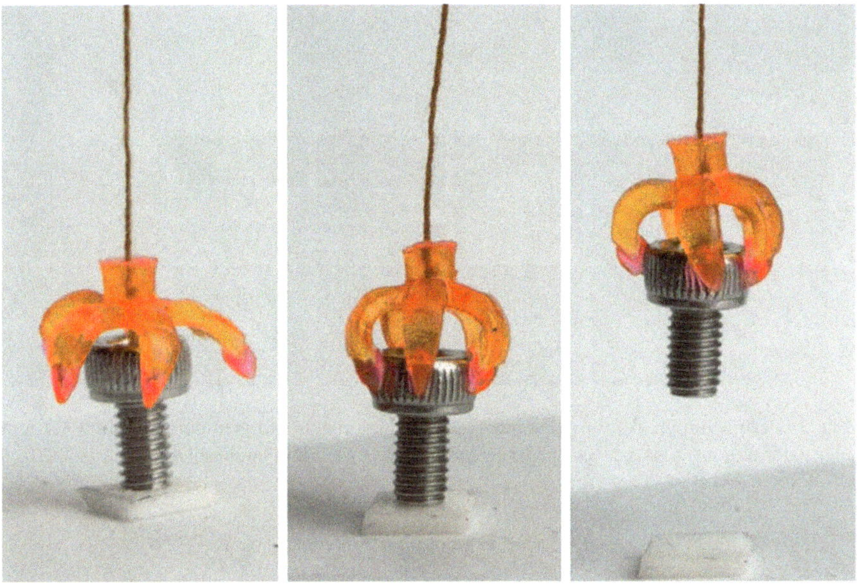

Fig. 4 Shape-memory polymers: transformation functions enabled by 4D printing. (Reproduced from [7], License CC BY)

programming of the desired shape by controlling their orientation. In response to temperature changes, these materials undergo expansion and transformation, adhering to the predefined code and thus offering a dynamic range of structural possibilities [8].

Furthermore, hydrogels, characterized by polymer chains primarily composed of water, constitute another notable category of 4D printing materials. Researchers investigated the versatility of a kind of 3D printable and shape-morphing hydrogel by designing and printing a highly complex orchid by preprogramming multiple shape-morphing domains. These hydrogels find significant relevance in photopolymerization processes and have garnered substantial interest from the medical sector due to their exceptional biocompatibility (Fig. 5) [9].

In multimaterial 4D printing processes, hydrogels are integrated with shape-memory polymers (SMPs) to create self-folding structures. These composite materials can incorporate components such as carbon or wood fibers, improving their properties and broadening the scope of various applications (Fig. 6) [10].

2 Research Objectives and Methods

The main objectives of this study are outlined as follows:

Investigate synergies. This study aims to investigate and analyze the potential synergistic relationship between the principles of biomimetics and the capabilities of 4D printing in creating innovative materials.

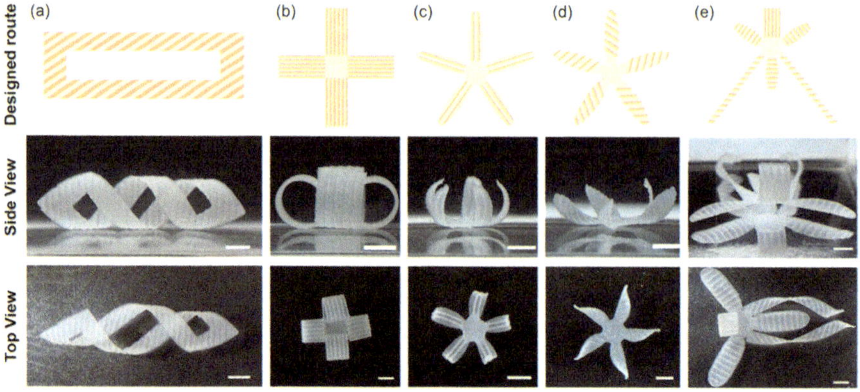

Fig. 5 Designed routes of the 4D-printed patterned hydrogels and their corresponding 3D complex deformations from the side and top views upon immersion in liquid solutions. (Adapted from [9], License CC BY)

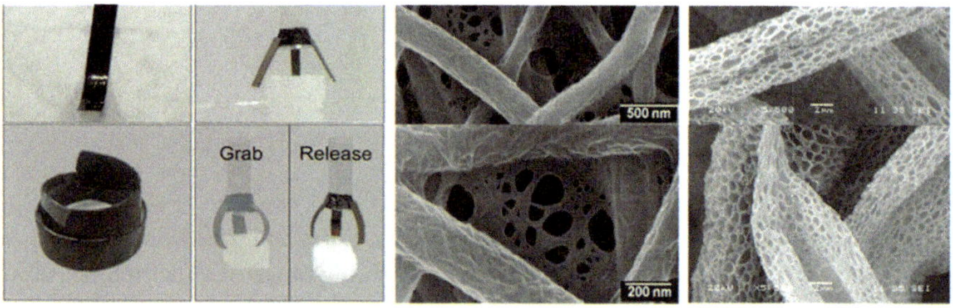

Fig. 6 Moisture-driven biomimetic cellulosic fiber and fiber-based materials robot, produced with balsa wood fibers. (Adapted from [10], License CC BY)

Examine design strategies. It also aims to explore the application of biomimetic design principles, drawing inspiration from nature's solutions, to steer the development of materials capable of dynamic adaptation, response, or transformation through the utilization of 4D printing techniques.

Demonstrate the intersection: This study seeks to showcase applications where the fusion of biomimetics and 4D printing has resulted in the creation of materials with enhanced properties, responsiveness, and functionalities.

Address challenges and opportunities. It seeks to uncover challenges and opportunities arising at the intersection of these two fields, discussing technical hurdles, design complexities, and potential breakthroughs.

This article aims to examine and outline the importance of the intersection of biomimicry and 4D printing in the engineering of advanced materials. The combination of these two approaches sets the stage for further innovation in the research and development of smart materials. These innovations not only draw from biological principles for design but also enable materials to actively respond and adapt to their surroundings through morphological changes and controlled structures.

To achieve this goal, it is essential to provide a comprehensive understanding of the foundational principles of biomimetics and 4D printing technology and highlight the significance of their convergence. This involves analyzing case studies and real-world examples related to the intersection of biomimetics and 4D printing for the development of innovative materials, which serves as concrete evidence to support insights into the diverse applications and outcomes of this synergy.

Additionally, the study highlights the significance of an interdisciplinary approach, featuring contributions from experts in biomimetics, materials science, and related fields to provide a broader perspective. Ultimately, the chapter aims to improve understanding of the progress in biomimetics for 4D printing and the development of bio-inspired materials with unparalleled adaptability and responsiveness.

3 The Intersection Between Nature and Technology

3.1 Foundational Biological Principles of Biomimetics

The foundational biological principles of biomimetics include the following.

Structures and models. Biomimicry is founded on the identification of parallels between natural features and processes and the challenges at hand. Scientists strive to comprehend how nature addresses analogous scenarios and then employ these principles in design. Hence, researchers from various disciplines study natural structures and patterns, like the lightweight and sturdy wing shape of butterflies or the adhesive capabilities of geckos. Their aim is to develop novel materials and technologies (Fig. 7) [11].

a) b) c) d)

Fig. 7 A butterfly (**a**) and biomimetic tissue that mimics the wings pattern (**b**), and a gecko (**c**) with Velcro, inspired by gecko skin (**d**). (Reproduce and adapted from Flickr, License CC0)

Evolutionary processes. Nature is a worthwhile endeavor, the product of evolutionary processes that have been ongoing for almost 4 billion years and have resulted in materials, structures, and mechanisms that are more effective than their human counterparts, especially when viewed from an environmental perspective [12].

Efficiency in design and sustainability. Solutions developed through biomimicry often seek to replicate the efficiency and sustainability of biological processes. Natural elements achieve optimal use of materials and resources, minimizing waste and energy consumption. By emulating nature's efficient designs, engineers can create materials that offer enhanced performance without unnecessary bulk or waste. For instance, cooling systems inspired by termite structures may be more energy-efficient [13].

Adaptation to the environment. Nature has perfected the art of adaptation to diverse environments over millions of years. Organisms have evolved to thrive in extreme conditions, such as deserts or deep-sea trenches. These adaptive strategies provide insights into material properties that can withstand harsh conditions, resist degradation, and maintain structural integrity over time [14].

Hierarchical structures. Biological structures often consist of intricate hierarchical arrangements, from nano- to macro-scales. These structures contribute to exceptional mechanical properties such as strength, flexibility, and toughness. Mimicking hierarchical designs can lead to composite materials with strength similar to bone structures or with customized properties that outperform traditional homogeneous materials [15].

Self-healing and regeneration. Many living organisms have the ability to self-heal and regenerate. The study of these mechanisms has inspired the development of materials that can autonomously repair damage, extending their life span and reducing maintenance needs. Self-healing materials offer great potential in industries where durability is crucial, such as aerospace and infrastructure [16].

Functional surfaces. Nature's designs often feature specialized surfaces optimized for specific functions. Examples include lotus leaves with self-cleaning properties and sharkskin with ribbed surfaces for drag reduction (Fig. 8) [17]. These bio-inspired surface patterns find practical use in the development of superhydrophobic, superoleophobic, and even antibacterial materials [18].

Responsive behavior. Living organisms exhibit responsive behaviors that allow them to interact with their environment. This concept has spurred research on materials capable of responding to external stimuli, such as changes in temperature, humidity, or light [19]. Biomimetic, responsive materials hold promise for diverse applications across fields such as textiles, architecture, and electronics.

Integration of form and function. In nature, form and function are intricately intertwined. Structures often serve multiple purposes, optimizing efficiency [20]. This principle guides the creation of multifunctional materials where a single material can perform multiple tasks, simplifying designs and reducing complexity.

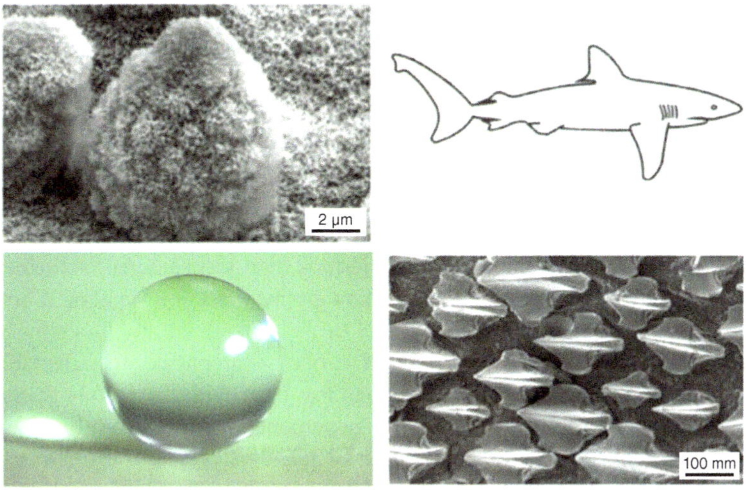

a b

Fig. 8 Two examples from nature: (**a**) lotus effect and (**b**) scale structure of shark reducing drag. (Reproduced from [17], License CC BY)

In summary, nature's designs are a database of solutions that have been optimized through countless iterations. By studying these designs and principles, scientists and engineers gain insights into the underlying mechanisms that drive nature's processes and provide strategies to replicate these insights into efficient materials with improved properties over those of traditional materials.

3.2 Principles of 4D Printing

Programming the controlled response into the material is crucial for achieving the intended transformation. The stimulus-responsive behavior depends on the selected stimuli that induce a change in shape and can manifest in various forms: thermo-responsiveness, the most prevalent stimulus in 4DP, along with photo-responsiveness, hygro-responsiveness, magneto-responsiveness, and chemo-responsiveness. Several studies have delved into significant key developments in this field, with the aim of exploring the various intersections between biomimicry and 4D printing technologies.

These studies categorize these intersections based on their functional mechanisms, such as self-actuating materials, nature-inspired shape transformations, shape-changing architectures, and multimaterial biofabrication for 4D printing [21]. Hence, closely connected to the foundational biological principles of biomimetics, at its core, 4D printing integrates the principles of smart materials, design and programming, and stimulus-responsive behavior to create structures that can shape-morph, self-assemble, self-repair, or adapt to various conditions.

3.3 The Intersection Between Biomimicry and 4D Printing

In a fundamental study conducted in 2015, researchers at Harvard's Wyss Institute for Biologically Inspired Engineering undertook a project centered on the development of engineered actuating structures drawing inspiration from natural botanical organisms [22].

Recognizing nature's profound influence, one must consider the adaptability of plants to environmental factors such as humidity, light, and temperature fluctuations [23]. These adaptations involve intricate movements that are driven by changes in internal turgor linked to the composition of plant tissue and the microstructural anisotropy of cell walls, as observed and investigated.

It is crucial to highlight two key concepts herein: shape-memory polymers and stimuli-responsive materials. Shape-memory polymers exhibit the remarkable capacity to retain a designated shape and subsequently revert to it when exposed to specific stimuli. In contrast, stimuli-responsive materials demonstrate the capability to react to alterations in their ambient conditions, such as variations in temperature or moisture, by modifying their inherent properties.

In the context of 3D and 4D printing and the creation of self-transforming structures, there is empirical evidence supporting the use of shape-memory polymers and stimuli-responsive materials. These materials are found to be suitable for constructing structures that can autonomously change their configuration or characteristics.

The 4D orchid In 2016, Jennifer Lewis and her research group at the Wyss Institute at Harvard University published the outcomes of their research [24].

Specifically, the team focused on the physical and biological qualities of cellulose, whose fibers make up much of a plant's structure, to print composite hydrogel architectures capable of reacting to different climate and weather conditions. For the experiment, the researchers introduced an innovative process involving the integration of a rotating nozzle system into a 3D printer.

This process was used to create an orchid composed of a composite hydrogel enriched with cellulose fibrils. This design aimed to facilitate anisotropic growth, enabling controlled growth in specific directions while restricting it in others (Fig. 9). This process involved the extrusion of a helical filament composed of four materials, similar to the helical patterns found in natural plants (Fig. 10) [25].

Mimicking the contraction processes observed in biological systems, researchers have termed this concept "living architecture," as hydrogel structure become alive in water [26]. One of the most notable aspects of this innovation lies in its predictability.

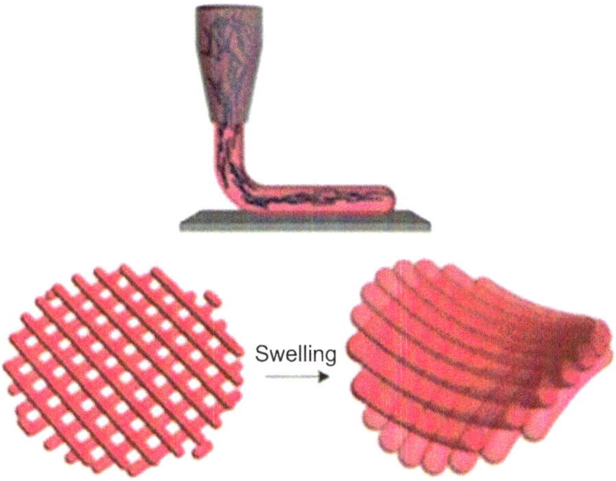

Fig. 9 Cellulose fibrils. Oriented structure. (Adapted from [25], License CC BY)

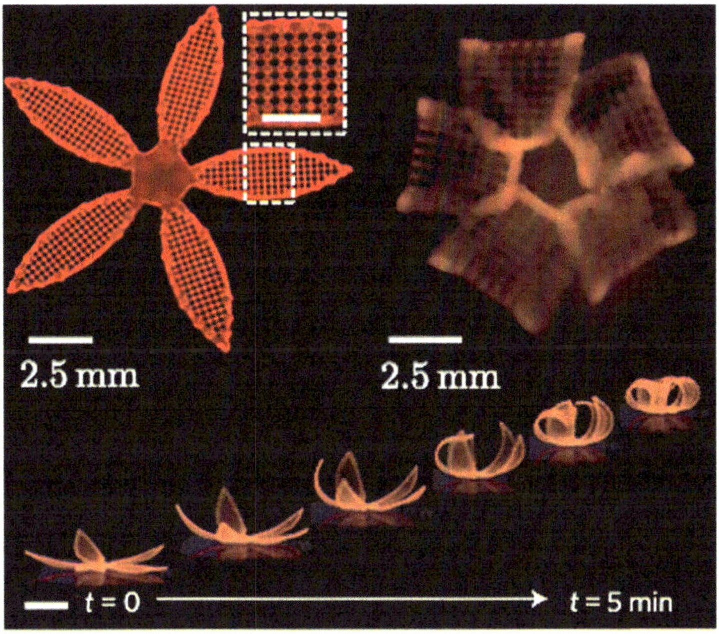

Fig. 10 Transformation of a hydrogel orchid enabled by 4D printing. (Adapted from [25], License CC-BY)

The hydrogel's characteristics allowed the researchers to formulate a precise mathematical model capable of predicting the dynamic morphology of the orchid with remarkable accuracy. In simpler terms, understanding and controlling the hydrogel material allows 4D-printed objects "to be taught" how to behave in various conditions.

Consequently, it becomes feasible to prescribe the extent and manner of an object's modification based on specific characteristics. While tested in water, this concept extends further, enabling the creation of materials that respond to varying levels of humidity or temperature. As a result, there is the potential to finely adapt the degree and method of transforming an object to be flexible for varying contexts [27].

This outcome extends beyond water-based testing and holds broader implications, including the creation of materials capable of responding to varying humidity levels and temperature changes.

Indeed, the hydrogel's ability to conduct electric charges adds an extra level of adaptability. Hence, the versatility of this technology goes from the microscopic to larger scales, enabling the development of small living devices and their use across various sectors.

The sector continued to evolve with a study by researchers from the Georgia Institute of Technology and the Singapore University of Technology and Design. They explored a range of shape-changing structures inspired by natural phenomena such as plant movements, demonstrating the potential of biomimetic principles in 4D printing [28]. A study conducted at the Swiss Federal Institute of Technology (ETH Zurich) in 2021 explored the integration of biomimetic principles with 4D printing to develop shape-changing architectures with hygroscopic characteristics.

Researchers investigated how responsive materials could be combined with structural design to respond to changes in humidity by expanding or contracting to achieve specific shapes. Such applications have wide-ranging potential, from adaptive architecture to biomedical devices that can change shape within the human body for enhanced functionality [29]. In this context, recent research conducted by Sonatkar et al. in 2022 delved into the field of multimaterial biofabrication for 4D printing, a process that involves shaping cellular structures through the transformation of 3D objects.

This study aimed to improve responsive structures by exploring various material combinations and stimuli to create versatile devices with potential applications in biomedicine [30]. Health professionals, especially in regions where access to 3D printing technology is limited, stand to greatly benefit from 4D printing, for creating dynamic structures and enabling the creation of more sophisticated implants and cutting-edge technologies.

4 Application Sectors and Future Impacts

The most significant application sectors benefiting from the convergence of biomimetic materials and 4D printing technologies concern the aerospace, healthcare, and architecture sectors. The adaptability adds a functional dimension, allowing materials to transition between states and reducing the need for continuous manual intervention.

One notable development comes from the Massachusetts Institute of Technology (MIT), where an innovative theory combines the principles of origami with 3D printing using shape-memory materials [31]. This theory takes inspiration from the Japanese art of origami, which involves creatively folding paper to create complex three-dimensional structures. It applies these origami principles to 3D printing, using materials capable of changing shape in response to stimuli (Fig. 11) [32, 33].

In essence, this theory involves the use of shape-memory materials that can be programmed to respond to specific conditions. This allows for the controlled creation of dynamic objects engineered to fold or change shape, achieved through a combination of geometric design and material properties [34].

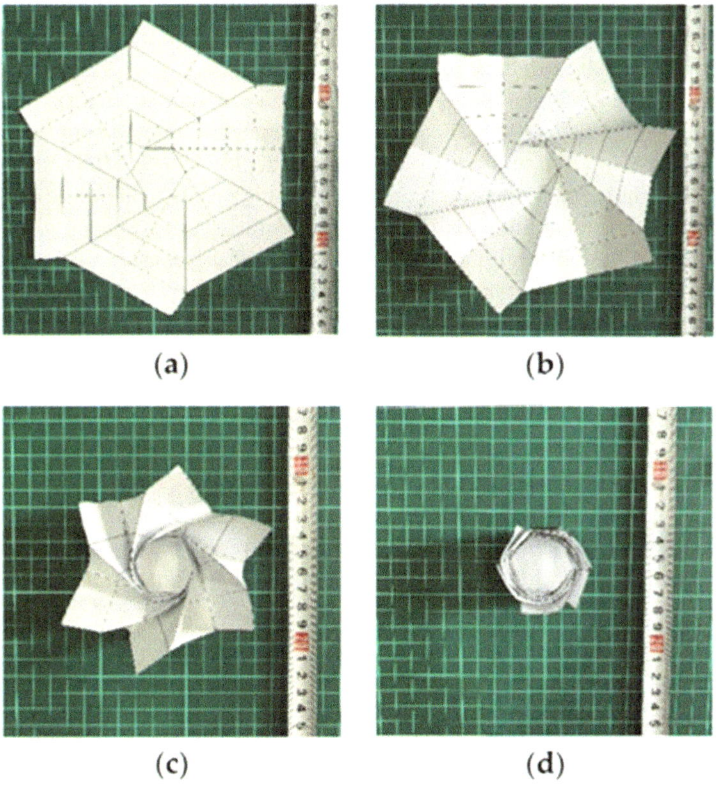

(a) (b) (c) (d)

Fig. 11 Folding sequences of the origami model of a pattern. (**a**) Completely deployed; (**b**, **c**) folding process 1; (**d**) folded. (Adapted from [32], Licensed CC BY)

4.1 *Aerospace*

4.1.1 Self-Deploying Structures Inspired by Origami and Plant Movements

In the field of aerospace engineering, a deep connection with nature emerges when creating intelligent structures for space exploration and satellites. An example of self-deploying aerospace structures inspired by origami and plant movements is the concept of deployable solar arrays used on spacecraft.

These solar arrays are designed to fold into a compact form during launch and then autonomously unfurl and position themselves in space once the spacecraft reaches its intended orbit [35]. The idea behind this concept is similar to how certain plants unfurl their leaves or flowers in response to environmental cues.

By drawing inspiration from origami and plant movements, solar arrays were developed with shape-memory alloy mechanisms that can be efficiently stowed for launch and then deployed to capture sunlight for power generation in space. This technology allows spacecraft to generate electricity for their missions, significantly enhancing their self-sufficiency and mission success.

Indeed, these intelligent materials provide the necessary energy to facilitate the unfolding process when required, and the stored energy within the unfolding mechanisms ensures precise and autonomous panel deployment [35]. This innovation reduces the requirement for complex and deployable mechanisms while eliminating the necessity for human intervention in the zero-gravity environment.

NASA's Jet Propulsion Laboratory in Pasadena, California, and Brigham Young University in Provo, Utah, have joined forces to create a prototype of a solar panel array that unfolds like origami, simplifying the deployment process (Fig. 12) [36]. These self-deployable solar panels are crafted from lightweight and flexible materials, such as thin-film solar cells and composite materials, with characteristics of durability, minimal weight, and resilience in the space environment. Initially compactly stored, these panels assume their ultimate configuration as the spacecraft or module reaches orbit, reducing their overall length and associated risks [37].

This technology has been exploited on various spacecraft, including communication satellites, scientific missions, and even Mars rovers, where self-deploying solar panels play a crucial role in ensuring the mission's accomplishment by providing a sustainable power source in the particular conditions of space.

The progression of self-deploying aerospace structures provides several benefits from efficiency and power generation to risk mitigation and the advancement of engineering knowledge [38, 39]. These adaptable structures, tailored to the unique demands of each mission, demonstrate their functioning within critical domains of exploration and aerospace engineering. They also deepen the understanding of biomimetic materials and their compatibility with 4D printing technologies.

Fig. 12 NASA's James Webb Space Telescope. (Reproduce from [36], License CC0)

4.2 *Medicine and Health*

4.2.1 Bacteria as Raw Material Suppliers for 4D Printing

Researchers at the University of Coimbra (UC) made considerable progress in enhancing the sustainability and ecological elements of 4D printing in 2021 by employing bacteria-produced cellulose as a raw material [40].

Their innovative strategy consists of exploiting the metabolic activity of bacteria in order to produce materials that adapt and transform over time when subjected to specific triggers.

Bacteria have evolved over millions of years, learning and adapting. Researchers, with help from the UC Environmental Microbiology Team, have chosen the best bacteria for making a versatile natural polymer that reacts to temperature changes [41]. In this process, the work of the bacteria produces pure cellulose, unlike that obtained from trees, which requires elaborate purification procedures.

Therefore, the celluloses obtained were mixed with polymers with different properties in order to produce biocomposites and develop filaments suitable for 4D printing. From this point of view, this innovative technology not only highlights the importance of sustainability but also proves cost-effective. Bacteria thrive on food waste to produce cellulose, and 97% of the material used remains integrated within the molded part, curbing waste production. Any residual waste can be reused to generate additional filaments, minimizing environmental impact [42].

Composite materials and bacterial cellulose are examples of biodegradability and biocompostability, demonstrating that biomanufacturing can be highly efficient in producing only what is actually needed, thus reducing the waste of resources and materials.

The ability to biofabricate tissues, organs, and even foods tailored to individual needs paves the way for personalized medical treatments and improved nutrition. Personalized medicine, a term that describes medical treatment tailored to the individual and responding to a patient's physiological changes over time, has grown in popularity since the Precision Medicine Initiative in 2015 [43].

Leveraging 4D printing technologies, biofabrication enables the creation of living tissues and artificial organs, which could revolutionize the field of regenerative medicine by enabling personalized organ transplantation and the self-repair of damaged tissue.

In summary, biomanufacturing offers a number of benefits that go beyond traditional industrial production, helping to meet people's needs in a sustainable and efficient way to improve human health and the environment. Achieving the reversibility of materials is one of the most significant challenges, not only in the Coimbra researchers' project but in biofabrication in general.

To address this challenge, researchers are exploring a number of approaches, including the use of smart materials, genetic engineering to create cells or organisms that produce reversible materials, and the design of material structures that can recover their original shape when subjected to certain conditions.

However, it is still an evolving research area and requires further efforts to achieve reliable reversibility in applications.

4.2.2 Bioinspired Soft Robotics for Medical Devices

The impact of 4D printing can be particularly significant in the field of soft robotics, which focuses on creating flexible and highly adaptable automata using deformable materials [44].

Soft robotics has the potential to bring about a profound transformation in the medical field due to its unique qualities and capabilities. Unlike traditional robotics, which often relies on rigid materials and mechanisms, soft robotics employs biocompatible materials that are well-tolerated by the human body.

This biocompatibility makes them suitable for use in medical devices and implants, without causing harm to patients or triggering immune responses (Fig. 13) [45]. One of the key advantages of soft robotics is its inherent flexibility and compliance. Soft robotic devices can bend, stretch, and conform to complex shapes, which is crucial when interacting with delicate human tissues and organs.

This flexibility reduces the damage or injury risk during medical procedures and surgeries. For instance, soft endoscopes can navigate through narrow and winding passages within the body with minimal trauma, resulting in shorter recovery times and fewer complications compared to traditional surgical approaches [46].

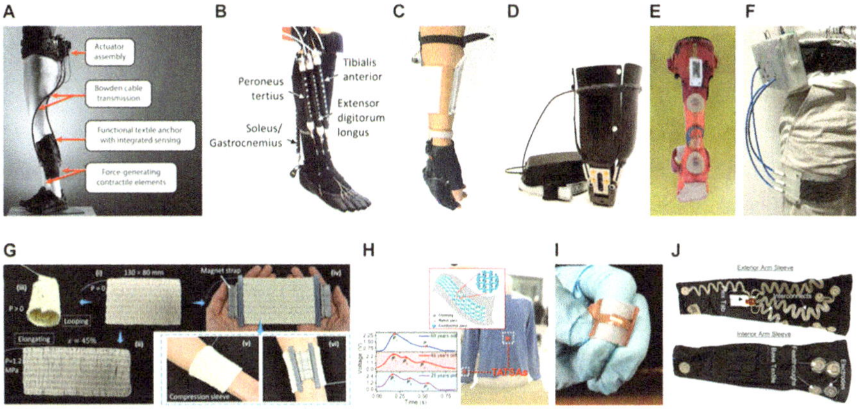

Fig. 13 Use of soft robotics in various medical applications. (Reproduced from [45], License CC BY)

Furthermore, implantable soft robots can be utilized for the targeted delivery of drugs, such as insulin when high blood sugar is detected, or other therapies inside the body. This approach minimizes side effects and enhances the efficacy of the treatment.

The customizability of soft robotic devices is particularly valuable in fields like prosthetics and orthotics. Since each patient may have unique anatomical requirements, these devices can be tailored to provide optimal comfort and functionality [45].

Consequently, these devices can adapt to changes in a patient's condition or anatomy, ensuring continuous comfort and support while also facilitating safer and more natural interactions between humans and machines.

In summary, soft robotics can radically transform the healthcare field by introducing devices and technologies that better adapt to the human body, enhance patient care, reduce invasiveness, and improve the overall quality of life for individuals with various medical conditions.

4.3 Architecture

4.3.1 A 4D-Printed House Hypothesis for Future Adaptations

In contemporary architecture, the search for innovative solutions to improve the energy efficiency of buildings and combine them with the evolving needs of the occupants is crucial [47].

This study proposes the hypothesis of a 4D-printed house, supported by biomimetic materials, capable of dynamically adapting to the environmental conditions and the needs of the occupants. The goal is to create a structure that not only consumes less energy but also changes according to specific user requests.

Methodology The methodological approach is based on the elaboration of the following phases:

1. Selection of a biological model to identify organisms or biological systems that present exceptional adaptations in terms of structure, resistance, and environmental sustainability, such as bee hives or plant tissues [12].
2. An in-depth analysis of the characteristics and strategies adopted by these organisms for environmental construction and resistance [12].
3. Adaptation of the biological model to 4D technology in order to transfer specified biological traits into design principles that can be printed in 4D and development of algorithms and computational models for the application of biomimetics in the 4D printing process, to ensure that nature-inspired concepts function as 3D printed objects change shape over time [48].
4. Material selection: to identify sustainable and 4D printing-compatible materials, such as bio-based composite materials or advanced polymers [49].
5. Evaluation of the thermal, structural, and environmental properties of the selected materials to ensure consistency with the biomimetic approach [50].

Expected Results The application of this approach is expected to produce significant outcomes, including the 3D printing of houses with efficient and resilient structures inspired by nature, enhancements in the environmental sustainability of residential construction, and the development of a replicable model for the large-scale application of this innovative technology.

Materials and Methods In this hypothesis, two types of bio-based shape-memory polymers are employed for the structure: cellulose-inspired hydrogels and bio-based composites (bioelastomers) integrating muscle-inspired polymers.

Cellulose-Inspired Hydrogel This material is designed to mimic the properties of plant cells, with the ability to absorb and release water in response to environmental changes [10]. This controls the movement of liquids and gases within the building, which improves ventilation and contributes to energy efficiency.

Muscle-Inspired Bioelastomers These materials mimic some of the properties and behaviors similar to natural muscle tissue, responding to stimuli such as heat, light, or electric fields by expanding, contracting, or generating mechanical force (Fig. 14) [51].

The material for the panels is thought to be made of a mesh structure of shape memory polymer that contains small pockets of air or liquid. During the day, when exposed to sunlight, the polymer may react by expanding slightly.

This expanding motion would open the air pockets or result in small amounts of liquid being pushed out. This process may improve air circulation within the structure, contributing to natural ventilation and cooling. On the other hand, at night or in cold conditions, the polymer may contract slightly, causing the air pockets to shrink in volume or push the liquid out [52]. This could increase the thermal insulation of the structure, helping to retain heat and improving energy efficiency.

Conceptual Framework Traditional construction is reimagined through the integration of advanced 3D printing technology and shape-shifting modules [53]. The core idea revolves around the utilization of 3D-printed modules, which not only establish the structural foundation of the house but also have the capability to adapt their configuration over time.

Shape Memory and Environmental Response Each module is designed and 3D printed, as an independent entity with self-contained functionality and made with materials specialized for their shape memory properties and response to specific

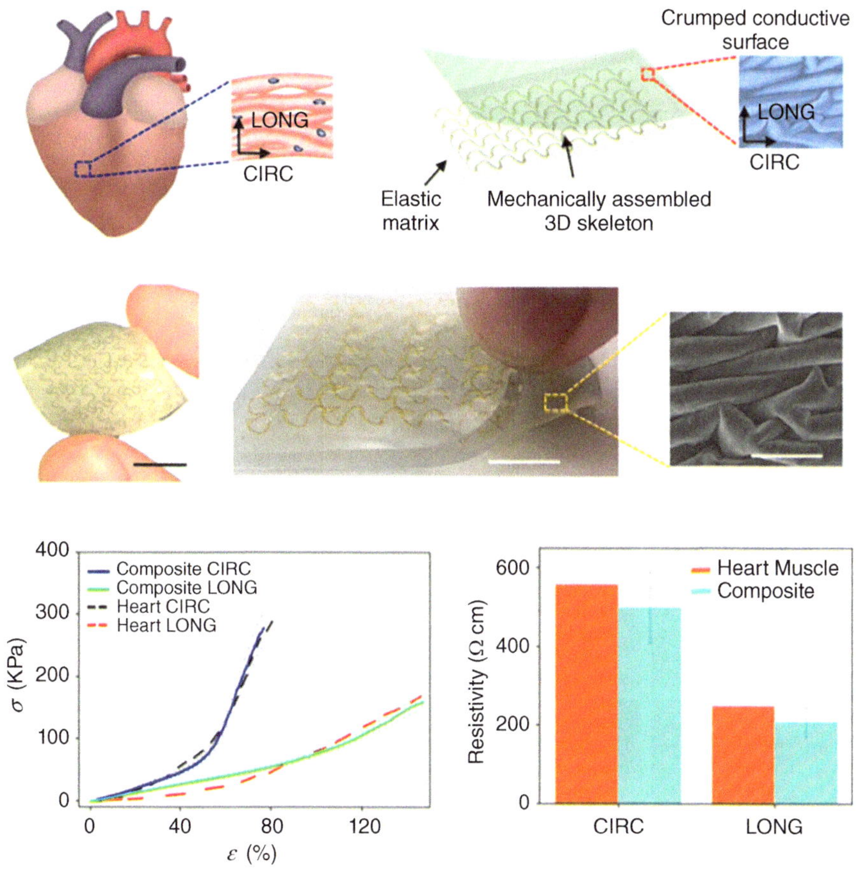

Fig. 14 Bio-inspired elastomer composites that behave like muscles, with mechanical and electrical anisotropies. (Adapted from [51], License CC BY)

environmental stimuli, including temperature fluctuations, humidity levels, or exposure to light [54, 55].

Technological Implementation The 3D printing process requires the use of advanced printers capable of intricately layering the chosen materials. The design specifications incorporate connectors, mechanisms, or joints that allow the modules to interconnect seamlessly [56].

Spatial Flexibility and User Experience Occupants have the option to change the spatial arrangement of their house, to create a configuration adapted to their changing demands by removing certain modular components and strategically relocating them.

4 Adaptive Modes It is also possible to classify four adaptation modes, each designed to meet specific structural and environmental needs, as described below.

1. Semipermeable Mode (Stretched Membranes) (Fig. 15)

Material. For the semipermeable mode, the house incorporates stretchable membranes composed of advanced polymer materials with elastic and breathable properties. These materials optimize ventilation and internal comfort, allowing controlled passage of air [57].

Method. During the day, when sunlight is strong, the stretched membranes respond by allowing the panels to stretch, curve, and provide shading, transitioning to the semipermeable mode. Further, the membrane creates a barrier that permits airflow while offering partial enclosure, reducing the amount of direct sunlight entering the building [58].

This helps in passive temperature regulation, optimizing natural ventilation and preventing excessive heat buildup within the building interior. Then, as the sun sets or weather conditions change, the membranes gradually return to their original state, allowing more natural light to enter.

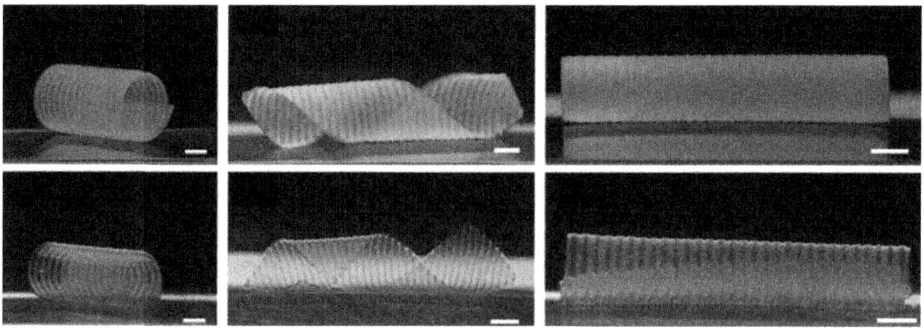

Fig. 15 3D shape deformations of 4D-printed simple patterned hydrogels with tube-curling structures, helix structures, and rolling structures in the liquid solution. (Adapted from [9], License CC BY)

2. Nesting Mode (Folded Plates/Walls) (Fig. 16) [59].

Material. Folding composite materials allow for the optimization of space when needed. Composite materials with foldable properties are chosen, allowing plates or walls to be bent without affecting structural stability. These materials draw inspiration from natural structures, such as insect wings or plant leaves, which can fold or unfurl in response to environmental changes [20].

Method: During the 4D printing process, these foldable materials are integrated into the modular components. When the nesting mode is activated, the plates or walls systematically fold using hinges or creases, compacting the structure efficiently [60]. This configuration can be useful for small living spaces or transportable units.

3. Deployable Mode (Released Plates/Unfolded Walls)

Material. These shape-memory materials are based on fiber-reinforced polymer composites (FRPs) and possess elasticity, enabling them to undergo repeated folding and unfolding without deformation (Fig. 17) [61].

Method: The shape-memory materials are embedded within the modular components. Activating the deployable mode triggers the release of stored energy within these materials, causing the plates or walls to unfold and expand [24]. This mode offers users the flexibility to create open and personalized spaces that cover diverse purposes and evolving needs.

4. Rigid mode (Locked Plates/Walls)

Material. The rigid mode necessitates materials recognized for their robustness and structural stability. Bio-based fiber-reinforced composites are used to ensure

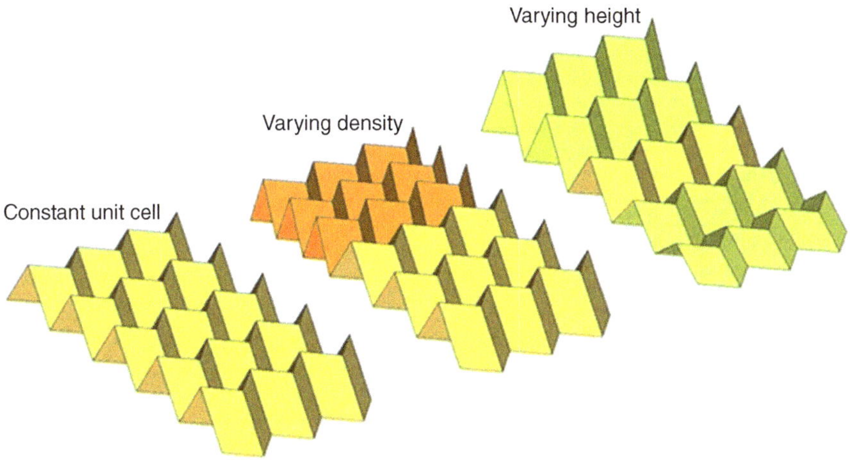

Fig. 16 Self-deployable structures inspired by origami: zigzag unit cell (from left), with varying densities and heights. (Reproduced from [59], License CC BY)

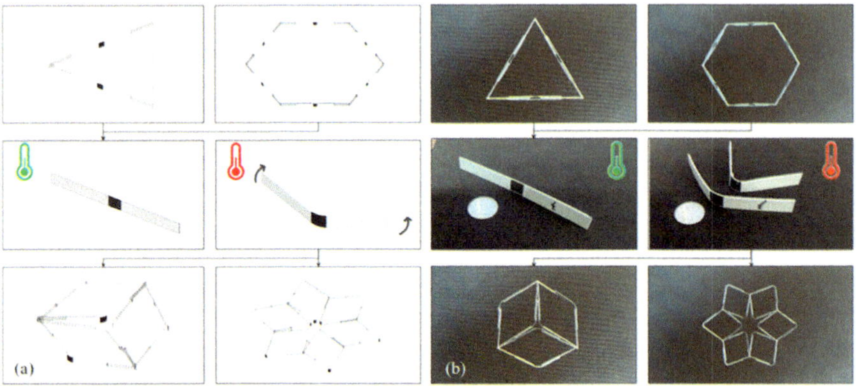

Fig. 17 Form controlled planar folding mechanism based on 4D printing technology. (a) Digital models, (b) prototype machines. (Reproduced from [61], License CC BY)

that the plates or walls can withstand external forces and maintain their shape without flexing [49, 62].

Method. The 4D printing process integrates these strong and rigid materials within the modular components. Switching to the rigid mode involves locking mechanisms that secure the plates or walls in position, conferring structural stability and integrity. This mode can provide benefits during adverse weather conditions or when the structure requires additional support.

Conclusion In summary, the hypothesis demonstrates that the materials and methods used for the adaptive modes of the 4D-printed modular house can match the specific requirements of each mode. The careful selection of materials and the precise integration of shape-memory, flexible, and rigid components allow the house to change configurations, providing space efficiency, ventilation, and stability as needed.

In practice, the printed house can have sections that flex, expand, or contract as needed, thereby increasing its resilience even under severe weather or earthquakes. Moreover, bio-inspired materials not only contribute to the structural efficiency of the 4D-printed house but also align with sustainable and eco-friendly building practices.

With the combination of smart biomimetic materials and 4D printing technologies, housing structures or building components can self-repair or adapt to their environment overstime, in line with the principles of living architecture (Fig. 18) [63].

Fig. 18 Biomimetic self-deployable structures with potential applications in building façades. (Reproduced by author)

5 Findings

All the material here explored offers a deep understanding of the results obtained from the intersection of biomimicry with 4D printing technology and the possibilities that the developments of continuous research and experimentation can provide. It is possible to generalize the main benefits as follows.

Innovative material development. By combining biomimetics concepts, which take inspiration from nature's efficient and time-tested designs, with the capabilities of 4D printing, emerging advances boost advancement in performing smart materials that can self-adapt for varied functions.

Revolutionizing design paradigms. This intersection challenges conventional design paradigms by enabling the creation of materials that blur the boundaries between static and dynamic, responsive, and passive. This has the potential to lay the groundwork for entirely novel paradigms in the fields of engineering and manufacturing.

Multifunctionality. The integration between biomimetics and 4D printing has the potential to create materials that exhibit multiple functionalities within a single structure, leading to streamlined designs and reduced material waste.

Sustainable solutions. The emulation of nature's efficiency through biomimetic design can contribute to more sustainable and environmentally friendly material solutions, while 4D printing's adaptability can lead to longer-lasting products and reduced consumption.

Cross-disciplinary collaboration. Exploring this intersection requires collaboration across diverse disciplines, including biology, materials science, engineering, and design. This multidisciplinary strategy promotes idea exchange and accelerates technical progress.

Future applications. The knowledge acquired through the exploration of the synergy between biomimetics and 4D printing is bringing about a revolution in various fields, ranging from aerospace (deployable structures), healthcare (smart implants), and architecture (adaptive structures) to robotics (shape-shifting components) [64].

6 Conclusions: Challenges and Prospects

The article demonstrates that, although they are still little explored due to the complexity and richness of our biological system, biomimetic materials promise to bring innovations to different fields of engineering through innovative technologies such as 4D printing.

The intersection between bio-inspired materials and 4D printing represents a very promising research field but presents several challenges to face. First and foremost, this research's multidisciplinary reinforcement is essential since it necessitates more collaborative efforts across biology, material science, engineering, and design. Success in general depends on integrating information from several fields.

The necessity for 4D printing to deposit materials layer by layer and program their future behavior demands progress in bio-inspired materials research, printing technologies, and design software. Since these materials might alter or degrade over time, it is also important to consider their stability and durability. The difficulty in assessing its long-term performance is a key contributor to concerns about issues such as safety, capacity to endure wear, pressure, or damage, and aesthetic appearance.

While the evolution of 3D printing toward advanced plug-and-play capabilities is becoming more evident, the same progression does not characterize 4D printing [66]. The latter mainly represents an experimental technology, limited to the realm of experts. In fact, the intrinsic complexity of 4D printing limits its access to a wider audience and, consequently, creates further obstacles to its diffusion and large-scale adoption beyond the confines of just research laboratories and technical sectors.

For instance, introducing the biomimetic discipline of 4D printing as a compulsory subject in all technical high schools and universities could be a beneficial practice aimed at spreading knowledge of its benefits and facilitating its democratization [66].

Large-scale production is another challenge to overcome. In the industrial sector, careful planning regarding investments in bio-inspired materials for 4D printing and tax incentives can encourage companies to invest in and exploit this technology, thus improving productivity and saving costs. On the other hand, ethical and regulatory issues need to be considered as new materials and technologies are developed. This includes evaluating the safety of materials and their impact on the environment.

Hence, the validation of practical applications is essential. Establishing the effectiveness of these emerging technologies in real-world scenarios requires concrete evidence of their functionality and safety, which entails continuous research, development initiatives, and increased institutional support to validate their viability.

References

1. Benyus, J. M. (1997). *Biomimicry: Innovation inspired by nature* (p. 320). Morrow.
2. ElDin, N. N., Abdou, A., & Abd ElGawad, I. (2016). Biomimetic potentials for building envelope adaptation in Egypt. *Procedia Environmental Sciences, 34*, 375–386.
3. Zhao, N., Wang, Z., Cai, C., Shen, H., Liang, F., Wang, D., & Xu, J. (2014). Bioinspired materials: From low to high dimensional structure. *Advanced Materials, 26*(41), 6994–7017.
4. Snell-Rood, E. (2016). Interdisciplinarity: Bring biologists into biomimetics. *Nature, 529*(7586), 277–278.
5. Tibbits, S. (2013). The emergence of 4D printingl TED talk.
6. Kotz, F., Helmer, D., & Rapp, B. E. (2020). Emerging technologies and materials for high-resolution 3D printing of microfluidic chips. *Microfluidics in Biotechnology*, 37–66.
7. Ge, Q., Sakhaei, A. H., Lee, H., Dunn, C. K., Fang, N. X., & Dunn, M. L. (2016). Multimaterial 4D printing with tailorable shape memory polymers. *Scientific Reports, 6*(1), 31110.
8. Guan, Z., Wang, L., & Bae, J. (2022). Advances in 4D printing of liquid crystalline elastomers: Materials, techniques, and applications. *Materials Horizons, 9*(7), 1825–1849.
9. Lai, J., Ye, X., Liu, J., Wang, C., Li, J., Wang, X., et al. (2021). 4D printing of highly printable and shape morphing hydrogels composed of alginate and methylcellulose. *Materials & Design, 205*, 109699.
10. Magalhães, M. I., & Almeida, A. P. (2023). Nature-inspired cellulose-based active materials: From 2D to 4D. *Applied Biosciences, 2*(1), 94–114.
11. Yang, Y., Song, X., Li, X., Chen, Z., Zhou, C., Zhou, Q., & Chen, Y. (2018). Recent progress in biomimetic additive manufacturing technology: From materials to functional structures. *Advanced Materials, 30*(36), 1706539.
12. Jamei, E., & Vrcelj, Z. (2021). Biomimicry and the built environment, learning from nature's solutions. *Applied Sciences, 11*(16), 7514.
13. Di Salvo, S. (2018). Advances in research for biomimetic materials. *Advanced Materials Research, 1149*, 28–40.
14. Chen, Y., Dang, B., Wang, C., Wang, Y., Yang, Y., Liu, M., et al. (2023). Intelligent designs from nature: Biomimetic applications in wood technology. *Progress in Materials Science, 101164*.
15. Luz, G. M., & Mano, J. F. (2010). Mineralized structures in nature: Examples and inspirations for the design of new composite materials and biomaterials. *Composites Science and Technology, 70*(13), 1777–1788.
16. Vates, U. K., Mishra, S., & Kanu, N. J. (2021). Biomimetic 4D printed materials: A state-of-the-art review on concepts, opportunities, and challenges. *Materials Today: Proceedings, 47*, 3313–3319.
17. Bhushan, B. (2011). Biomimetics inspired surfaces for drag reduction and oleophobicity/philicity. *Beilstein Journal of Nanotechnology, 2*(1), 66–84.
18. Barthlott, W., Mail, M., & Neinhuis, C. (2016). Superhydrophobic hierarchically structured surfaces in biology: Evolution, structural principles and biomimetic applications. *Philosophical Transactions of the Royal Society A: Mathematical, Physical and Engineering Sciences, 374*(2073), 20160191.
19. Wang, J., Wang, Z., Song, Z., Ren, L., Liu, Q., & Ren, L. (2019). Biomimetic shape-color double-responsive 4D printing. *Advanced Materials Technologies, 4*(9), 1900293.

20. Khalid, M. Y., Arif, Z. U., & Ahmed, W. (2022). 4D printing: Technological and manufacturing renaissance. *Macromolecular Materials and Engineering, 307*(8), 2200003.
21. Yun Lu Tee & Phuong Tran. (2021). On bioinspired 4D printing: Materials, design and potential applications. *Australian Journal of Mechanical Engineering, 19*(5), 642–652. https://doi.org/10.1080/14484846.2021.1988434
22. Hardin, J. O., Ober, T. J., Valentine, A. D., & Lewis, J. A. (2015). Microfluidic printheads for multimaterial 3D printing of viscoelastic inks. *Advanced Materials, 27*(21), 3279–3284.
23. Ryan, K. R., Down, M. P., & Banks, C. E. (2021). Future of additive manufacturing: Overview of 4D and 3D printed smart and advanced materials and their applications. *Chemical Engineering Journal, 403*, 126162.
24. Sydney Gladman, A., Matsumoto, E. A., Nuzzo, R. G., Mahadevan, L., & Lewis, J. A. (2016). Biomimetic 4D printing. *Nature Materials, 15*(4), 413–418.
25. Han, I. K., Chung, T., Han, J., & Kim, Y. S. (2019). Nanocomposite hydrogel actuators hybridized with various dimensional nanomaterials for stimuli responsiveness enhancement. *Nano Convergence, 6*, 1–21.
26. Jang, T. S., Jung, H. D., Pan, H. M., Han, W. T., Chen, S., & Song, J. (2018). 3D printing of hydrogel composite systems: Recent advances in technology for tissue engineering. *International Journal of Bioprinting, 4*(1).
27. Taylor, J. M., Luan, H., Lewis, J. A., Rogers, J. A., Nuzzo, R. G., & Braun, P. V. (2022). Biomimetic and biologically compliant soft architectures via 3D and 4D assembly methods: A perspective. *Advanced Materials, 34*(16), 2108391.
28. Advances in 4D printing: Materials and applications. *Advanced Functional Materials, 29*(2), 1805290.
29. Grönquist, P., Panchadcharam, P., Wood, D., Menges, A., Rüggeberg, M., & Wittel, F. K. (2020). Computational analysis of hygromorphic self-shaping wood gridshell structures. *Royal Society Open Science, 7*(7), 192210.
30. Sonatkar, J., Kandasubramanian, B., & Ismail, S. O. (2022). 4D printing: Pragmatic progression in biofabrication. *European Polymer Journal, 169*, 111128.
31. Mao, Y., Yu, K., Isakov, M. S., Wu, J., Dunn, M. L., & Jerry Qi, H. (2015). Sequential self-folding structures by 3D printed digital shape memory polymers. *Scientific Reports, 5*(1), 13616.
32. Lin, Q., Jia, W., Wu, H., Kueh, A. B., Wang, Y., Wang, K., & Cai, J. (2021). Wrapping deployment simulation analysis of leaf-inspired membrane structures. *Aerospace, 8*(8), 218.
33. Alshahrani, H. A. (2021). Review of 4D printing materials and reinforced composites: Behaviors, applications and challenges. *Journal of Science: Advanced Materials and Devices, 6*(2), 167–185.
34. Behl, M., & Lendlein, A. (2007). Shape-memory polymers. *Materials Today, 10*(4), 20–28.
35. Xin, X., Liu, L., Liu, Y., & Leng, J. (2020). Origami-inspired self-deployment 4D printed honeycomb sandwich structure with large shape transformation. *Smart Materials and Structures, 29*(6), 065015.
36. Image from https://www.flickr.com/photos/nasawebbtelescope/50221539843/
37. Wang, X., He, Y., Liu, Y., & Leng, J. (2022). Advances in shape memory polymers: Remote actuation, multi-stimuli control, 4D printing and prospective applications. *Materials Science and Engineering: R: Reports, 151*, 100702.
38. Joshi, S., Rawat, K., Karunakaran, C., Rajamohan, V., Mathew, A. T., Koziol, K., Thakur, V. K., & Balan, A. S. S. (2020). 4D printing of materials for the future: Opportunities and challenges. *Applied Materials Today, 18*, 100490.
39. Liu, Z. Q., Qiu, H., Li, X., & Yang, S. L. (2017). Review of large spacecraft deployable membrane antenna structures. *Chinese Journal of Mechanical Engineering, 30*(6), 1447–1459.
40. Kanaan, A. F., & Piedade, A. P. (2022). Electro-responsive polymer-based platforms for electrostimulation of cells. *Materials Advances, 3*(5), 2337–2353.
41. Piedade, A. P., & Pinho, A. C. (2022). 4D-printed stimuli-responsive hydrogels modeling and fabrication. In *Smart materials in additive manufacturing* (pp. 151–192). Elsevier.

42. Pinho, A. C., & Piedade, A. P. (2023). Stimuli-responsive smart materials for additive manufacturing. *Nanotechnology-Based Additive Manufacturing: Product Design, Properties and Applications, 1*, 249–276.
43. Green, S., Prainsack, B., & Sabatello, M. (2023). Precision medicine and the problem of structural injustice. *Medicine, Health Care and Philosophy*, 1–18.
44. Whitesides, G. M. (2018). Soft robotics. *Angewandte Chemie International Edition, 57*(16), 4258–4273.
45. Paternò, L., & Lorenzon, L. (2023). Soft robotics in wearable and implantable medical applications: Translational challenges and future outlooks. *Frontiers in Robotics and AI, 10*, 1075634.
46. Ahmed, A., Arya, S., Gupta, V., Furukawa, H., & Khosla, A. (2021). 4D printing: Fundamentals, materials, applications and challenges. *Polymer, 228*, 123926.
47. Demoly, F., Dunn, M. L., Wood, K. L., Qi, H. J., & Andre, J. C. (2021). The status, barriers, challenges, and future in design for 4D printing. *Materials & Design, 212*, 110193.
48. López, M., Rubio, R., Martín, S., Croxford, B., & Jackson, R. (2015). Active materials for adaptive architectural envelopes based on plant adaptation principles. *Journal of Facade Design and Engineering, 3*(1), 27–38.
49. AL-Oqla, F. M., Hayajneh, M. T., & Nawafleh, N. (2023). Advanced synthetic and biobased composite materials in sustainable applications: A comprehensive review. *Emergent Materials, 6*, 1–18.
50. Zhang, Z., Liu, G., Li, Z., Zhang, W., & Meng, Q. (2023). Flexible tactile sensors with biomimetic microstructures: Mechanisms, fabrication, and applications. *Advances in Colloid and Interface Science*, 102988.
51. Ling, Y., Pang, W., Liu, J., Page, M., Xu, Y., Zhao, G., et al. (2022). Bioinspired elastomer composites with programmed mechanical and electrical anisotropies. *Nature Communications, 13*(1), 524.
52. Jayawardena, I., Turunen, P., Garms, B. C., Rowan, A., Corrie, S., & Grøndahl, L. (2023). Evaluation of techniques used for visualisation of hydrogel morphology and determination of pore size distributions. *Materials Advances, 4*(2), 669–682.
53. Yao, X., Chen, H., Qin, H., & Cong, H. P. (2023). Nanocomposite hydrogel actuators with ordered structures: From nanoscale control to macroscale deformations. Small. *Methods, 2300414*.
54. Borra, N. D., Neigapula, V. S. N., Sanivada, U. K., & Fangueiro, R. (2023). Tailoring the shape memory properties of silica nanoparticle infused photopolymer composites for 4D printing applications: A Taguchi analysis. *European Polymer Journal, 194*, 112174.
55. Zhao, W., Li, N., Liu, L., Leng, J., & Liu, Y. (2023). Mechanical behaviors and applications of shape memory polymer and its composites. *Applied Physics Reviews, 10*(1).
56. Žujović, M., Obradović, R., Rakonjac, I., & Milošević, J. (2022). 3D printing technologies in architectural design and construction: A systematic literature review. *Buildings, 12*(9), 1319.
57. Fu, P., Li, H., Gong, J., Fan, Z., Smith, A. T., Shen, K., & Sun, L. (2022). 4D printing of polymers: Techniques, materials, and prospects. *Progress in Polymer Science, 126*, 101506.
58. Smith, S. I. (2017). Superporous intelligent hydrogels for environmentally adaptive building skins. *MRS Advances, 2*, 2481–2488.
59. Fischer, S. (2015). Realistic Fe simulation of foldcore sandwich structures. *International Journal of Mechanical and Materials Engineering, 10*(1), 1–11.
60. Zhang, Z., Demir, K. G., & Gu, G. X. (2019). Developments in 4D-printing: A review on current smart materials, technologies, and applications. *International Journal of Smart and Nano Materials, 10*(3), 205–224.
61. Zhang, W., Ge, Z., & Li, D. (2023). Design and research of form controlled planar folding mechanism based on 4D printing technology. *Chinese Journal of Mechanical Engineering, 36*(1), 1–13.
62. Koch, S. M., Grönquist, P., Monney, C., Burgert, I., & Frangi, A. (2022). Densified delignified wood as bio-based fiber reinforcement for stiffness increase of timber structures. *Composites Part A: Applied Science and Manufacturing, 163*, 107220.

63. Pittas, P. *Exploring self-repairing materials and their application towards sustainable design*. Obtenido de: https://unswcode. org/wpcontent/uploads/2018/12/Exploring_Self-Repairing_Materials_and_their_Applications_Towards_Sustainable_Design. pdf
64. Mallakpour, S., Tabesh, F., & Hussain, C. M. (2021). 3D and 4D printing: From innovation to evolution. *Advances in Colloid and Interface Science, 294*, 102482.
65. Joharji, L., Mishra, R. B., Alam, F., Tytov, S., Al-Modaf, F., & El-Atab, N. (2022). 4D printing: A detailed review of materials, techniques, and applications. *Microelectronic Engineering*, 111874.
66. Vatanparast, S., Boschetto, A., Bottini, L., & Gaudenzi, P. (2023). New trends in 4D printing: A critical review. *Applied Sciences, 13*(13), 7744.

Santina Di Salvo, Architect, Ph.D. in Architecture. Associate Professor at Polis University of Tirana, she focuses on biomimetic materials research for resilient and sustainable architecture. She was the curator and coordinator of the project "Perspectives of Resilient Architecture" promoted by the Italian Cultural Institute of Melbourne and taught sustainable architecture topics in Italy and abroad. She is author of patents and publications on issues on Architectural Technology in relevant Journals.

Innovative Trends in Academic Pavilions: Exploring Processes in Architecture

Elton Cristovão da Silva Lima, Luís Miguel Cotrim Mateus, and Amilton José Vieira Arruda

1 Introduction

The last decade has seen considerable growth in the construction of full-scale study pavilions, increasing exponentially in recent years. This is due to the popularization of access to prototyping and digital fabrication laboratories in the university context, and the fact that the pavilion serves as a subject of exercise in project disciplines and workshops that seek to explore parametric design, form conception, generative algorithms, and digital fabrication [10]. When this type of experience derives inspiration from nature, it is possible to adopt and manipulate a series of parameterized constraints that make the design process more flexible, thereby achieving a high degree of innovation. Furthermore, it is possible to establish a relationship between biomimicry, digital fabrication, and architecture [8].

Although the field of biomimicry has its roots in observations and experimentation that go back centuries, it only gained widespread recognition more recently, in 1997, when the American natural science writer Janine Benyus released her influential book *Biomimicry: Innovation Inspired by Nature*. Benyus defined biomimicry as a discipline that seeks inspiration from nature, studying its ideas and imitating its designs and processes to solve human challenges, aligned with the principles of sustainability and resource efficiency [1]. Since then, this approach has spread to various fields of research and industries, where companies from a variety of niches have adopted this methodology/philosophy to improve products and services. This

E. C. da Silva Lima (✉) · L. M. C. Mateus
Lisbon School of Architecture, University of Lisbon, Lisbon, Portugal
e-mail: elton.lima@edu.ulisboa.pt; lmmateus@fa.ulisboa.pt

A. J. V. Arruda
BioDesign Lab, Federal University of Pernambuco, Recife, PE, Brazil
e-mail: amilton.arruda@ufpe.br

A. J. V. Arruda, F. L. Palombini (eds.), *Biomimetics, Biodesign and Bionics*,
Environmental Footprints and Eco-design of Products and Processes,
https://doi.org/10.1007/978-3-031-51311-4_8

has not only resulted in remarkable technological advances but has also stimulated a growing awareness of socioeconomic and environmental issues. By allowing nature to serve as a mentor, biomimicry not only promotes innovation but also rekindles the importance of considering socioeconomic and environmental impacts in its solutions, reviving valuable ideals of sustainability and responsibility.

As for significant technological development, this has caused design theories and processes to be restructured. The approach defined as parametric design, for example, is based on defining topological relationships between the parts of an architectural object before dimensioning. Generally, the dimensions are not fixed values and can be selected from numerical ranges related to the measurements of building components, allowing for greater flexibility and variety of forms [7]. Therefore, the architectural shape arises from well-defined rules and restrictions that are coherent with the parameters previously established and added to digital design tools capable of associating the manipulation of form with programming. Biomimicry associated with parametric design makes up an emerging design methodology whose viability needs to be analyzed and tested according to the specifics of each project [6].

One promising application of biomimicry and parametric design methods centers on the development of ephemeral pavilions. These pavilions, due to their temporary nature and few functional program restrictions, play the role of a true "creative laboratory" in architecture. They offer an ideal environment for experimenting with a wide range of elements, from innovative materials and construction techniques to advanced design methods and theoretical concepts. The fact that these pavilions are temporary means that the unintended consequences of such experiments are not permanently perpetuated in a building [3].

This flexibility and freedom provided by the ephemeral pavilions offer an ideal platform for the creative application of biomimicry and parametric design, allowing architects to explore new frontiers of efficiency, aesthetics, and sustainability in contemporary architecture. These pavilions not only demonstrate the potential of biomimicry and parametric design but also serve as catalysts for innovation, influencing the evolution of architecture.

The combination of biomimicry and the pavilion design process represents an innovative approach that seeks inspiration from nature to guide the development of specific shapes and functions in architectural design. The pavilion, once considered primarily as an exhibition element, has evolved to become a recurring and versatile architectural typology, playing a key role in introducing new forms of interaction with the urban landscape [13]. This scenario has encouraged architecture faculties and research groups around the world to get involved in this design challenge, exploring complex geometries through the application of the latest digital technologies.

The use of these digital tools not only allows for creative exploration but also establishes a fundamental integration between theory and practice. This interdisciplinary and multidisciplinary approach, promoted by the exercise of designing pavilions inspired by nature, encourages a more holistic view of architectural design. It also brings the design process closer to reality through practical experimentation, in line with the "learning by doing" methodology [10]. In this way,

Fig. 1 Examples of two editions of the Serpentine Gallery's annual pavilion program. The 2013 Serpentine Pavilion, designed by the Japanese architect Sou Fujimoto (**a**), and the 2016 Serpentine Pavilion, designed by the Danish firm Bjarke Ingels Group (BIG) (**b**)Source: (**a**) photo by David Hawgood, CC BY-SA 2.0, via Wikimedia Commons. Available at: https://commons.wikimedia.org/wiki/File:Serpentine_Gallery_Pavilion_2013_-_geograph.org.uk_-_3556424.jpg (Accessed 16 September 2023); (**b**) photo by David Hawgood (2016), CC BY-SA 2.0, via Wikimedia Commons. Available at: https://commons.wikimedia.org/wiki/File:Serpentine_Pavilion_2016_view_towards_gallery_-_geograph.org.uk_-_5001785.jpg (Accessed 16 September 2023)

biomimicry and parametric design combine to expand the creative potential of pavilion architecture, creating innovative solutions that harmoniously merge the influences of nature and the potential of computational tools.

The British institution Architectural Association (AA) played a pioneering role by taking inspiration from the Serpentine Gallery[1] (Fig. 1) and launching an annual program to develop ephemeral pavilions, the Summer Pavilion Programme [11]. This innovative program paved the way for the exploitation of this type of temporary structure. The success and influence of the AA's program soon spread, inspiring other notable institutions to follow suit. Globally renowned academic and research organizations such as MIT[2] (United States), ETH Zurich[3] (Switzerland), University of Stuttgart (Germany), and IAAC[4] (Spain), among others, began exhibiting temporary pavilions resulting from workshops, academic disciplines, or annual programs,

[1] Serpentine Gallery: It is an important British institution located in Kensington Gardens in London, United Kingdom. The gallery was founded in 1970, and it is known for its commitment to contemporary art and innovative architecture. One of its most important events is the Serpentine Pavilion, an annual project that invites architects from different parts of the world to develop highly creative ephemeral structures to exhibit in its gardens [9]. This influential program has played an important role in promoting experimental architecture and collaboration between art, architecture, and engineering.

[2] MIT: Massachusetts Institute of Technology, Cambridge, Unites States

[3] ETH Zürich: *Eidgenössische Technische Hochschule Zürich* – Swiss Federal Institute of Technology in Zurich, Switzerland

[4] IAAC: Institute for Advanced Architecture of Catalonia, Barcelona, Spain

representing a remarkable convergence of architectural innovation and education, and contributing significantly to exploring new boundaries of ephemeral architecture.

In the Portuguese-Brazilian context, the construction of ephemeral pavilions is still little explored, but most of the experiments, like the pavilions developed by the AA, are the result of workshops. In 2015, the Federal University of Rio de Janeiro (UFRJ) presented The Butterfly Gallery as the result of a workshop that sought to demonstrate strategies for approaching ephemeral architecture through theoretical reflection and design experimentation [4]. Later, in 2017, the university presented the Tornado Pavilion, focusing on the issue of temporary structures and regulated surfaces [5]. In addition to the experiences at UFRJ, it is possible to highlight the small pavilion developed in the course "*Prototipagem e Fabricação Digital* – Prototyping and Digital Fabrication" taught by Professor Gabriela Celani at FEC/UNICAMP[5] and the Constructive Geometry Pavilion (*Pavilhão Geometria Construtiva*), which was originally designed and manufactured at the DFL (*Laboratório de Fabricação Digital* – Digital Fabrication Laboratory) at the School of Architecture of the University of Porto, in Portugal, in 2012 and was later reproduced at FAU/USP[6] under the coordination of Professor Paulo Fonseca [10].

Two of the most emblematic ephemeral academic pavilion programs are those of the AA and the University of Stuttgart.

1.1 Architectural Association (AA), London, England

The Summer Pavilion Programme was launched by the AA School in London between 2005 and 2009, led by tutors Martin Self and Charles Walker. The program was a proposal in which second- and third-year students would design and produce ephemeral pavilions in order to develop their skills and discover their architectural potential through experimenting with wood construction. The pavilions were produced at the AA's Hooke Park campus and received assistance from the Fine Forest timber company and engineering support from Arup [11].

In all, the AA's Summer Pavilion series was exhibited in four editions: 2006 (Fractal Pavilion), 2007 (Bad Hair Pavilion, Fig. 2a), 2008 (Swoosh Pavilion, Fig. 2b) and 2009 (Driftwood Pavilion). The series became a success and was marked by the experience of collaborative work and developed in several phases: initial concept, development of ideas, tests, mock-ups, financing, planning, and construction.

[5] FEC/UNICAMP: *Faculdade de Engenharia Civil, Arquitetura e Urbanismo da Universidade Estadual de Campinas* – School of Civil Engineering, Architecture, and Urbanism of the State University of Campinas, Brazil

[6] FAU/USP: *Faculdade de Arquitetura e Urbanismo da Universidade de São Paulo* – School of Architecture and Urbanism of the University of São Paulo, São Paulo, Brazil

Fig. 2 Bad Hair Pavilion (**a**) and Swoosh Pavilion (**b**). Both pavilions were developed by the AA's Summer Pavilion Programme in England in 2007 and 2008, respectivelySource: (**a**) photo by Andy Potter, CC BY-SA 2.0, via Wikimedia Commons. Available at: https://commons.wikimedia.org/wiki/File:Bad_Hair_Day_Pavilion_-_geograph.org.uk_-_2556564.jpg (Accessed 16 September 2023); (**b**) photo by Ronaldccwong, public domain, via Wikimedia Commons. Available at: https://commons.wikimedia.org/wiki/File:Inter2pavilion.JPG (Accessed 16 September 2023)

1.2 University of Stuttgart, Germany

In this context of building and exhibiting pavilions as a way of learning, testing skills, and as an experimental laboratory, the University of Stuttgart in Germany has also been contributing to the development of these ephemeral structures. In 2010, the ICD (Institute for Computational Design) and ITKE (*Institut für Tragkonstruktionen und Konstruktives Entwerfen* – Institute of Building Structures and Structural Design), both at the University of Stuttgart, began developing temporary research pavilions. The designed structures demonstrate the latest developments in computational design, robotic production processes in architecture, and material simulation [2].

The ICD/ITKE academic pavilions are a remarkable project by the University of Stuttgart and represent a truly biomimetic approach to experimental architecture. By mimicking biological processes, the team responsible for these pavilions began working directly with biologists, which demonstrates a deep commitment to incorporating principles from nature into architectural design [12] (Fig. 3). This interdisciplinary collaboration has enabled the exploration of innovative solutions that go beyond the visual imitation of nature, seeking to understand and adapt biological processes to build efficient architectural structures.

The German institution's pavilions have become a notable annual project, having started in 2010 and with their most recent edition taking place during the 2020–2021 academic year. Located in the heart of the University of Stuttgart campus, these structures were conceived not only as architectural installations but also as an extension of the university's central square. This strategic positioning not only transforms the urban landscape of the campus but also emphasizes the importance of the

Fig. 3 Research pavilions 2011 (**a**) and 2016–2017 (**b**) developed by IC/ITKE at the University of Stuttgart in Germany. Both pavilions were developed using biological inspirations, mimicking the shapes and functions of the sea biscuit skeleton (**a**) and apple leaf miner moth (**b**)Source: (**a**) photo by MSeses, CC BY-SA 4.0, via Wikimedia Commons. Available at: https://commons.wikimedia.org/wiki/File:ICD_Research_Pavilion_2011_Stuttgart_01.jpg (Accessed 16 September 2023); (**b**) photo by MSeses, CCBY-SA 4.0, via Wikimedia Commons. Available at: https://commons.wikimedia.org/wiki/File:ICD_Research_Pavilion_2016-17_Stuttgart_01.jpg#metadata (Accessed 16 September 2023)

interaction between architecture, nature, and academic life, creating a rich environment for teaching and research.

These pavilions represent in an exceptional way how biomimicry and innovation in digital fabrication and robotics can come together to create advanced architectural structures that not only incorporate the principles of nature but also promote interdisciplinary research and the experience of the built environment.

2 Methodology

The experiment developed outlined the survey of various references through what was known as a "Reference Map" (Fig. 4), where the online platform MyMaps was used to spatialize the academic pavilions according to the location of the universities that developed them. This type of representation made it possible to see more clearly the distribution of similar research carried out around the world.

At each point representing a reference, it was possible to insert several pieces of information pertinent to the example, such as year, technique, and bio-inspiration (if any), among others (Fig. 5). Based on this diversity of information entered, the platform allows filters to be made, and, as a result, it is possible to reach different conclusions more quickly and intuitively than simply analyzing a database developed using tables. Despite this, the map was developed through the implementation of a large spreadsheet of data filled in after bibliographical research.

In addition, it is important to emphasize the need to insert a "Location" column, in which the address corresponding to the location of each point on the map must be entered. The images of each reference had to be inserted individually, within each

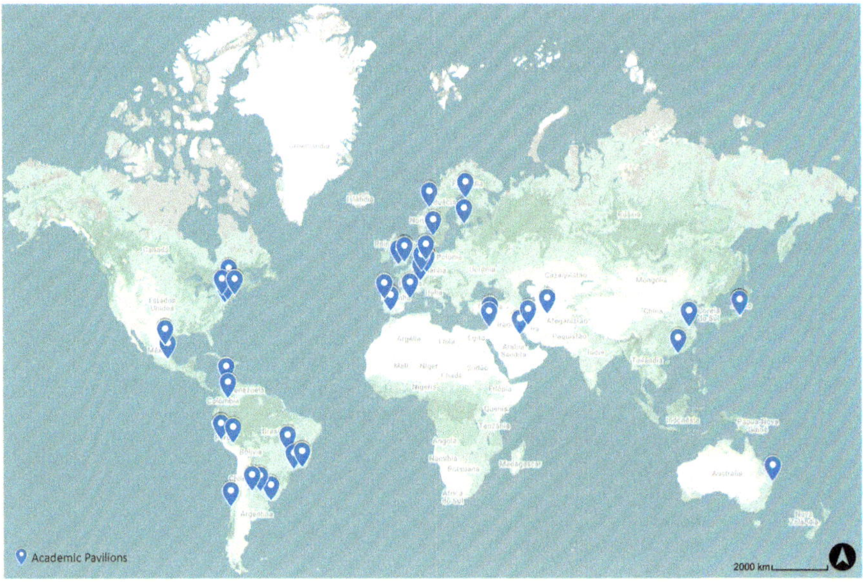

Fig. 4 Reference Map – Academic Pavilions. (Source: authors, 2023)

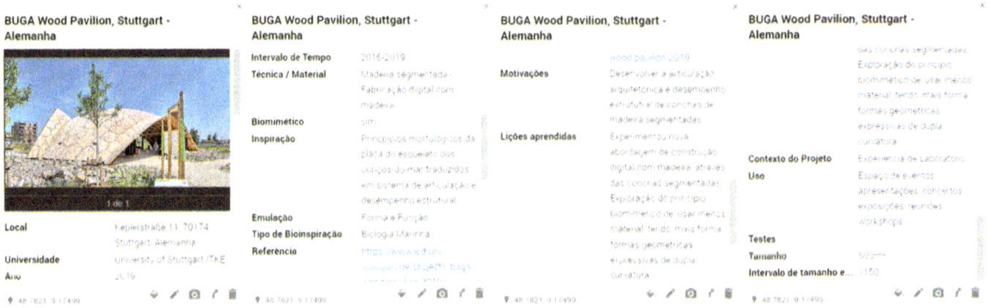

Fig. 5 Information panel inserted in the Reference Map. (Source: authors, 2023; image of the BUGA Wood Pavilion retrieved from Wikimedia Commons, photo by Joachim Köhler, CC BY-SA 4.0, available at: https://commons.wikimedia.org/wiki/File:Mannheim-BUGA-Spinelli-Park-Bionischer-Holzpavillon-IMG_5838-x3B-02-06-2023.jpg (Accessed 18 September 2023))

information panel, after importing the spreadsheet and spatially distributing the points on the map.

Based on the spreadsheet previously developed with the various parameters collected to analyze the pavilions, a disjunctive table (Table 1) was constructed, as shown below, with the aim of grouping categories and making it easier to analyze the information. The following table shows the categories used to develop the disjunctive table and the strategy used to group subcategories.

Table 1 Grouping of categories and subcategories used to analyze the academic pavilions studied

Categories	Subcategories	Grouping strategy
Location	Europe Americas Asia Oceania Africa	Grouping by continent
Year of construction	2005–2009 2010–2012 2013–2015 2016–2019 2020≥	Time period interval
Predominant material	Plastic Wood CTCR (Cardboard, Textile, Canvas, Rope) Concrete Metal Fibre Glass/Polycarbonate Bamboo	*Grouping of similar types of materials *The same pavilion may have more than one category marked
Size (m^2)	<30 30–60 60–150 >150	*Range of built areas *Some pavilions have estimated areas
Project context	Annual Programme Curricular units Workshops Academic work Research project Uninformed	Separated according to project/construction development context
Utility	Temporary installation Event space Rest area Habitable module Classroom	Grouping of similar uses
Biomimetic	Yes No	Not all pavilions are biomimetically inspired
Type of bio-inspiration	Mineral Natural structure Botany Marine biology	Grouping the types of inspirational elements
Emulation	Shape Function Shape and function Uninformed	Separation according to the type of emulation

(continued)

Table 1 (continued)

Categories	Subcategories	Grouping strategy
Motivations	Academic exercise Exploring architectural potential Exploring new techniques Testing/rethinking techniques Exhibition space University use Explore construction process Design development Experimentation	These categories had long and very specific descriptions of each project, so they were grouped into general categories that could cover several pavilions in a generalized way
Lessons learned	Workflow Design/construction methods Material handling Computational/ automation techniques Development of new techniques Experimenting with parametric structures	
Future works	Techniques for rationalizing computer geometries Materialization and assembly techniques Biomimetic exploration Geometrical exploration Construction/ manufacturing techniques Computer simulations Simulations of physical models	

Source: authors, 2023

3 General Analysis of the Pavilions

Sixty-one temporary pavilions developed in academic contexts around the world from 2005 to 2022 were selected (Table 2). From this sample, some conclusions were drawn that can be analyzed using the following graphs.

Most of the pavilions were developed between 2016 and 2019; however, it is possible to notice a considerable increase in practice from 2013 onward (Fig. 6). These are mainly the result of workshops and annual programs developed by universities (Fig. 7).

Table 2 List of the 61 pavilions analyzed separated by country and university where they were developed

ID	Country	University	Pavilion's name
01	Argentina	Universidad Nacional de Córdoba	Tesis Uno en Uno
02	Argentina	Universidad del Litoral	SSFS Pavilion
03	Australia	Bond University	Cellular Tessellation Pavilion
04	Brazil	UFRJ – Faculdade de Arquitetura e Urbanismo	The Butterfly Gallery
05	Brazil	UFRJ – Faculdade de Arquitetura e Urbanismo	Pavilhão Tornado
06	Brazil	Escola da Cidade	Contemplay
07	Brazil	Universidade Federal de Goiás	Sem Nome
08	Canada	Université McGill	Contemplay
09	Chile	Universidad del Bío-Bío	Pabellón BioDuna
10	China	Design School of Nanjing University of the Arts	Flora Pavilion
11	Colombia	Universidad Nacional de Colombia	The Coccon Pavilion
12	Colombia	Universidad del Norte	Molusco Pavilion
13	England	AA Architectural Association	Fractal Pavilion
14	England	AA Architectural Association	Bad Hair Pavilion
15	England	AA Architectural Association	Swoosh Pavilion
16	England	AA Architectural Association	Driftwood Pavilion
17	Finland	University of Oulu-School of Architecture	Hila Pavilion
18	Finland	Aalto University	Säie Pavilion
19	Germany	Detmolder Schule für Architektur und Innenarchitektur	Boxel Pavilion
20	Germany	Universität des Saarlandes	Bowoos Bionic Pavilion
21	Germany	University of Stuttgart ITKE	ICD Sewn Timber Shell
22	Germany	University of Stuttgart ITKE	ICD Aggregate Pavilion
23	Germany	University of Stuttgart ITKE	BUGA Fibre Pavilion
24	Germany	University of Stuttgart ITKE	BUGA Wood Pavilion
25	Germany	University of Stuttgart ITKE	ICD/ITKE Research Pavilion 2010
26	Germany	University of Stuttgart ITKE	ICD/ITKE Research Pavilion 2011
27	Germany	University of Stuttgart ITKE	ICD/ITKE Research Pavilion 2012
28	Germany	University of Stuttgart ITKE	ICD/ITKE Research Pavilion 2013–2014
29	Germany	University of Stuttgart ITKE	ICD/ITKE Research Pavilion 2014–2015
30	Germany	University of Stuttgart ITKE	ICD/ITKE Research Pavilion 2015–2016
31	Germany	University of Stuttgart ITKE	ICD/ITKE Research Pavilion 2016–2017
32	Germany	University of Stuttgart ITKE	ICD/ITKE Research Pavilion 2018–2019

(continued)

Table 2 (continued)

ID	Country	University	Pavilion's name
33	Germany	University of Stuttgart ITKE	ICD/ITKE Research Pavilion 2020–2021
34	Hong Kong	The Chinese University of Hong Kong School of Architecture	ZCB Bamboo Pavilion
35	Iran	Khayyam University	Pipe Pavilion
36	Iran	Islamic Azad University of Najafabad	Kooshk Pavilion
37	Japan	The University of Tokyo	Digital Fabrication Lab Research Pavilion
38	Japan	The University of Tokyo	Toca Pavilion
39	Japan	The University of Tokyo	Pop-up Shelter
40	Kuwait	Kuwait University	OV Pavilion
41	Lebanon	American University of Beirut	MARAH Pavilion
42	Mexico	Instituto Tecnológico e de Estudos Superiores de Monterrey	Parametric Pavilion
43	Mexico	Universidad de Monterrey	BICHO VI
44	Mexico	Universidad Nacional Autónoma de México	Cactus Pavilion
45	Norway	Norwegian University of Science and Technology	Rindal Star Cube
46	Palestine	Birzeit University	Waffle Pavilion
47	Peru	Pontificia Universidad Católica del Peru	NYMBÚ
48	Peru	Pontificia Universidad Católica del Peru	Tubotella
49	Peru	Pontificia Universidad Católica del Peru	Pempén
50	Portugal	Faculdade de Arquitectura da Universidade do Porto	Constructive Geometry Pavilion
51	Spain	Institut d'Arquitectura Avançada de Catalunya	Heliocell
52	Spain	Universidad de Sevilla	The Caterpillar Gallery
53	Spain	Universidad de Sevilla	Globoide Pavilion
54	Sweden	Chalmers Tekniska Högskola	The Crack Pavilion
55	Switzerland	École Polytechnique Fédérale de Lausanne	HOUSE 1
56	Uruguay	Universidad de la República	Dieste Pavilion
57	United States	MIT School of Architecture and Planning	Kerf Pavilion
58	United States	MIT School of Architecture and Planning	Unblocked
59	United States	Cornell University College of Architecture, Art, and Planning	Jello Pavilion
60	United States	Graduate School of Architecture, Planning and Preservation-Columbia University	Kerby
61	United States	University of Pennsylvania	PennDesign Pavilion

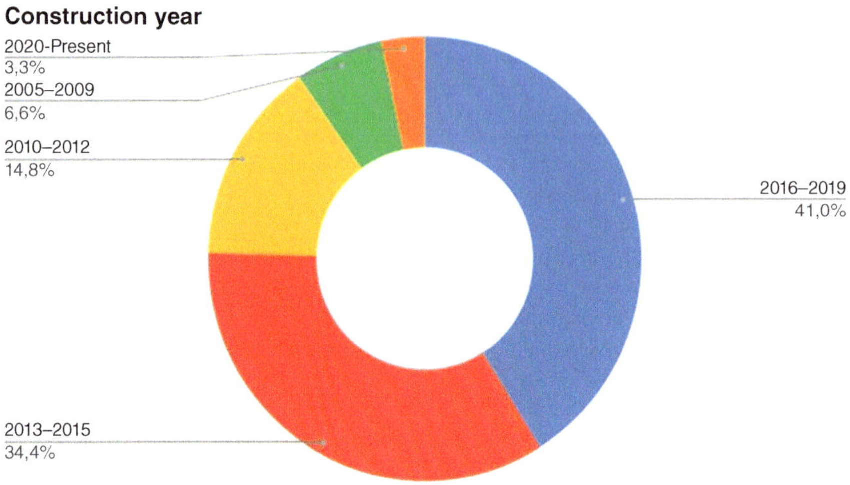

Fig. 6 Graph relating to the time interval for construction of the pavilions. (Source: authors, 2023)

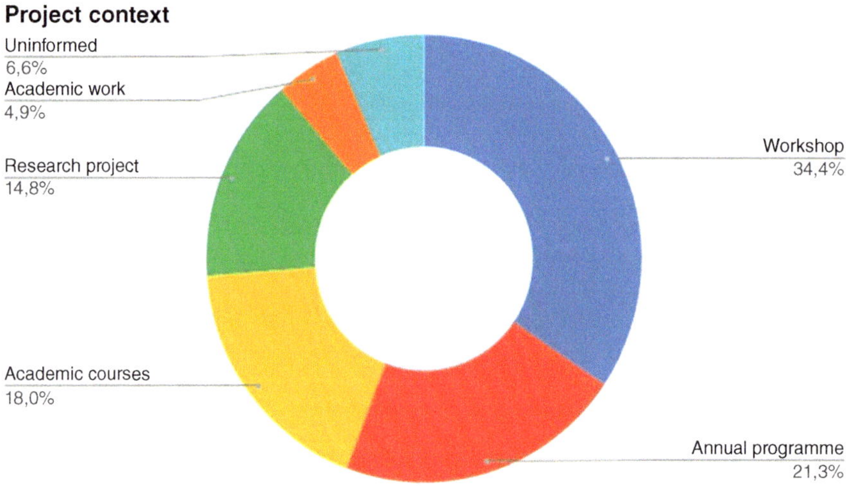

Fig. 7 Graph of the production context of the pavilions. (Source: authors, 2023)

Due to the ephemeral and multifunctional nature of the pavilion, the vast majority of the sample functioned as a temporary installation on campus or in a public space, where people could interact with the structure, but without an established use (Fig. 8).

Unfortunately, it was not possible to make a detailed analysis of the tests conducted during the design and manufacturing process, but it was noted that structural and material tests and simulations are commonly conducted (Fig. 9). Similarly with the size analysis, where 40 out of the 61 pavilions did not make this information

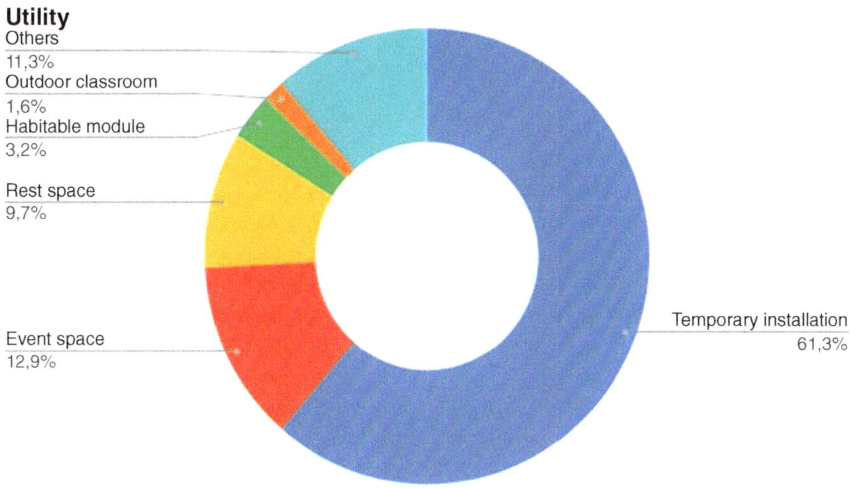

Fig. 8 Graph of utility. (Source: authors, 2023)

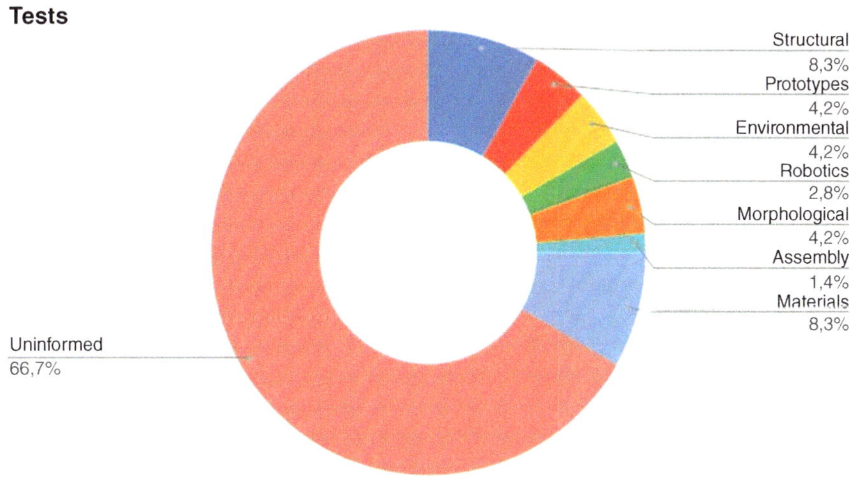

Fig. 9 Graph of the types of tests conducted on the pavilions. (Source: authors, 2023)

clear, however, an estimate of the area of these pavilions was made based on the images found, and it was realized that a large part of the pavilions studied fall within the range of up to $30m^2$ (Fig. 10). This area estimate was approximated according to some scale/reference present in the available images of the pavilions.

It is noticeable that a considerable number of the pavilions were developed by the University of Stuttgart in Germany. This is due to the university's traditional annual program, which seeks to combine research into materials, biomimicry, and

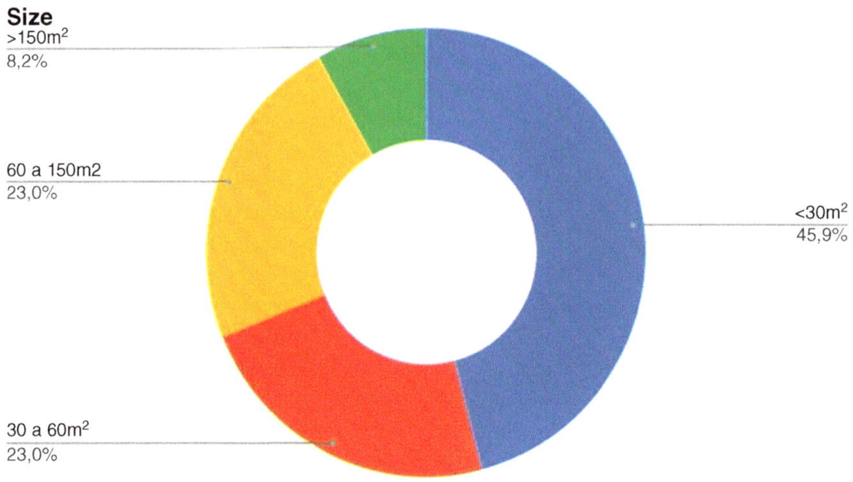

Fig. 10 Graph of the range of estimated pavilion areas in m². (Source: authors, 2023)

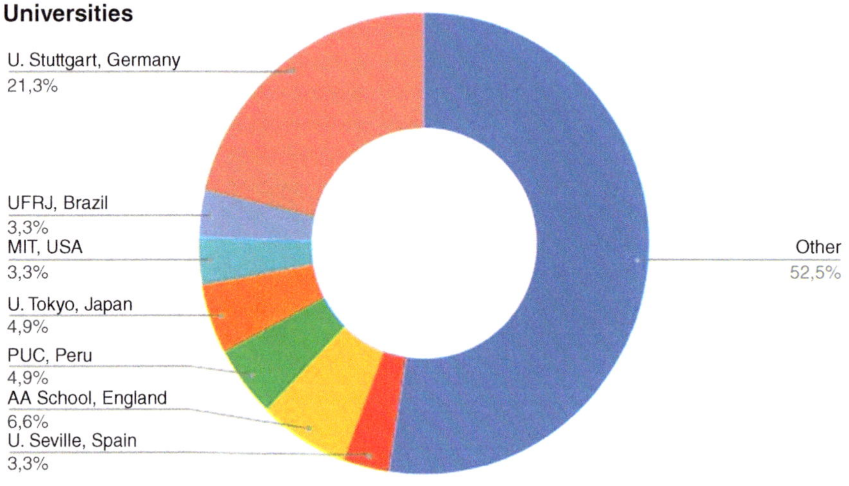

Fig. 11 Graph of the most recurring universities in the sample studied. (Source: authors, 2023)

recent advances in computational design techniques (Fig. 11). In general, there is a strong European influence in the development of these experiments (Fig. 12).

In terms of materiality, most of the pavilions were developed using wood and bamboo construction techniques. The majority use a combination of materials that are related to each other and seek to employ local techniques and materials. It's important to emphasize that this category refers to the predominance of materials in the pavilion and not the detailed materials present in the structure's constructive composition. Other experiments seek to develop and investigate new techniques and

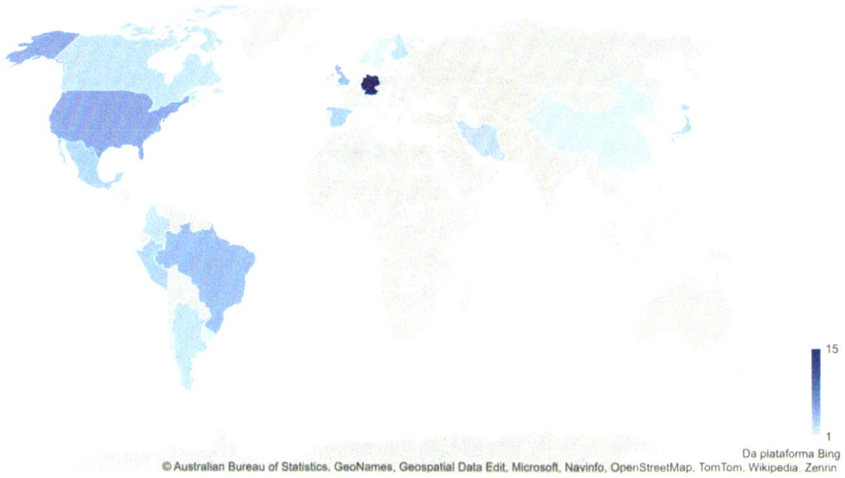

Fig. 12 Graph of the location of the studied experiments. (Source: authors, 2023, through Bing Platform)

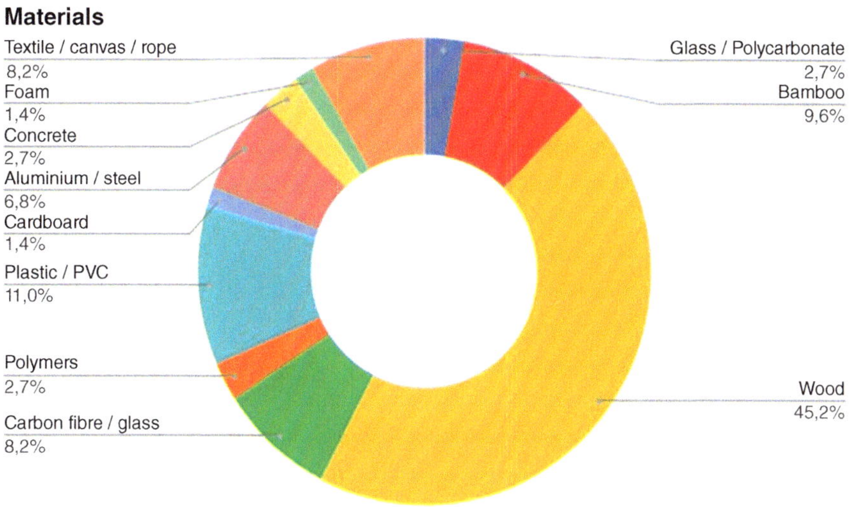

Fig. 13 Graph of the materials used to build the pavilions studied. (Source: authors, 2023)

materials and end up using elements that are not so common in architecture, such as fiberglass, carbon fiber, linen, foam, and textile materials (Fig. 13).

From the 61 units studied, only 19 were developed following biomimetic precepts (Fig. 14). The majority, in turn, are based on research into natural elements from the animal kingdom, botany, and marine biology (Fig. 15). In addition, many of the biomimetic pavilions seek to emulate both the shape and the function of the inspirational element, with the aim of bringing natural functionality to the architectural element (Fig. 16).

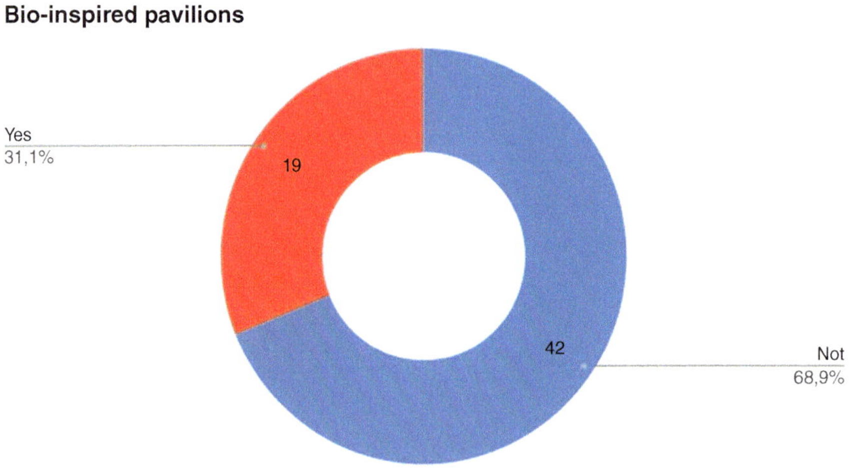

Fig. 14 Graph of identification of biomimetic pavilions

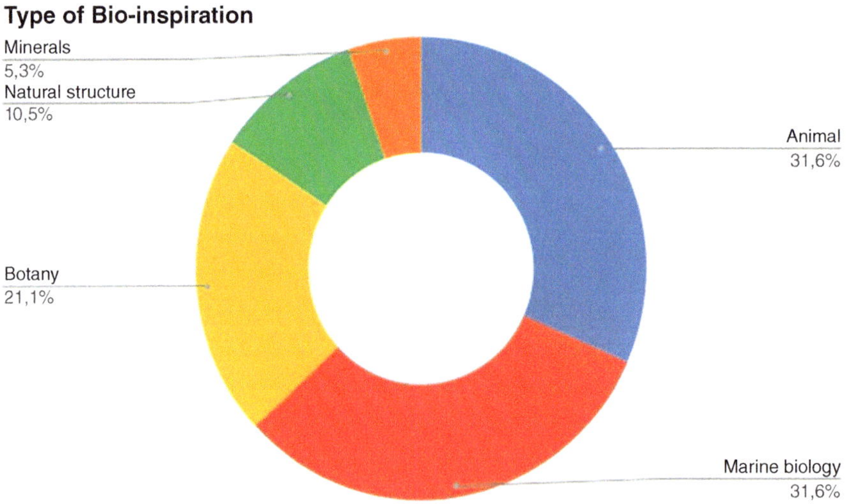

Fig. 15 Graph of the type of bio-inspiration used in the pavilion design. Graph based on the analysis of the 19 pavilions shown in Fig. 14. (Source: authors, 2023)

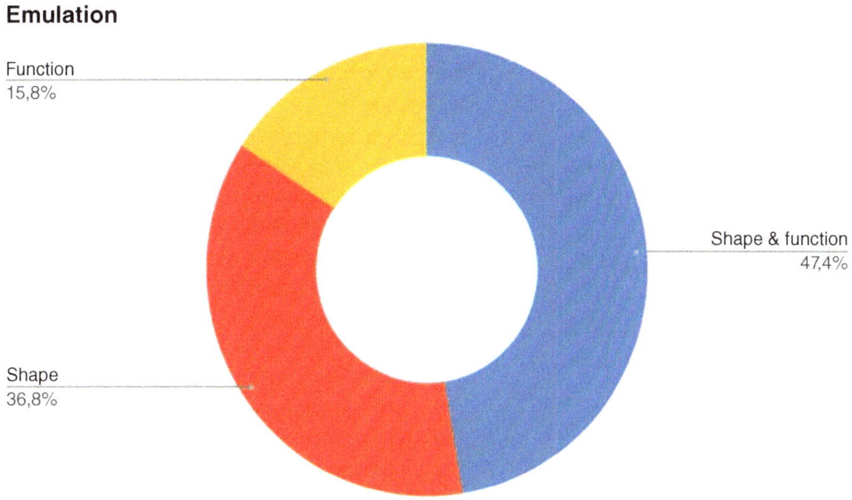

Fig. 16 Graph of the type of emulation used in the design of the pavilions. Graph based on the analysis of the 19 pavilions shown in Fig. 14. (Source: authors, 2023)

4 Motivations

The relevance of the pavilion typology as a school building site is clear. The pavilions function as the results of a laboratory of experimentation, the fruit of years of research, or they can even be the goal of a practical academic exercise.

The pavilions studied seek to explore the relationship between thinking and making architecture, investigate the architectural potential of temporary experimental constructions, and approach the design process from the first drawings to manufacture, resulting in a collaborative academic exercise (human-human, human-machine). Some seek to rescue and rethink construction techniques, others, the cultural bias of a site through contemporary remodeling, by applying technological and scientific principles. They seek technological innovation and the application of topics covered in the classroom, mainly related to geometry.

They function as a means of presenting advances in the field of materials-orientated computational design, simulation, and production processes. Thus, it is possible to approach unconventional structures with foldable, adaptable, interactive systems, controlled by the user or the environment, and unusual materials such as fiberglass, carbon fiber, linen fiber, polymers, and textiles.

Some examples take advantage of the limited approach to parametric modeling in architecture courses to develop these design exercises and explore different types of techniques and materials. The vast majority investigate the potential of digital tools (parametric design, digital fabrication techniques, and robotic manufacturing) in the design process and seek to encourage students and researchers to use them.

The pavilions, as a temporary, artistic, and experimental construction, provide dynamic experiences and create democratic and multipurpose public spaces through flexible, light, demountable, practical structures with poetic appeal. They function as a stage for parties or as a piece or part of an exhibition circuit. Some examples also seek to promote local tourism, create a transitional space within a natural environment, and be a haven of rest and well-being.

By combining biomimicry with the experimental exercise of pavilions, it is possible to integrate the performative capacity of biological structures into architectural practice. In addition, the aim is to raise awareness about the exploitation of natural resources, discuss ideals of sustainability and recycling, and explore biodegradable, renewable, and local materials. These biomimetic experiments also make it possible to investigate the possible relationship between bio-inspired design and new digital fabrication and robotic processes.

In some cases, their experimental character is aimed at developing construction systems, techniques, and innovative shapes, as well as bringing their findings to the market. In doing so, they seek to strengthen ties between academia and industry through collaborations and partnerships.

Finally, in general, these experiments seek to combine architecture and technology to improve people's living conditions by reacting dynamically to natural, social, and constructive circumstances.

5 Lessons Learnt

Analyzing the pavilions made it possible to verify the importance of a full-scale educational construction experience that can be adapted to different circumstances, as well as building lightweight systems, creating spaces with a great social and programmatic impact. The possibilities for learning through this type of experience are wide-ranging and cover different aspects. However, in most cases, discussions are restricted to the design process, manufacture, simulation, and materials, rather than the space and quality of the environment created through the pavilion.

Using parametric design makes it possible to develop algorithmic thinking and analytical reasoning, but it requires prior knowledge of the systems you want to design, material properties, and geometric knowledge. This makes students realize the tool's potential, but also its restrictions and difficulties.

Parametric modeling makes it possible to explore different geometries, twisted, sculpted, undulating, with variable shapes, giving the idea of movement, fluidity, and even applying the same algorithm to different surfaces. As the design flow is very particular, it will sometimes be necessary to rationalize, plan, and prepare the parametric model for manufacture.

This design method also allows integration with common construction techniques, such as concrete, for example, and it is not necessarily mandatory to follow the entire design process using digital tools (although these do optimize the entire design and construction process).

Some of the advantages of parametric modeling are the possibility of working with form-finding, geometric optimization, creating more expressive shapes that are rarely used in construction, and performing real-time simulations using plug-ins available for each type of analysis required (environmental, structural, loads, etc.).

In addition, there is the possibility of combining the flexibility of parametric modeling with the foresight of a CNC milling machine. Digital manufacturing makes mass customized construction possible, through additive and subtractive manufacturing techniques, for example.

The experimental nuance of the works made it possible to work with mobile structures to optimize lighting and ventilation and bending, twisting, and overlapping wood, combined with different materials such as metal loops, foam, and ropes.

Through the study of materials, it was realized that translucent PVC[7] affects the perception of the pavilion depending on the incidence of sunlight and the angle of view.

Furthermore, the use of solar panel skins that change according to the position of the sun, the reuse of materials in order to transform by-products into usable materials, and the handling of unusual materials such as Mylar©,[8] foam, and burned wood were explored. Finally, it was noted that there are materials that can restrict design freedom, such as fiber composites.

The pavilions with all their prominence are capable of generating feelings according to the space, so they can provide a sense of surprise, curiosity, closure, openness, and hospitality.

In another dimension of analysis, this type of design exercise allows for the integration of parametric modelling and digital fabrication with biomimicry, enabling the exploration of natural ideas and patterns, as well as the search for sustainability. One example explores the association of artisanal design techniques with the use of natural resources, organic materials, and digital fabrication.

Other cases dealt with recyclable materials and used everyday materials (recycled bamboo chopsticks) to develop a self-sustainable construction, related to the local context, combining nature and technology to develop cutting-edge environmental and energy technologies to deal with the ecological, economic, and sociocultural challenges that the built environment demands.

6 Possible Future Work

Based on the analysis of the 61 pavilions studied, in some cases, there is a need to exploit the potential of digital tools to develop bolder shapes and surfaces and thus extract the maximum usability from parametric design tools and digital fabrication and/or robotics.

[7] PVC, or polyvinyl chloride, is a synthetic plastic polymer valued for its versatility, durability, and resistance to moisture, chemicals, and fire.

[8] Mylar© is also known as BoPET (biaxially oriented polyethylene terephthalate) and is a type of polyester film known for its high tensile strength, transparency, and insulating properties.

It should be noted that just as important as the product is the process of obtaining it. The process of form finding through the exploration of different geometric shape patterns found in nature and the generation of multiple pavilion possibilities would be interesting work to develop and analyze together with optimization tools.

Another factor not addressed is the structure's impact on the installation site, as an activator of space and as a point of attraction for people.

It is noticeable that the presentations, at least those that have been made available on the Internet, do not detail the process of the object, encompassing design, construction, and assembly. Some pavilions have retractable or adaptable structures, but the approach of dynamic architecture that changes according to the environment has been little explored.

Some materials, such as Mylar© and some fiber composites, do not allow us to explore design freedom due to some limitations that have not been disclosed, so it would be interesting to investigate these challenges along with comparing the limitations of digital fabrication and robotics techniques.

As for the field of biomimicry, there is a need to investigate techniques for translating biological knowledge more objectively into the field of architecture. There is also a need to explore biomimicry beyond the shape and function of natural elements, extracting all possible potential from nature in a sustainable way. The exploration of sound in a natural context is a possible avenue of research, as it extracts unique characteristics that are imperceptible to most people, but it would represent a side of nature that has not yet been explored, even in other sciences.

As for construction and assembly techniques, it is inferred that the use of a 3D scanner would facilitate the process of assembling structures with multiple and different parts. Addressing techniques for folding and compacting structures, techniques for planning curved shapes and digital models in a computer environment, additive manufacturing techniques related to biological materials, possibilities for joining parts, techniques for folding flat steel plates, and wood construction techniques and their respective limitations are all topics that could be addressed in the set of samples studied.

Regarding materials, we realize that it would be interesting to explore the limits of material workability; to approach local and conventional materials such as the soil itself; to explore the economy of the material; and to give it as much resistance as possible. The manufacture of lightweight structures with fibers (linen, carbon, and glass) is still underdeveloped and could be further explored, as could foam, textiles, expandable materials, and translucent and inflatable materials.

As for the computer environment used to develop the project, it was realized that some pavilions could have explored structural simulation using parametric digital tools or used Galapagos as a way of working with generative algorithms to find more efficient shapes.

7 Biomimicry of the Pavilions

Although only 19 out of the 61 pavilions analyzed were designed following biomimetic precepts (Table 3), it is clear that the pavilions studied represent a remarkable expression of the intersection between architecture, technology, and nature, revealing a notable trend toward biomimicry. Thus, among the various facets explored in these projects, biomimicry emerges as a particularly relevant and inspiring approach.

Table 3 Identification of biomimetic pavilions with specification of the type of bio-inspiration, inspiration, and type of emulation. The ID is based on the construction of Table 2

ID	Pavilion's name	Inspiration	Type of bio-inspiration	Emulation
10	Flora Pavilion	Flower	Botany	Shape
12	Molusco Pavilion	Mollusk *Bolinus brandaris*	Marine biology	Shape
18	Saië Pavilion	Forest, due to joints and overlaps	Botany	Shape
20	Bowoos Bionic Pavilion	Marine plankton shells (diatoms)	Marine biology	Shape
22	ICD Aggregate Pavilion	Granularity of sand, gravel, snow; erosion process	Mineral	Shape
23	BUGA Fibre Pavilion	Fiber compounds: cellulose, chitin, collagen – transfer to the structural system	Natural structure	Function
24	BUGA Wood Pavilion	Morphological principles of the sea urchin skeleton plate translated into articulation system and structural performance	Marine biology	Shape and function
26	ICD/ITKE Research Pavilion 2011	Sea biscuit skeleton (*Clypeaster reticulatus*) – Load-bearing capacity and geometric arrangement	Marine biology	Shape and function
27	ICD/ITKE Research Pavilion 2012	Exoskeleton of the lobster (*Homarus americanus*) – Material anisotropy[a] and functional morphology of arthropods[b]	Marine biology	Shape and function
28	ICD/ITKE Research Pavilion 2013–2014	Wings and abdomen of beetles – Performance of these lightweight structures is based on geometric morphology	Animal	Shape and function
29	ICD/ITKE Research Pavilion 2014–2015	Diving bell spider (*Argyroneta aquatica*) – Basic behavior patterns and web design rules	Animal	Shape and function
30	ICD/ITKE Research Pavilion 2015–2016	Urchin and starfish (*Clypeasteroida*)/ morphological principles and behavioral principles of growth	Marine biology	Shape
31	ICD/ITKE Research Pavilion 2016–2017	Apple leaf miner moth (*Lyonetia clerkella, Leucoptera erythrinella*) – Morphological and functional principles, and construction logic of natural light structures	Animal	Shape and function

(continued)

Table 3 (continued)

ID	Pavilion's name	Inspiration	Type of bio-inspiration	Emulation
32	ICD/ITKE Research Pavilion 2018–2019	Ladybird (*Coleoptera coccinellidae*) – Wing fold patterns and elasticity	Animal	Shape and function
33	ICD/ITKE Research Pavilion 2020–2021	Saguaro cactus (*Carnegia gigantea*) and spiny cactus (*Opuntia* sp.) – Mechanical properties of the cactus' cross-linked fiber structures	Botany	Function
40	OV Pavilion	Snail – Shape and capabilities of the animal	Animal	Shape and function
44	Cactus Pavilion	Cactus (*Echinocactus platyacanthus*)	Botany	Shape
47	NYMBÚ	Nest and arboreal aspect – Birdwatching experience	Natural structure	Shape and function
49	Pempén	Butterfly wing movement	Animal	Function

Source: authors, 2023

[a]Anisotropy: it is a property of a material or system that displays distinct physical characteristics or behaviors when measured in different directions. It means that material properties, such as conductivity, elasticity, or optical behavior, vary depending on the measurement direction rather than being uniform in all directions.

[b]Arthropods: it is a diverse group of invertebrates' animals characterized by jointed legs and a segmented body covered by an external skeleton made of chitin, encompassing insects, spiders, and crustaceans.

In the context of academic pavilions, biomimicry manifests itself as a bridge between biological knowledge and architectural practice. These pavilions seek to emulate the shape of natural elements and in some cases also to understand and transpose the intrinsic function of these elements into architecture.

Furthermore, these pavilions embrace the challenge of taking biomimicry beyond mere shape, delving into broader, more holistic territory. They seek to extract the full potential of nature, exploiting not only the visual appearance but also less perceptible characteristics, such as certain functions of natural elements.

In addition to biomimicry, these pavilions also stand out for their exploration of parametric design techniques and digital fabrication. Combining these tools with biomimetic concepts allows natural shapes to be integrated with cutting-edge technologies, resulting in structures that not only imitate nature but also function efficiently and sustainably, in some cases using materials such as wood, bamboo, and fibers. This includes the use of biodegradable, recyclable, and renewable materials, as well as cladding systems that respond to the environment.

The biomimetic inspirations for these pavilions are diverse. For instance, some of them have turned to marine biology and animal anatomy, while others have delved into geometric patterns found in nature. As for the inspirations, it is possible to verify the following.

Marine Biology: Six out of the 19 biomimetic pavilions were inspired by marine organisms, such as urchins and starfishes, to create organic shapes and textures that mimic the natural patterns of these living beings.

Plant Structures (Botany): Four pavilions focused on plant nature, incorporating the shapes and mechanical principles of plant structures to create bio-inspired designs.

Animal Anatomy: Biomimicry also extended to animal anatomy. Six pavilions were inspired by features of bird wings, insect skeletons, and even behavioral patterns.

Natural structures: Two biomimetic pavilions explored the shapes and functions of natural elements, such as the chemical compounds chitin and cellulose, to transfer knowledge to the development of structural systems.

Minerals: One of the pavilions drew inspiration from natural elements such as sand particles, granularity, and snowflake structures.

These inspirations not only borrowed visual elements from nature but also sought to understand and apply the underlying principles that make these natural structures efficient and adaptable to the environment. The result is a diverse collection of pavilions that not only stand out visually but also explore innovative architectural solutions.

8 Conclusion

Based on an in-depth analysis of the 61 temporary pavilions developed in an academic context between 2005 and 2022, it is possible to extract valuable insights into the impact of this typology on contemporary architecture and related interdisciplinary fields. These pavilions, often conceived as the results of full-scale experimentation laboratories, play a significant role in the training of future architects and researchers.

Firstly, the academic pavilions work as an ideal training ground, allowing students and researchers to experiment and apply their architectural skills and theories discussed in the classroom to concrete projects. They represent the transition between theoretical thinking and actual practice, addressing a wide range of topics from geometry and materials to innovative construction techniques. In addition, these projects often emphasize collaboration, whether between students and professors or with experts from other fields, promoting multidisciplinary thinking.

The pavilions also serve as catalysts for innovation, encouraging the exploration of new design and manufacturing processes. Many of them make use of parametric design and digital fabrication tools, maximizing the possibilities offered by these technologies. This not only prepares future architects to deal with the demands of a constantly evolving industry but also contributes to significant advances in construction methods and the development and use of new materials.

In addition, these temporary pavilions have a remarkable impact on the spaces where they are installed. They act as environment activators, transforming public areas into places of interaction and experimentation. These ephemeral structures can generate a variety of feelings and experiences, from surprise to contemplation, creating democratic and multifunctional public spaces.

Despite being used in less than half of all the pavilions studied, the analysis also revealed the importance of using biomimicry in this context. Biomimetic pavilions not only imitate shape but also seek to incorporate principles and functionalities from nature. This approach not only enriches the aesthetics of the structures but also raises awareness about the responsible exploitation of natural resources and environmental sustainability.

By spatializing the pavilions studied on a map, it was possible to see more clearly the distribution of similar research around the world, noting the predominance of certain locations to the detriment of others. It can be seen that most of the pavilions studied are located in Europe and South America. However, the absence of points on certain parts of the map does not mean that there are no academic pavilions but rather that they are not publicized on digital platforms.

Finally, it is clear that these academic pavilions represent more than just temporary architectural projects. They are living laboratories of innovation, where learning and experimentation intertwine. These ephemeral structures play a crucial role in the evolution of architecture and in the training of highly qualified professionals, with experience not only in hypothetical projects loaded with theoretical concepts but also with practical on-site experience. Therefore, the continued exploration and promotion of this architectural typology is fundamental to the advancement of the discipline and the development of creative solutions to the challenges of the built environment.

References

1. Benyus, J. M. (2006). *Biomimicry: Innovation inspired by nature* (p. 320). Harper Perennial.
2. Fleischmann, M., & Menges, A. (2011). ICD/ITKE research pavilion: A case study of multidisciplinary computational design. In C. Gengnagel, A. Kilian, N. Palz, & F. Scheurer (Eds.), *Computational design modeling* (pp. 239–248). Springer.
3. Jodidio, P. (2011). *Ten years serpentine gallery pavilions* (p. 332). Taschen.
4. LabIT-PROURB. (2015). *Butterfly Gallery*. Available at http://intervencoestemporarias.com.br/intervencao/butterfly-gallery/. Accessed 5 Sept 2023.
5. LabIT-PROURB. (2017). *Pavilhão Tornado*. Available at http://intervencoestemporarias.com.br/intervencao/inauguracao-do-pavilhao-tornado/. Accessed 5 Sept 2023.
6. da Silva Lima, E. C., Matsunaga, C., & e Mendes, L. T. (2021) Exploração do processo de projeto bio-paramétrico baseado na solução para um pavilhão efêmero. *Gestão & Tecnologia de Projetos, 17*(1), 65–79. https://doi.org/10.11606/gtp.v17i1.183658.
7. Monedero, J. (2000). Parametric design: A review and some experiences. *Automation in Construction, 9*(4), 369–377.
8. Oliveira Junior, J. A. (2012). *Biomimética e processo de Fabricação Digital: aplicações na produção da Arquitetura*. Paper presented at the XVI SIGraDi (pp. 352–325). Fortaleza: Brazil.
9. Pereira, S. C. M. (2013). *Pavilhões de Verão da Serpentine Gallery – o espaço arquitetônico em exposição*. Master thesis, Universidade da Beira Interior, Covilhã.

10. Quintella, I., Ferreira, I., & Florêncio (Ed.). (2016). *Making pavilions: Os pavilhões temporários no contexto das faculdades de arquitetura e urbanismo*. Paper presented at the XX SIGraDi (pp. 318–325). Buenos Aires: Argentina.
11. Self, M., & Walker, C. (2010). *Making Pavilions* (p. 184). Architectural Association Publications.
12. Soares, W. S., & Quintella, I. (2018). O campo ampliado da arquitetura: Os pavilhões temporário e sua contribuição para os novos paradigmas no ensino e pesquisa em arquitetura na era digital. Paper presented at the 70ª Reunião Anual da SBPC. Universidade Federal de Alagoas, Maceió: Brazil.
13. UNStudio. (n.d.). Interview: Ben van Berkel – Pavilions. [online] Available at https://www.unstudio.com/en/page/7577/interview-ben-van-berkel-pavilions

Elton Cristovão da Silva Lima, He graduated in Architecture and Urbanism by the Federal University of Pernambuco (UFPE) – Brazil in 2020 and received a recognition of degree and the title of master's in Architecture by the University of Lisbon in 2022. He is currently a PhD candidate in Architecture at the University of Lisbon (Specialty in Design and Computing) with a double degree partnership with the BioDesign Lab at UFPE.

Luís Miguel Cotrim Mateus, Assistant Professor at the Faculty of Architecture of the University of Lisbon. He teaches the courses of "Descriptive Geometry," "Generative and Geometric Modelling," and "3D Digitization, Virtual Reality and Animation." He is a member of the Research Centre for Architecture, Urbanism and Design (CIAUD) and develops research in the "Design and computation Group" and in "ArcHC3D". Moreover, he is the head of the 3D Scanning Lab.

Amilton José Vieira Arruda, He graduated in Industrial Design Product Project by UFPE, Master in Design and Bionics by IED in Milan, Ph.D. in Ricerca in Disegno Industriale by University of Milan Politecnico, postdoctoral in Design and Bionics at IADE European University UNIDCOM Lisbon, and postdoctoral in Bionics at University Luigi Vanvitelli Naples – IT. He is currently an associate professor IV at the UFPE Design Course and coordinates the BioDesign Lab.

Levels of Access to Biomimetics

Carla Langella

1 Introduction

In the history of design, nature has always been considered an example to imitate in order to create products that are more efficient, effective, and acceptable to users [1]. Throughout history, humans have empirically observed the functionalities of organisms and how they respond to the primary needs of survival and reproduction.

For centuries, architects, engineers, and designers have sought to mimic the characteristics and functionalities of biological systems that perform similar functions, such as emulating the aerodynamic shape of birds for designing aircraft or replicating the branching structures of trees for designing roofs and canopies [2]. This type of bio-inspiration has often been limited to predominantly formal or mechanistic observations, based on a rarely in-depth understanding of biology.

In recent decades, the field of design has benefited from a considerably deeper and more comprehensive comprehension of the fundamental principles, processes, and rationale that underlie biological systems. This enhanced understanding has been made possible by the utilization of increasingly advanced investigative tools, such as electron microscopy and micro-tomography [3]. These technologies allow for the examination of natural structures and the generation of three-dimensional models with an exceptional level of detail, reaching down to submicrometer scales. Certain techniques even enable nondestructive analysis, permitting observations in vivo and over time. Consequently, it is feasible to swiftly acquire intricate morphological, structural, and dynamic insights, intricately connected to functional, physiological, and metabolic aspects. These advancements yield results that

C. Langella (✉)
Department of Architecture, University of Naples Federico II, Naples, Italy
e-mail: carla.langella@unina.it

A. J. V. Arruda, F. L. Palombini (eds.), *Biomimetics, Biodesign and Bionics*, Environmental Footprints and Eco-design of Products and Processes,
https://doi.org/10.1007/978-3-031-51311-4_9

facilitate the precise understanding of the intricate mechanisms governing the development and functionality of living organisms.

Biology is thus able to know and understand in-depth natural structures and their ability to respond and react to stimuli, self-organize, or self-repair [4]. This new knowledge leads biomimetic designers to try to reproduce more complex properties such as the adaptive capacity of biological systems that are able to conform their structures, functionality, and behavior to certain conditions and situations in their environment.

Contemporary biomimetic design cannot ignore the latest scientific results that biology is progressively achieving. The designer who chooses to draw inspiration from nature must necessarily be able to consult the relevant biological scientific journals and collaborate with biologists in a participatory, conscious, and rigorous manner [5]. Biomimetics is a design approach that must be handled in a strictly scientific manner, with methodological consistency.

In biomimetic design, two approaches are possible. One can adopt a bottom-up approach, which entails commencing with a biological model to identify potential opportunities for biomimetic application in design. Alternatively, and more commonly, one can employ a top-down approach, searching in nature for solutions to design problems in artificial objects [6]. In both cases, it is crucial to clearly specify the biological species serving as the nature model, the particular biological trait being referenced, and the perspective from which this trait is observed and replicated.

The bio-inspired approach allows for various levels of access to nature's information for transfer. Consequently, it can draw inspiration from nature through observation from different viewpoints, commencing with the most intuitive and accessible aspects, such as formal and external characteristics, and extending to more complex aspects involving self-organizing and learning capabilities.

The beauty of nature encompasses its inherent intelligence, which is intricately woven into its very fabric. This intelligence is constructed upon elements such as ribs, overlapping layers, hierarchical structures, porosity, dynamics, processes, and intricate details that can be observed at various dimensional scales. Remarkably, each of these facets often serves multiple functions.

For instance, shell riblets fulfil roles associated with hydrodynamics, camouflage, and stability requirements [7]. Consequently, nature emerges as a generous teacher of multifunctionality, a quality of increasing significance in contemporary design. It empowers artifacts to adapt to evolving and intricate demands and contexts by offering a multitude of solutions.

Furthermore, when one observes an organism across a spectrum of scales, ranging from the macroscopic to the microscopic, from the entire body to individual molecules, using microscopes with varying degrees of magnification, a fascinating revelation unfolds. Different forms, movements, and structures manifest and are organized and interconnected in a hierarchical manner across these diverse scales [8].

The diverse hierarchy and multifunctionality inherent in every aspect of nature underscore the imperative need to clarify and explicitly define, during the design process, which function or functions are intended to be transferred to the project. Simply indicating the part or detail being referenced is insufficient. It is essential to elucidate the relationship between the chosen biological organism or its specific component and the intended level of access with respect to functional, physiological, and metabolic aspects.

The mere mention of drawing inspiration from, for instance, the wing of a particular insect does not suffice. One must specify whether the focus is on replicating the aesthetic texture, the mechanical structure, the generative constructional logic, the dynamics of movement, or the sensory capacity. Different levels of access to the same biological element yield vastly different characteristics and functionalities in biomimetic artifacts.

Within this chapter, a categorization of potential levels of access is presented, serving as both a reference code for the critical examination of case studies involving bio-inspired products and a framework to assist designers in consciously selecting the perspective through which to emulate nature.

2 Aesthetic-Morphological Level

The first level of access involves the aesthetic-morphological perspective, wherein the design draws upon aesthetic and geometric characteristics such as shapes, patterns, colors, shades, surface textures, ornamentation, porosity, streaks, opacity, and transparencies. These attributes can be observed at their full scale or under magnification with a microscope. While in nature form always serves a functional or process-driven purpose, the transfer of aesthetic characteristics in biomimetics generally lacks a functional or phenomenological rationale and typically serves solely as an aesthetic, communicative, and at times, evocative element.

When geometric morphologies are enlarged or reduced, they maintain their proportional relationships and overall form. Consequently, during the transition from the biological realm to the artificial dimension, aesthetic bio-inspiration retains its validity, even when translated across vastly different scales [9].

Regrettably, many biomimetic design products predominantly interpret biological inspiration from a formal standpoint, often missing out on the potential for functional innovation that this design paradigm offers.

The process of transferring the geometric morphology of biological elements to artifacts begins with the identification of the geometries and dimensions of the biological system. Advanced scanning technologies, such as X-ray micro-tomography, are employed for this purpose, generating two- or three-dimensional models that can readily serve as nature models for study [10]. The ongoing advancement of scanning techniques and technologies in the field of biology has significantly expanded the opportunities to analyze intricate details for inspiration. This allows access to geometries existing in nature at very small scales, which were previously impossible to detect and observe. When conducting such a survey with a biomimetic intent, the level of detail in the scan and the choice of what to capture depend on the specific part of the organism intended for replication. This necessitates collaboration between the designer and biologist. The transfer of aesthetic-morphological attributes may encompass elements of a biological system such as a particular texture, ornamentation, the shape of a specific detail, or the entire overall form.

Following the stages of surveying or scanning, digital modeling ensues. During this phase, design and biology work in concert to align with the design objectives, leading to the actual design stage. This stage involves abstractions and formal, structural, and functional reinterpretations.

The organic form should not be replicated in the project automatically without any form of interpretation or decoding. It must be adapted to the designer's language and tailored to the specifics of the cultural, social, economic, and production context within which the project is situated.

Regarding the production phase, design inspired by biological geometries encounters a new opportunity with the advancements in digital fabrication technologies. These technologies enable the realization of increasingly complex forms, thanks to the resemblance between 3D printing processes and biological growth processes [11]. Digital fabrication technologies allow for the creation of intricate patterns, weaves, and layers without the need for extensive post-production interventions. Presently, digital fabrication technologies are undergoing a significant shift from their initial role in prototyping to broader applications in large-scale production, largely due to the emergence of industrial 3D printers [12]. This paradigm shift opens up new avenues for the integration of biomimetics in the field of industrial design. It provides designers with a wealth of possibilities to experiment with unique, highly distinctive, and original expressive languages that can deviate from the flattering often seen in the contemporary market, which tends to favor neutral and minimalist design languages [13].

The transmission of patterns and forms from nature broadens expressive opportunities, offering a vocabulary rooted in morphological structures, geometric configurations, symmetries, as well as interruptions and variations that the human eye perceives as linked to its biological foundation.

The concept of biophilia, initially coined by Edward O. Wilson in 1984 [14], encapsulates the inherent tendency of humans to establish an emotional connection with other natural systems.

Researchers across diverse fields, including neuroscience, psychology, and ecology, such as Orians and Heerwagen in 1995 [15], Mealey and Theis in 1995 [16], Kawabata and Zeki in 2004 [17], Ryan in 2014 [18], and Hettinger in 2017 [19], have delved into the primal instinct of the human mind, which inclines toward natural forms associated with elements that do not pose threats to the survival of the species or offer advantages as shelters or sources of nourishment.

Nonetheless, the principle put forth by Buckminster Fuller in the 1950s remains pertinent: one should not merely mimic the outward appearance of nature but rather strive to comprehend the underlying principles that define it [20].

In the Hybrid Design Lab, the aesthetic-morphological level of nature is investigated especially in cases where the reference to nature serves to convey a particularly ethically important meaning or message. Biomimetic design is an effective means of communicating principles and values related to environmental sustainability.

In the photovoltaic lamp Fotosintetica (Fig. 1), through the morphological reference of a diatom, used as the lamp's porous structure, and the use of waste green algae to make the diffuser's bioplastic, the possibility of using photovoltaic energy as a strategy to reduce environmental impact is highlighted, even in small lamps as well as in buildings. The lamp prompts people to reflect on how nature draws energy from the sun to fuel the life of microalgae and macroalgae. Each alga produces oxygen through photosynthesis and contributes to the oxygen production of the entire planet. Similarly, each individual lamp can make a contribution to reducing the consumption of oil resources. Fotosintetica offers a low-intensity ambient light to dimly illuminate a room for relaxation. During the day, it must be placed by users near a window so that the photovoltaic panel charges. In the evening, thanks to its lightness, it can be easily moved to where it is needed. This concept is intended to make people realize that sustainability requires commitment and work on a daily basis. We can no longer just push a button unconsciously to get light. Instead, we must engage with small gestures to reduce the environmental impact of our activities. Moving a lamp from one part of the house to another is a small effort that is worth making and that the designer must facilitate so that it becomes a healthy habit.

Fig. 1 Fotosintetica. Design: Martina Ponticelli, Claudia Portanova, Lucio Prisco Junios, Valentino Scarpati, Simona Suarino. Research team: Carla Langella, Gabriele Pontillo, Valentina Perricone

3 Generative-Constructional Level

The generative-constructional level pertains to the transfer of constructive strategies and logics employed by nature. This includes methods related to the assembly of discrete components, growth patterns, and diffusion processes utilized in the creation of natural textures and structures [21]. It encompasses how organisms are generated and develop during their initial life stages, how parts and elements are regenerated throughout their life span, and how living structures with protective functions like nests, cocoons, and shells are formed. Additionally, it encompasses the organization of structures with predatory functions, such as spider webs; the construction of structures designed to attract breeding partners, as seen in the nesting behaviors of certain birds; and the development of materials that facilitate movement and locomotion, such as burrs [22, 23].

Designers should be mindful that organisms not only construct for protective purposes. An illustrative example is the ptilonorhynchids (Ptilonorhynchidae), commonly known as garden birds. These birds, belonging to the Passeriformes order, are renowned for the intricately detailed structures constructed by males, primarily for the purpose of attracting potential female mates.

They craft these structures using a diverse array of elements they find in their environment, including colored objects, both natural and artificial, such as fruit, berries, mushrooms, bits of tin foil, and fragments of plastic. In some instances, they even employ natural pigments to add color to their creations. Interestingly, many males in this bird family exhibit brightly colored plumage, some even iridescent, but it has been observed that males with more conspicuous colors tend to build simpler structures, while those with less conspicuous plumage create more complex and visually prominent structures, as if compensating for their less striking appearance [24].

As the field of biology increasingly integrates computational and processing tools, it becomes possible to develop mathematical models of biological growth processes and logic. Research is revealing that numerous biological structures, regardless of scale or function, share common mathematical principles, such as the Voronoi diagram, which possess significant constructional value [25].

In imitating the living and protective structures produced by biological organisms, the designer observes and summarizes the biological strategies aimed at ensuring protection, adaptability to external thermo-hygrometric conditions, flexibility with respect to changing situations, resistance and defense against external aggression, and effective anchorage to substrates.

In the project of the traveling museum pavilion Hemppy Campania (Fig. 2), developed in the Hybrid Design Lab in cooperation with Assocanapa, to promote the ancient culture of hemp production as a material of the past and a possible sustainable future, the reference of the bioclimatic functioning of the caterpillar's tent was transferred to ensure thermal well-being in all seasons. Biological studies [26] have shown, by means of electronic monitoring, that lasiocampids of the genera *Malacosoma* and *Eriogaster* construct their tents by laying overlapping layers of

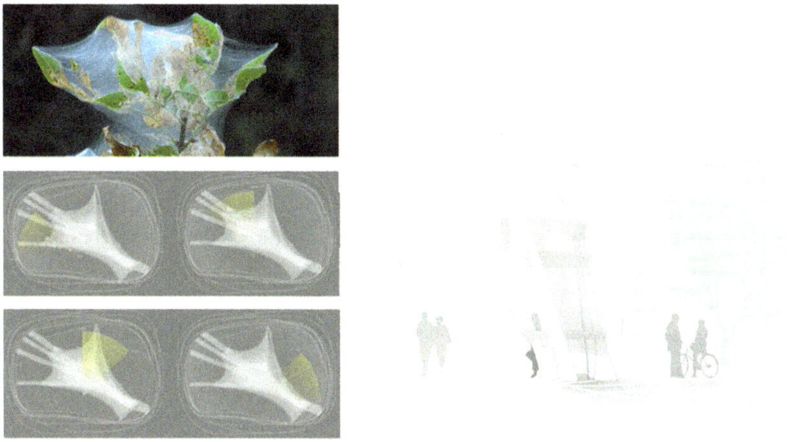

Fig. 2 Hemppy Campania. Design: Laura Guarino. research Group: Carla Langella and Carlo Santulli

silk separated by air-filled spaces and modulating the layers according to environmental exposure, irradiation, and wind-exposed parts. This creates a beneficial microclimate for the organisms. When the nest is exposed to solar radiation, the layered structure of the tent creates a thermally heterogeneous and insulated environment within which the caterpillars thermally regulate themselves by moving between warmer and cooler parts. In the bioinspired, bioclimatic museum concept, the sustainable, weather- and light-resistant hemp fabric, treated with natural waxes to increase waterproofing, is structured in layers with the same logic and strategically positioned according to climatic orientation. The structure is supported by a modular, dry-assembled concrete and hemp support system.

4 Structural Level

The field of design is increasingly drawn to the application of structural optimization principles and adaptation to specific loading conditions observed in natural structures.

In biomimetics, the structural level involves the transfer of strategies and solutions rooted in biomechanics, which examines how stresses are distributed within biological structures and, consequently, how materials can be optimized. Within this domain, designers adopt biomechanical strategies and replicate cellular, porous, ribbed, folded, and layered structures found across the animal, plant, and mineral kingdoms. These structures, shaped by centuries of biological evolution, offer a wealth of solutions applicable to diverse fields.

However, it's crucial to note that biomechanics is an exceptionally complex field that demands expertise in both structural engineering and biology [27]. These are skills that may not always be readily accessible to designers. At times, the challenge of fully comprehending biomechanical principles can lead to misunderstandings, resulting in the transfer of structural logic based on incorrect analogies between the biological reference system and the applied product.

Structural optimization offers the potential to minimize the consumption of materials and energy throughout the life cycle of products, resulting in a significant reduction in environmental impact. Not all forms of biomimetics are inherently sustainable, but structural biomimetics is, by its very nature, sustainable because it draws inspiration from nature's inherent drive to conserve material and energy resources [28].

At the microscopic level, an increasingly detailed understanding of the microstructures found in biological tissues, combined with digital simulation systems and virtual structural calculations, allows us to gain a deeper appreciation of the biomechanical principles underlying the beauty observed in nature. This beauty is characterized by features like ribbing, porosity, and texture, all governed by the logic of structural optimization, which seeks to minimize the use of materials and energy, ultimately regulating the processes of life itself. biomimetic designers now have access to specialized software and plug-ins explicitly designed for topological optimization, mirroring the processes found in nature.

The Protex (Fig. 3), developed by Lucia Paciolla in the Hybrid Design Lab, draws inspiration from the optimized, light, and strong mechanical structure of the Xylem cactus roots to propose a customized and comfortable cover for lower limb prostheses for amputees. This biomedical aid performs the task of covering the prosthesis with a structure consisting of two shells that can be easily assembled and disassembled, thanks to the presence of small magnets in the edges that come into contact. It is designed by parametric digital modeling and manufactured via 3D

Fig. 3 Protex. Design: Lucia Paciolla, coordinated by Carla Langella

printing so that users can choose textures, colors, and styles according to taste, context of use, and type of outfit. Structural optimization plays a fundamental role in this project because it allows the prosthesis to not add additional weight to avoid unbalancing the two limbs. For this reason, the level of structural mechanical access is what guided the nature model's research.

5 Material Level

The material level involves the transfer of materials and the underlying principles governing how these materials are assembled and structured in nature. This approach does not involve scaling up; instead, it focuses on directly applying biological materials and their logic to products while staying within the realm of materials. Durable material structures found in nature, such as spider's webs or mother of pearl, are carefully studied to replicate some of their unique properties in relevant contexts. A prime example of this approach is the hierarchical organization seen in structures like shells. These natural examples teach us that alternating layers of hard and elastic materials can result in enhanced mechanical performance and a reduction in the propagation of damage or injuries [29]. Furthermore, emulating the chemical composition or performance characteristics of natural materials proves valuable in achieving environmental compatibility and biodegradability. This ensures that, at the end of their useful life, artifacts can reintegrate into ecosystems without causing harm.

Bio-inspiration also plays a crucial role in addressing specific design challenges, such as developing adhesives capable of effectively bonding elements in aquatic environments, even when exposed to water. One area of advanced research that draws inspiration from natural models is the field of biomaterials designed for use in the human body. These biomaterials must rely on biological substances to be integrated into the body without triggering rejection reactions. Biomaterials encompass substances, distinct from drugs, utilized to treat, regenerate, replace, or enhance tissues, organs, or bodily functions.

Some biomaterials are employed to create implants with the goal of restoring or regenerating damaged or deteriorated parts of the human body, such as bone and dental implants. One prevalent method of applying biomaterials involves the use of scaffolds—porous structures crafted from biodegradable materials that are implanted into the body to provide a framework for gradual cell growth, facilitating the replacement of the degrading biomaterial. The discipline dedicated to studying these processes is known as tissue engineering [30]. Tissue engineering seeks to summarize the principles underlying the generation and regeneration of biological tissues, primarily for medical applications, using biofabrication and bioproduction technologies.

Various biomaterials serve specialized functions by replacing specific organs within the human body, including intraocular lenses, heart valves, catheters, and stimulating electrodes. Others find use in manufacturing surgical equipment, devices for controlled drug release, and internal diagnostic tools, such as probes employed in gastroscopy. Over the centuries, biomaterials used for grafting have undergone significant advancements, driven by progress in materials science and changes in application paradigms, particularly due to the evolution of nanotechnology.

The realm of biomaterials, once confined primarily to the field of medicine, is increasingly intersecting with the field of design. Utilizing tissue engineering techniques, it is now possible to create in vitro replicas of animal skin, offering an ethical alternative to leather and hides traditionally used in the production of accessories, furniture, and vehicles [31]. These same techniques can also produce replicas of human skin for testing cosmetics and drugs, providing platforms for research and development.

These emerging design approaches bring forth significant ethical considerations, as they challenge the distinction between what is natural and what is artificial. Moreover, ethical standards are currently lacking in this hybrid scenario. Biogenerated materials represent a shift from the traditional production paradigm, which involves assembling materials and components, toward a new paradigm where production is conceived as a form of growth [32].

Within this context, biodesign comes to the forefront, utilizing biology in the manufacturing processes of artifacts, particularly at the material level. These materials are generated, consumed, and reused by nature through circular logic and processes, minimizing the consumption of materials and energy.

Biodesign proposes the integration of living biological systems into the creation of artifacts to enhance their compatibility with the ecosystem. This approach seeks to contribute to the achievement of Sustainable Development Goals (SDGs). In the development of their designs, biodesigners frequently employ synthetic biology techniques to manipulate biological matter, sometimes even delving into genetic manipulation of organisms to attain properties and performance characteristics aligned with the design requirements.

A concrete example of this biodesign approach is mycelium, the vegetative component of fungi composed of a network of interconnected filamentous cells. Mycelium has emerged as a potential substitute for plastics due to its capacity to grow on various substrates and adapt its morphological characteristics, including resistance, elasticity, uniformity, and water repellency, depending on the material it colonizes. The US company Ecovative was among the pioneers in leveraging the versatility of mycelium. They initially used it as an alternative to polystyrene in packaging and later extended its applications to the production of entirely organic tables and chairs [33].

Another prevalent resource in contemporary biodesign is bacterial products, particularly bacterial nanocellulose. In material production processes, bacteria can serve as catalysts for metabolic processes such as fermentation. One well-known example is the utilization of bacterial cellulose, which was first introduced by Brown in 1886 using the *Acetobacter xylinum* strain. Bacterial cellulose is characterized by its three-dimensional structure of porous fibrils, ranging in size from 10 to 100 nanometers [34].

Over the past decade, there has been a significant focus on bacterial cellulose obtained through the fermentation of kombucha, largely due to the dissemination of research results from the BioCouture project conducted by Suzanne Lee at Central Saint Martins College of Art and Design. This fabric, commonly referred to as scoby, can be easily produced at home and has found widespread use in DIY and artisanal fashion circles as a vegan alternative to leather [35].

In the realm of biology, researchers have identified numerous strains of both wild and recombinant bacteria that excel in cellulose production. Additionally, various carbon sources as growth substrates have been explored, including the utilization of various types of waste materials. Bacteria can be employed in their natural "wild-type" state as they exist in nature, or they can be genetically engineered to enhance their functional performance and productivity. Notably, bacterial cellulose differs from plant cellulose in that it is purer, lacking the impurities typically found in vegetable pulp, such as hemicellulose, lignin, or pectin.

In addition to its higher purity, bacterial cellulose exhibits superior mechanical properties, increased crystallinity, and improved water-holding capacity. These qualities are making it increasingly attractive in various industries, including cosmetics, biomedicine, and food packaging. Bacterial cellulose is sometimes used in the form of biocomposites, incorporating organic or inorganic fillers to enhance its mechanical, electrical, and thermal properties [36].

By cultivating mycelium or bacterial cellulose in molds crafted with digital manufacturing systems, designers can explore intricate three-dimensional morphologies. Particularly noteworthy are the flexible silicone molds that allow for the creation of highly detailed shapes. This opportunity presents the design community with a broad spectrum of new and original expressive possibilities, resulting in innovative forms of ergonomics and usability. One promising avenue to explore involves using these materials to craft casings or interfaces for electronic and digital devices, which typically have a significant environmental footprint due to their short lifecycles resulting from rapid technological advancements.

Interactive artifacts integrated into the soft and malleable morphologies of mycelium or the organic tactility of bacterial cellulose take on a more natural, organic, and notably distinctive character. Despite using molds, each artifact created through biodesign techniques possesses its own unique aesthetic qualities, featuring variations in color, consistency, and mechanical properties. These variations, in addition to the intended shape and the properties of the mold, arise from environmental factors like humidity levels, temperature, and nutrient availability. This scenario offers a glimpse of a new generation of electronic or digital technological artifacts that are one of a kind. They depart from the paradigm of perfection, impersonality, and standardization and instead embrace irregularity, differentiation, and distinctiveness. Moreover, they emphasize ecological sustainability as a central aspect of their design and production.

When biodesign generates materials in which microorganisms remain alive and active, these materials are referred to as living materials. Developing such materials necessitates a deep understanding of the molecular mechanisms governing the assembly and functions of biological materials.

Biodesign can also be employed to establish sustainable decorative processes, addressing the environmental issue that conventional decorative finishes often involve paints and materials with a significant ecological footprint. The minimal-sustainable approach seeks to create single-material artifacts to avoid the separation of components from various materials during disposal and aims to minimize decorations to reduce their environmental impact. However, this approach can limit the expressive potential of designers.

Biodesign offers a response to these challenges by applying treatments to common materials such as ceramics, stone, or wood, enabling controlled growth of microorganisms like moss or mold on surfaces. These treatments result in visible traces that either stabilize or evolve over time, creating dynamic decorations. This approach paves the way for innovative methods of bio-ornamentation characterized by fractal, branching, and textured morphologies shaped by the growth processes of the organisms involved [37].

In this process, decoration emerges from a synergistic and cooperative relationship between humans and organisms. Living decoration can be designed and executed by manipulating environmental parameters like temperature, humidity, and the chemical-physical treatment of the substrate. Additionally, it involves the strategic positioning and manipulation of nutrients. In some cases, when experiments are conducted in collaboration with microbiologists, genetic manipulations may also be involved, such as introducing chromatic modifications or imbuing organisms with photoluminescent properties.

In the Hybrid Design Lab, designer and artist Marco Fiume has developed the project Bio Ornamentum (Fig. 4) which proposes a new form of experimental decoration made from live organisms of the plasmodial myxomycete species known as *Physarum polycephalum*, to be used on ceramic vessels. The decoration, characterized by its fractal and branched morphology, is generated by collaboration between

Fig. 4 Bio Ornamentum. Design: Marco Fiume, coordinated by Carla Langella

the designer and organism to contrast the high environmental impact of decorative finishes for ceramics made with polluting and toxic paints and materials. The living-colored decorative patterns have been obtained on black ceramic vases by modulating the environmental thermo-hygrometric parameters and the positioning of colored nourishment with food pigments.

Living components like bacteria, fungi, and cells can be seamlessly integrated with synthetic components to create composites with programmable and responsive functions. An illustrative example of this approach is the development of wound healing gels containing spores of genetically modified bacteria capable of producing antibiotics in response to the presence of pathogenic bacteria [38]. Furthermore, engineered bacteria and fungi can be employed to construct cellulose-based bacterial biosensors for hormone detection [39]. Notably, regenerative building materials that incorporate living components have been created. These materials are composed of a hardened sand-hydrogel matrix formed through biomineralization processes [40].

6 Dynamic Level

The dynamic access level to biological models pertains to understanding the principles governing the movements of natural systems, with a primary focus on the animal kingdom, although botanical movements, such as those of carnivorous plants or the gradual growth of plant parts like roots, leaves, or the geometry of falling seeds, are also observed.

For instance, researchers delve into the intricate processes underlying the movement and fine control of tentacles in octopuses, the locomotion of snakes, and the complex leg structures of insects that enable them to perform high jumps, rapid sprints, or precise maneuvers.

This approach holds particular significance in the field of robotics, leading to the emergence of a specific discipline known as biorobotics. This discipline focuses on identifying the fundamental principles underlying biological functions and translating them into technological solutions that can be realized with the tools and technologies available, closely emulating natural models. Studies often focus on omni-dimensional structures like the arms of octopuses or the trunks of elephants, both of which exhibit complex movements and the ability to integrate various modes of grasping, locomotion, and exploration. To study dynamic functions, researchers employ sophisticated cameras capable of capturing as many movement fragments as possible. Biomimetic robotics is the primary application area for these systems [41].

In the Hybrid Design Lab, several biomimetic projects refer to movement and dynamism. Tune (Fig. 5), designed by Valentina Mello, is a wearable device for knee physiotherapy inspired by the relationship between movement and sound of the rattlesnake. Various research have shown that the motor areas of the brain can be activated by rhythmic sound sequences and more complex acoustic stimuli. Tune is

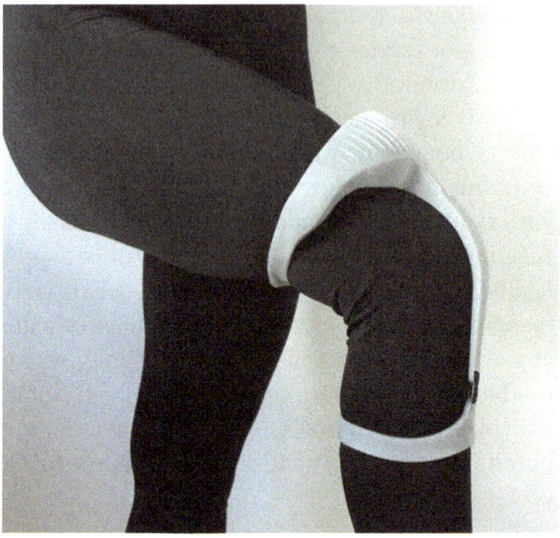

Fig. 5 Tunes. Design: Valentina Mello, coordinated by Carla Langella

based on the biofeedback potential of movement sonification, i.e., the conversion of certain physical parameters of body movement deemed relevant (force, speed, acceleration) into synthetic sound parameters. The device acts as a tool guide during the rehabilitation session by translating, thanks to the presence of movement sensors and sound actuators, the movement of the knee into sounds of various types: an alarm sound if the patient is positioning the knee incorrectly, a continuous sound in real time describing the movement made by the knee (sonification), and various sounds of different intensities indicating the various angles assumed during the bending of the knee (sound feedback).

ActiveAid (Fig. 6), designed by Teresa Iavarone, is a walking stick for the elderly inspired by the ability of the horse's foot to adapt to different ground textures such as soil, pavement, or sand during its movement. This ability is attributable to the lobed morphology of the muscular part inside the hoof, which deforms in different ways depending on the weight of the animal, the type of movement, and the mechanical characteristics of the ground. The handle is equipped with a safety call system and a small notification monitor connected to the telephone. It can be unscrewed and, thanks to an internal elastic band, used to train the hands and arms.

The growth movement of the climbing plant *Parthenocissus tricuspidata* [42], also known as Boston ivy, inspired Climbing Light (Fig. 7), designed by Emanuela De Conciliis, a variable configuration lighting system with 3D-printed wall hooks, on which a wire led can be hooked to form different figures. The lamp was designed for juvenile prisons, as a tool for creative expression and manifestation of the emotions of young prisoners. By configuring the lamp in a different way each time they wish, they can characterize and personalize their common living spaces by expressing their emotions. The lamp thus also becomes a tool for interpreting children's moods by analyzing the hidden message in the coffered forms through the psychology of forms.

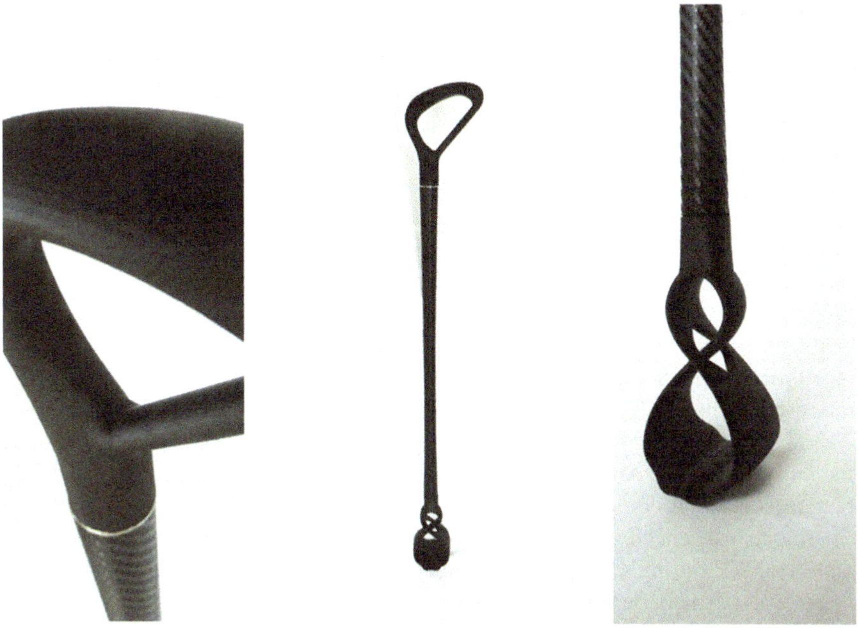

Fig. 6 ActiveAid. Design: Teresa Iavarone, coordinated by Carla Langella and Eugenio Ruocco

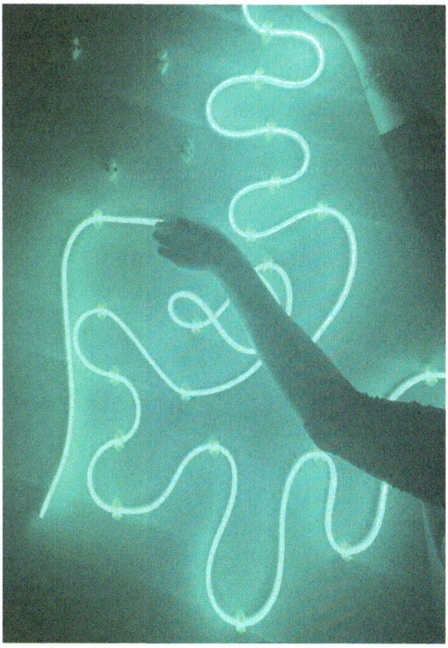

Fig. 7 Climbing light. Design: Emanuela De Conciliis, coordinated by Carla Langella and Antonio Bosco

7 Physiological Level

Projects grounded in the physiological level of biomimetics draw inspiration from the essential processes within biological systems, including metabolic processes and mechanisms that enable organisms to regulate their body temperature, particularly in challenging environmental conditions.

For instance, in fields like sportswear and architecture, biomimetics takes inspiration from the thermoregulation strategies observed in animals like penguins and other amphibious creatures living in extremely cold climates. These strategies are built upon the ability to insulate the body while submerged in water and to gather heat from solar radiation when out of the water [43].

Hotring (Figs. 8 and 9), designed by Matteo Galeotti and carried out in collaboration between the Hybrid Design Lab and the IUAV University of Venice, is a wearable device inspired by the thermogenic capacity of the plant *Symplocarpus foetidus* [44] of the Araceae family, which is able to warm itself up in winter by piercing snow in order to protect itself from the cold of its surroundings. Thermogenesis enables this herbaceous perennial plant to maintain its temperature at around 20° during the winter when the air temperature is below zero. This happens because the tissue of the inner part of the plant, the spadix, produces endogenous heat thanks to a plant physiological process.

The project consists of a pendant that allows people to keep warm while practicing outdoor sports such as skiing and climbing in very cold climates. Referring back to the nature model, the concept of a heating core capable of transferring heat to the body in order to maintain a stable body temperature has been reintroduced. The pendant is equipped with a photovoltaic panel, a rechargeable battery, and a heating element, components that are encased in a case that is thermally insulated on one side and capable of conducting heat on the other.

8 Communication Level

The access level related to communication involves the transfer and reinterpretation of communication models derived from biology into communication protocols for use in artefacts. In nature, organisms primarily communicate for purposes such as recognizing other organisms and making themselves recognizable (parents and offspring, predators, positioning in herd hierarchies), signaling aggression, self-defense, and attracting potential mates [45]. The primary modes of animal communication encompass chemical signaling using pheromones and other substances [46], auditory communication relying on sound signals [47], tactile communication, and visual communication encompassing colors, shapes, patterns, and movements [48].

All organisms, or genotypes, strive to enhance their fitness, denoting the difference in reproductive and survival capabilities one individual possesses compared to

Figs. 8 and 9 Hotring. Design: Matteo Galeotti, coordinated by Carla Langella, Francesca Toso, Valentina Perricone

another of the same species. Communication, across its various dimensions, plays a pivotal role in pursuing this objective [49].

Nature often employs highly intricate visual strategies based on communicative hierarchies of colors and shapes. These strategies are used to emphasize, highlight, or conceal parts of the body, convey messages, and indicate conditions, intentions, gender differences, and hierarchical positions. These natural communication codes offer designers a vast reservoir of unique communication cues to explore and apply creatively.

In the realm of visual communication design, there is an abundance of intricate and strategically rich references, particularly in the domain of mimicry and optical deception that employ shapes and colors reminiscent of other organisms [50]. The primary types of mimicry include cryptic mimicry and phaneric mimicry [51].

In cryptic mimicry, an organism blends seamlessly with its surrounding environment by adopting colors, shapes, and patterns that render it nearly invisible to predators or prey. In contrast, phaneric mimicry involves organisms that are clearly visible but are designed to be mistaken for other organisms. Phaneric mimicry can be Batesian when a harmless organism, which could potentially be prey, mimics a dangerous, toxic, unpleasant, or otherwise undesirable organism to deter predators [52]. For instance, the owl's head butterfly, *Caligo idomeneus*, features eye-like ovals on its wings to confuse predators into thinking they resemble the eyes of owls [53]. Mimicry isn't limited to animals; it extends to plants as well [54]. Some orchid species, for instance, mimic the bodies of certain female wasps to attract pollinators [55].

In Müllerian mimicry, two or more phylogenetically distinct species, all of which are unpalatable, mimic each other and share visual characteristics like colors, markings, and warning morphologies [56]. This collaborative strategy benefits all the involved species because it reduces the time predators need to learn to recognize a single type of visual warning signal instead of distinct ones for each species. The outcome is a decrease in the number of individuals from each species that must be sacrificed for predators to learn the lesson. For instance, certain species of Lepidoptera in the *Zygaena* genus, the homopteran *Cercopis sanguinea*, and the Cleridae beetle *Trichodes apiarius* all share the same red and black aposematic colors [57].

This type of strategy could find applications in co-branding systems, allowing different brands to effectively convey a common message through shared codes, even if they operate in vastly different sectors or product areas.

Beyond visual communication, design can also draw inspiration from various forms of communication that rely on different physical-chemical channels operating at different dimensional levels, ranging from the scale of pheromones to the intracellular level. The biological and natural sciences are continually advancing their understanding of the intricacies and manifestations of these mechanisms, enabling designers to translate them into conceptual transitions.

Molecular biology offers intriguing sources of inspiration for conceiving new forms of communication, such as the enzymatic mode, which relies on sequences for combining amino acids to create diverse proteins capable of performing specific functions. Even highly complex sequences function as keys that can only engage with the specific sequences they are predestined for, activating processes. Synthetic biology heavily relies on deciphering and controlling these mechanisms [58], which could be translated into new forms of communication conveyed through design.

In the Hybrid Design Lab, different ways of communicating nature are studied and transferred to the design of artifacts. Chroma (Fig. 10) is a cardiotocographic band for pregnant women to monitor the fetus' heartbeat and uterine contractions, which also has the function of a containment band to support the weight of the belly. To distract the mother-to-be from the stress of fetal monitoring, a communication strategy inspired by how cephalopods, such as octopuses and squids, change color to blend into their environment and communicate effectively with each other was employed. The design of the belly tissue derives specifically from the observation of chromatophores, specialized cells responsible for the variable coloration of cephalopods, located in the tegument and containing numerous pigment granules. These are surrounded by muscles that, as they contract and relax, cause the cell size to vary and the pigments to become visible, altering the color of the skin almost instantaneously. This nature model led to the design of a belly pan consisting of three overlapping layers of elastic fabric of different colors on which holes are drilled. The growth of the abdomen as the pregnancy progresses causes the material used to stretch and the underlying layers of various colors to gradually emerge, thus creating a pleasing visual effect and leading the pregnant woman to focus on changing her body and communicating her condition to the outside world, rather than on the health of the baby.

Drawing inspiration from the communication between plants and insects based on color codes [59], a color communicating table has been developed by Osorio and Vorobyev. Mox is a table for bars and restaurants that optimizes communication between customers and service staff. Through the use of a color code that allows different colors to be selected and emitted from a linear LED source located underneath the table projecting them onto the floor, customers can communicate whether they are waiting to order, the dishes they have already ordered, the bill, or whether they do not need the staff. The focus on interpersonal communication is also directed at the analogue relationship between the people occupying the table.

By decorating the tabletop with a texture made of thermochromic inks, people are induced to focus their attention on the color variations resulting from the hot dishes or their hands and to interact analogically, putting aside their smartphones (Fig. 11).

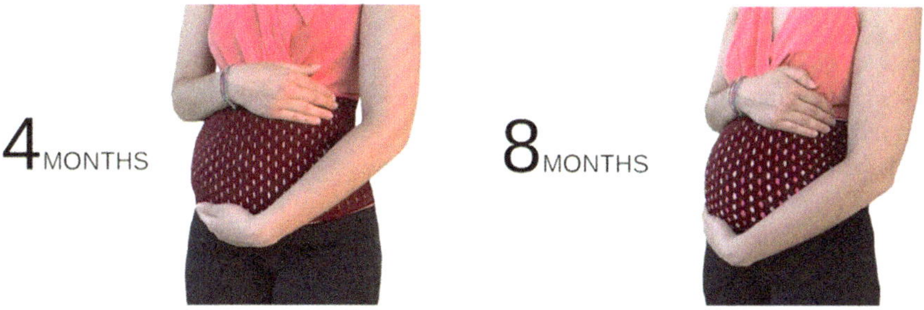

Fig. 10 Chroma. Design: Flavia Mastroberardino, Roberta Gragnano, Daria Cermola, coordinated by Carla Langella and Valentina Perricone

Fig. 11 Mox. Design: Alessia Serino, Arianna Moxedano, Giovanna Pannone, Alessio Monti, Annarita Palladino, Giulia Svetti. Research team: Carla Langella, Valentina Perricone, Gabriele Pontillo

9 Behavior and Organization Level

The most intricate of the access levels pertains to the transfer of behavioral and organizational principles that biological systems employ to fulfil their needs for protection, attraction, defense, and aggression. These principles encompass forms of collective organization, such as those seen in social insects structured into super-organisms [60].

This level of access also encompasses human entomology, an area of investigation teeming with ideas, inspirations, and models useful for projects aiming to explore opportunities to influence user behavior through products and services, promoting healthier or socially more desirable behavioral models [61]. Design can mediate and modulate relationships among individuals, particularly in today's complex and ever-changing relational dynamics increasingly shaped by digital tools. It can encourage empathy, prompting individuals to make eye contact, perceive each other's presence, stimulate physiological reactions associated with proximity and physical contact in personal contexts, or aid in decoding and recoding signals between individuals to facilitate dialogue and understanding in professional settings [62]. This type of intervention necessitates knowledge of the principles and the most recent advancements in the field of human entomology, integrated with psychology, anthropology, and sociology. It can also draw parallels between human behavior and that of animals or imitate mechanisms related to relational and behavioral aspects observed in nature.

Fig. 12 Deeplamp. Design: Simona Sbriglia. Research team: Carla Langella, Valentina Perricone, Gabriele Pontillo, Nicola Corsetto

Additionally, this level encompasses aspects related to the intricate organization of living systems, including their complex system qualities as self-organization, self-assembly, self-adaptation, resilience, redundancy, cyclicity, reactivity, cooperation, and synergy [63].

In the Hybrid Design Lab, the biological concepts of symbiotic and parasitic behavior [64] are experimented with through objects in which different elements cooperate with each other to achieve mutual or unique advantages. Deeplamp is a photovoltaic outdoor lamp with an LED source that integrates the technological innovation of 3D printing and solar energy with the efficiency of nature. The production of photovoltaic energy is facilitated by inserting a marine sponge, *Euplectella*, into the lamp, which is able to focus solar radiation downwards, as if it were an optical fiber. The result is a hybrid lamp in which the natural component is grafted into the artificial one to improve its efficiency. The bio-inspired appearance makes it different from other similar products (Fig. 12).

10 Conclusions

The conceptual framework proposed in this chapter constitutes both a design and interpretative tool for biomimetic design. It could be useful in guiding the process of collaboration between designers, biologists, and technologists to support them in selecting the level of access to nature providing a common context that can lead to coherent biological research and design activities projected in a clear and explicit direction.

It could be also useful for students to learn how to apply the biomimetic approach in a scientifically rigorous manner. With such a methodological grid, students are facilitated to design biomimetic products without risking naivety in the identification and interpretation of nature models.

Finally, it can be a conceptual frame of reference for design historians and critics because it makes it easier to catalogue biomimetic products in a scientific way.

References

1. Gruber, P. (2011). *Biomimetics in architecture: Architecture of life and buildings* (pp. 418–426). Springer.
2. Bar-Cohen, Y. (2005). *Biomimetics: Biologically inspired technologies*. CRC Press.
3. Langella, C., & Perricone, V. (2019). Hybrid biomimetic design for sustainable development through multiple perspectives. *GRID-Architecture Planning and Design Journal, 2*(2), 44–76.
4. Misteli, T. (2001). The concept of self-organization in cellular architecture. *The Journal of Cell Biology, 155*(2), 181.
5. Langella, C., Arruda, A. J., & Di Bartolo, C. (2022). Revising Biomimetics: Opportunities and ambiguities in the bioinspired design approach. *DIID Disegno Industriale Industrial Design, 78*.
6. Speck, T., Speck, O., Beheshti, N., & McIntosh, A. C. (2008). Process sequences in biomimetic research. *Design and Nature IV, 114*, 3–1.
7. Baumeister, D., Tocke, R., Dwyer, J., Ritter, S., & Benyus, J. (2013). *Biomimicry resource handbook: A seed bank of knowledge and best practices* (Vol. 3, p. 8). Biomimicry.
8. Ball, P. (2016). *Patterns in nature: Why the natural world looks the way it does*. University of Chicago Press.
9. Perricone, V., Santulli, C., Rendina, F., & Langella, C. (2021). Organismal design and biomimetics: A problem of scale. *Biomimetics, 6*(4), 56.
10. De Tommasi, E., De Luca, A. C., Lavanga, L., et al. (2014). Biologically enabled sub-diffractive focusing. *Optics Express, 22*(22), 27214–27227.
11. Sane, S. P. (2016). Bioinspiration and biomimicry: What can engineers learn from biologists? *Journal of Applied Science and Engineering, 19*(1), 1–6.
12. Jandyal, A., Chaturvedi, I., Wazir, I., Raina, A., & Haq, M. I. U. (2022). 3D printing–A review of processes, materials and applications in industry 4.0. *Sustainable Operations and Computers, 3*, 33–42.
13. Chirazi, J. E. (2022). Commercialization journeys: Bringing biomimetic innovation to the market. In *Biomimicry for materials, design and habitats* (pp. 393–436). Elsevier.
14. Kellert, S. R., & Wilson, E. O. (Eds.). (1995). *The biophilia hypothesis*. Island Press.
15. Orians, G. H., & Heerwagen, J. H. (1992). Evolved responses to landscapes. In J. H. Barkow, L. Cosmides, & J. Tooby (Eds.), *The adapted mind: Evolutionary psychology and the generation of culture* (pp. 555–579). Oxford University Press.
16. Mealey, L., & Theis, P. (1995). The relationship between mood and preferences among natural landscapes: An evolutionary perspective. *Ethology and Sociobiology, 16*(3), 247–256.
17. Kawabata, H., & Zeki, S. (2004). Neural correlates of beauty. *Journal of Neurophysiology, 91*(4), 1699–1705.
18. Browning, W. D., Ryan, C. O., & Clancy, J. O. (2014). *14 Patterns of biophilic design*. Terrapin Bright Green llc.

19. Hettinger, N. (2017). Evaluating positive aesthetics. *Journal of Aesthetic Education, 51*(3), 26–41.
20. Laidre, M. E. (2021). Animal architecture. *Current Biology, 31*(22), R1458–R1464.
21. Perricone, V., Grun, T. B., Marmo, F., Langella, C., & Carnevali, M. D. C. (2020). Constructional design of echinoid endoskeleton: Main structural components and their potential for biomimetic applications. *Bioinspiration & Biomimetics, 16*(1), 011001.
22. Hansell, M. (2007). *Built by animals: The natural history of animal architecture*. Oxford University Press.
23. Fuller, R. B. (1982). *Synergetics*. Macmillan Publication.
24. Hansell, M. H. (2005). *Animal architecture*. Oxford University Press.
25. Perricone, V., Grun, T. B., Rendina, F., Marmo, F., Candia, C. M. D., Kowalewski, M., Langella, C., & Micheletti, A. (2022). Hexagonal Voronoi pattern detected in the microstructural design of the echinoid skeleton. *Journal of the Royal Society Interface, 19*(193), 20220226.
26. Fitzgerald, T. D., Miller, S., & Smith, M. (2012). Thermal properties of the tent of early instar colonies of the eastern tent caterpillar, Malacosoma americanum (Lepidoptera: Lasiocampidae). *Journal of Thermal Biology, 37*(8), 615–624.
27. Knudson, D. V., & Knudson, D. (2007). *Fundamentals of biomechanics* (Vol. 183). Springer.
28. Aziz, M. S. (2016). Biomimicry as an approach for bio-inspired structure with the aid of computation. *Alexandria Engineering Journal, 55*(1), 707–714.
29. Barthelat, F., Tang, H., Zavattieri, P. D., Li, C. M., & Espinosa, H. D. (2007). On the mechanics of mother-of-pearl: A key feature in the material hierarchical structure. *Journal of the Mechanics and Physics of Solids, 55*(2), 306–337.
30. Burdick, J. A., & Mauck, R. L. (Eds.). (2010). *Biomaterials for tissue engineering applications: A review of the past and future trends*. SpringerWien.
31. Palmiero, C., Imparato, G., Urciuolo, F., & Netti, P. (2010). Engineered dermal equivalent tissue in vitro by assembly of microtissue precursors. *Acta Biomaterialia, 6*(7), 2548–2553.
32. Langella, C. (2023). Biodesign as a strategy of autonomy in the emergency scenario. *Cuadernos del Centro de Estudios de Diseño y Comunicación, 190*, 29–47. ISSN 1668-0227.
33. Brownell, B. (Ed.). (2010). *Transmaterial 3: A catalog of materials that redefine our physical environment*. Princeton Architectural Press.
34. Brown, A. J. (1886). An acetic ferment which forms cellulose. *Journal of the Chemical Society, 49*, 432–439.
35. Llanos, M. (2012). *Suzanne Lee and her celluloid clothing*. Trendland. Retrieved June 20, 2012, from http://trendland.com/suzanne-lee-and her celluloid-clothing/
36. Langella, C. (2023). Bacterioscape. *For nature/with nature: New sustainable design scenarios*, Springer Series.
37. Langella, C., & Fiume, M. (2023). Biodesign: The project that instructs nature. *Cuadernos del Centro de Estudios de Diseño y Comunicación, 190*, 17–28. ISSN 1668-0227.
38. González, L. M., Mukhitov, N., & Voigt, C. A. (2020). Resilient living materials built by printing bacterial spores. *Nature Chemical Biology, 16*(2), 126–133.
39. Gilbert, C., Tang, T. C., Ott, W., Dorr, B. A., Shaw, W. M., Sun, G. L., et al. (2021). Living materials with programmable functionalities grown from engineered microbial co-cultures. *Nature Materials, 20*(5), 691–700.
40. Heveran, C. M., Williams, S. L., Qiu, J., Artier, J., Hubler, M. H., Cook, S. M., et al. (2020). Biomineralization and successive regeneration of engineered living building materials. *Matter, 2*(2), 481–494.
41. Sequeira, A. A., Usman, A., Tharakan, O. P., & Ali, M. Z. (2016). Biologically inspired robots into a new dimension-A review. *International Journal of Automation, Mechatronics & Robotics, 3*(1), 106–116.
42. He, T., Li, Z., & Deng, W. (2011). Biological adhesion of Parthenocissus tricuspidata. *Archives of Biological Sciences, 63*(2), 393–398.
43. Badarnah, L., & Fernandez, J. E. (2015). Morphological configurations inspired by nature for thermal insulation materials. *Proceedings of IASS Annual Symposia, 2015*(4), 1–12. International Association for Shell and Spatial Structures (IASS).

44. Takahashi, K., Ito, T., Onda, Y., Endo, T., Chiba, S., Ito, K., & Osada, H. (2007). Modeling of the thermoregulation system in the skunk cabbage: Symplocarpus foetidus. *Physical Review E, 76*(3), 031918.[44].
45. Bradbury, J. W., & Vehrencamp, S. L. (1998). *Principles of animal communication* (Vol. 132). Sinauer Associates.
46. Wood, W. F. (1983). Chemical ecology: Chemical communication in nature. *Journal of Chemical Education, 60*(7), 531.
47. Hopp, S. L., Owren, M. J., & Evans, C. S. (Eds.). (2012). *Animal acoustic communication: Sound analysis and research methods*. Springer.
48. Osorio, D., & Vorobyev, M. (2008). A review of the evolution of animal colour vision and visual communication signals. *Vision Research, 48*(20), 2042–2051.
49. Patricelli, G. L., & Hebets, E. A. (2016). New dimensions in animal communication: The case for complexity. *Current Opinion in Behavioral Sciences, 12*, 80–89.
50. Quicke, D. L. J. (2017). *Mimicry, crypsis, masquerade and other adaptive resemblances*. Wiley.
51. Pasteur, G. (1982). A classificatory review of mimicry systems. *Annual Review of Ecology and Systematics, 13*(1), 169–199.
52. Mappes, J., & Alatalo, R. V. (1997). Effects of novelty and gregariousness in survival of aposematic prey. *Behavioral Ecology, 8*(2), 174–177.
53. Quesnel, V., & Stradling, D. J. (2012). Evidence for the function of the eye-spots in the butterfly genera Caligo and Eryphanis (Lepidoptera: Nymphalidae: Morphinae: Brassolini). *Living World, Journal of the Trinidad and Tobago Field Naturalists' Club.*
54. Launchbaugh, K. L., & Provenza, F. D. (1993). Can plants practice mimicry to avoid grazing by mammalian herbivores? *Oikos*, 501–504.
55. Stökl, J., Brodmann, J., Dafni, A., Ayasse, M., & Hansson, B. S. (2011). Smells like aphids: Orchid flowers mimic aphid alarm pheromones to attract hoverflies for pollination. *Proceedings of the Royal Society B: Biological Sciences, 278*(1709), 1216–1222.
56. Sherratt, T. N. (2008). The evolution of Müllerian mimicry. *Naturwissenschaften, 95*(8), 681–695.
57. Raška, J., & Pekár, S. (2019). Do ladybird spiders really mimic ladybird beetles? *Biological Journal of the Linnean Society, 126*(1), 168–177.
58. Llopis-Lorente, A., Díez, P., Sánchez, A., Marcos, M. D., Sancenón, F., Martínez-Ruiz, P., et al. (2018). Toward chemical communication between nanodevices. *Nano Today, 18*, 8–11.
59. Cuthill, I. C., Allen, W. L., Arbuckle, K., Caspers, B., Chaplin, G., Hauber, M. E., et al. (2017). The biology of color. *Science, 357*(6350), eaan0221.
60. Sasaki, T., & Pratt, S. C. (2018). The psychology of superorganisms: Collective decision making by insect societies. *Annual Review of Entomology, 63*, 259–275.
61. Harwood R. F., & James M. T. (1979). *Entomology in human and animal health* (7th Edn.). Macmillan Publishing Co. Inc
62. Kempe, K. J., & Shaltout, H. A. (2011). Non-verbal communication of compassion: Measuring psychophysiologic effects. *BMC Complementary and Alternative Medicine, 11*(1), 1–9.
63. Kelso, J. S. (2007). *Understanding complex systems*. Springer.
64. Lewin, R. A. (1982). Symbiosis and parasitism—Definitions and evaluations. *Bioscience, 32*(4), 254–260.

Carla Langella Architect, Associate Professor of Industrial Design at the Department of Architecture of University of Naples Federico II, currently teaching User Experience Design and Industrial Design. She has founded and coordinates since 2006 the Hybrid Design Lab, a project and research laboratory dedicated to mutual relations between design and biosciences with a particular interest in biomimetics, design-driven materials, design for health, and design of cultural experiences mediated by digital technologies.

Biodesign as a Tool to Achieve Sustainable Construction Through Additive Manufacturing

Maria De Los Angeles Ortega Del Rosario, Carmen Castaño, and Miguel Chen Austin

1 Introduction and General Outlook

Biological systems (BS) result from millions of years of evolution. This evolutionary development has generally led BS into function-related models that optimize their performance [1], in most cases as adaptative mechanisms to cope with the environment and to enhance material, energy, and water uses [2, 3]. Therefore, outstanding properties can be observed in biological materials forming skins, envelopes, and skeletal structures, among many others [4–6].

Reproducing structures inspired by BS or using living and nonliving biomaterials can exhibit many challenges. First, in several cases, most structures found in nature are not the result of a fabrication process but rather a growing process, resulting in highly hierarchical and intricate structures that link multiple scales to develop their shapes and functions fully. Second, most traditional manufacturing methods

M. De Los Angeles Ortega Del Rosario
Research Group in Design, Manufacturing and Materials (DM+M), Faculty of Mechanical Engineering, Universidad Tecnológica de Panamá, Ciudad de Panamá, Panamá

Sistema Nacional de Investigación (SNI), Clayton Panama, Panama
e-mail: maria.ortega@utp.ac.pa

C. Castaño
Faculty of Industrial Engineering, Universidad Tecnológica de Panamá, Ciudad de Panamá, Panamá
e-mail: carmen.castano@utp.ac.pa

M. C. Austin (✉)
Research Group in Energy and Comfort in Bioclimatic Buildings (ECEB), Faculty of Mechanical Engineering, Universidad Tecnológica de Panamá, Ciudad de Panamá, Panamá

Sistema Nacional de Investigación (SNI), Clayton Panama, Panama
e-mail: miguel.chen@utp.ac.pa

A. J. V. Arruda, F. L. Palombini (eds.), *Biomimetics, Biodesign and Bionics*, Environmental Footprints and Eco-design of Products and Processes,
https://doi.org/10.1007/978-3-031-51311-4_10

fail to reproduce the shapes and the inter-scale relationships that give rise to the properties generally sought when reproducing these BS since biological architectures and geometries are highly complex. Third, there is a need to develop design methods that simultaneously cover the micro- to macroscale performance of the BS, the complex geometry, and the suitable fabrication method, all while ensuring the mechanical performance expected from the product.

The use of digital tools and methods to deal with problems and challenges in several commercial and industrial sectors, including the architecture, engineering, and construction industry (AEC), has become common in the last decade [7, 8] since they provide opportunities to obtain well-tailored parts and structures capable of being adapted to specific functions [9]. Tools such as additive manufacturing (AM) and robotic manufacturing, integrated with techniques such as artificial intelligence (AI), augmented reality, virtual data connectivity, machine learning (ML), and automation, among others, have created a landscape with product design possibilities that would not be possible with traditional manufacturing techniques [8, 10].

AM first appeared in 1987 as a patent for a stereolithography process for 3D systems [11, 12]. Nowadays, AM processes are standardized into seven categories: material extrusion, binder injection, powder bed fusion, light curing, material injection, direct energy deposition, sheet laminations [13, 14] based on several technologies.

Lately, AM has gained significant attention as a fabrication technique to reproduce bioinspired or biodesigned structures [1, 4]. AM is well-fitted to be coupled with digital design and manufacturing, which usually consists of material deposition layer by layer. Thus, AM offers great flexibility to work with multimaterial, multilayered, and multiscale designs. Besides, this technique offers benefits over traditional methods, such as rapid production, bespoke parts, and reduced material waste, among many others. However, the most attractive feature of AM is that complex shapes can be generated with geometric freedom and a high degree of detail [15], achieving a high degree of automation and reducing the need for human intervention in processes, costs, investment times, and associated risks [16].

The AEC, one of the largest industries worldwide, is one typically characterized by traditional and, in some cases, somewhat manual activities due to several complexities, interdisciplinarity, individuality, and diversity of its products [17–20]. Consequently, its level of digitalization lags behind that of other industrial sectors, and it is more labor accident-prone.

In contrast to other industrial sectors, in the AEC, most companies are small- and medium-sized enterprises (SME) due to ease of market entry and low initial capital [21–23], which result in a vast sector with different levels of technical competencies, technological capabilities, and economic revenues. All these factors result in critical barriers to the development of this sector, leading to a slow pace of digitalization [20, 24]. Despite these challenges, digital development in the construction sector has been steady during recent years [25]. This is largely because digitalization and automation have proven to enhance productivity [26].

Nevertheless, the shift toward digitalization brings about multiple challenges. Factors such as population growth, depth of scientific and technical understanding, competency levels, supply chain oversight, and environmental consequences must all be addressed to realize the full spectrum of what is often called Construction 4.0 [21, 27–29]. The AEC sector adopts many digital tools and techniques. Among them, one can find building information modeling (BIM), AI and ML, Internet of Things (IoT), time-sensitive network (TSN), digital twins (DT), 5G, cloud management, geographical information system (GIS), virtual and augmented reality, unmanned aerial vehicles (UAVs), robotics, and computer numerical control (CNC), such as AM [17, 22, 25, 27, 30–32]. Thus, by adopting digital tools, data acquisition, and visualization, this sector now faces the opportunity to reduce waste, water, and energy consumption; promote environmental safety; increase product precision, resilience, and human safety; and achieve sustainability [17, 22, 33]. The latter is a key aspect if one considers that, currently, the construction sector is responsible for 40% of solid waste generation, 12% of drinking water depletion, 36% of global energy demand, 37% of energy-related CO_2 emissions, and 38% of greenhouse gas emissions [34] adding that the concrete manufacturing sector is responsible for 8% of anthropogenic CO_2 emissions [4].

Consequently, it is unsurprising that AM has garnered increasing attention recently due to its previously mentioned features, such as adapting to intricate geometries, providing freedom in crafting bespoke shapes, employing functional and topological optimization, and minimizing material waste and usage, which are just a few of its many advantages [35, 36].

AM encompasses various fabrication methods that could replicate the porous hierarchical structures and growth mechanisms in biological systems. The layer-by-layer deposition in AM bears similarities to these natural processes. Shakib et al. [37] stated that there are common features between AM and biological synthesis, such as the supply of building blocks, orientation-controlled mechanisms, programming codes for the macroscopic shape, and assembly of materials and bottom-up layers' deposition.

The use of AM with biomimetic, biomimicry, bionics, bioinspired design, and biomaterials aligned with nature-based design (NbD) has been widely covered in literature in recent years [1, 4, 6, 9, 37–41]. Regarding standardization, the ISO 18458:2015 standard [42] already offers a structured approach to biomimetics. It aids developers and designers in incorporating biomimetic methodologies into their initiatives, as shown in Fig. 1, covering a wide range, from shape-like-mimic to process-mimic, to inspire structural products and new materials.

Beyond leveraging nature for process enhancement and the development of highly functional structures, it has also been tapped as a source for novel materials. Natural-based materials or biomaterials have been tested to produce sustainable, high-performance structures as a strategy to reduce greenhouse gas emissions and fossil fuel dependency [40, 43–45].

Despite the great efforts and results set by several authors [9, 39–41], the approach to effective design is still minimal and variable, depending on the field. In this chapter, the authors cover selected studies to establish some design guidelines

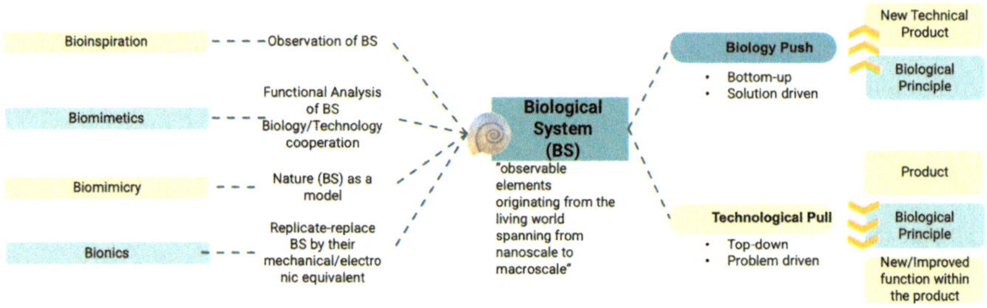

Fig. 1 General framework of nature-based design in line with the definitions from the ISO 2015:18458 standard [42]

and development opportunities based on the current challenges. To gain insight into the interplay between AM and NbD, the authors raised the research question: How have the principles of AM been integrated with NbD methodologies to hold potential value for the AEC industry? Scopus and Web of Science were used as scientific databases. The inclusion criteria covered articles, book chapters, conference papers, technical notes, and review papers of the last 5 years (2018–actuality) for building construction and architecture-related studies. Nevertheless, the exploration was broadened to include works that have been cited as foundational in understanding the connection between these two design and manufacturing approaches.

2 Geometry-Based Approaches

Geometry has been critical when studying AM because this fabrication technique has a virtually infinite capacity for generating bespoke shapes and combinations while improving and controlling the mechanical performance of the structures. Advances in AM, digital design, and digital manufacturing, as well as simulation tools, have significantly enhanced the development of the materials known as architectured materials. Since structure leads to properties, the microstructure can be tailored into desired mechanical responses while ensuring the accomplishment of complex geometries and structures at the macroscale [46, 47].

Different sources of inspiration have been analyzed to obtain architectured materials with superior mechanical performances for the AEC industry. This is sometimes based on traditional geometries, seeking to adapt naturally to the existing market [48, 49]. The most recent studies explore the modifications of both the internal and external geometries [50–52] for nonstructural parts and the geometrical curves in large-scale structures, such as walls, facades, and others [24, 53, 54]. In this sense, structures with cellular geometries, lattices, and other bioinspired geometries found in nature have attracted attention.

2.1 Cellular Geometries and Materials

Cellular materials are characterized by a porous microstructure composed of solid-void networks. They are geometrically based structures in topological space frequently found in nature [55]. These can be divided into open cell or interconnecting and closed cell or non-connecting typologies. Fabrication of complex cellular geometries has been challenging until very recently, with the commercial development of AM techniques. One can fully profit from the advantages of using AM when designing and fabricating with such complex geometries [56].

Cellular structures, like the honeycomb design, have garnered significant interest for their material-saving potential. Their use extends beyond AM to various techniques and disciplines, aiming for lightweight components with impressive mechanical properties or as infill structures to bolster parts during the printing process. They can be found in space as cubes, prisms, dodecahedra, rhombuses, and elongated dodecahedra. However, not all cellular structures are suited for application in all AM techniques [50, 55].

Beyond their natural occurrence, cellular materials' appeal in additive manufacturing also stems from the fact that they can frequently be represented through mathematical formulations [55]. This may ease its inclusion in analytical models and simulations and, to furthermore, establish design guidelines. For instance, open cell modeling in most cases is well-characterized and even standardized [57–59].

Porosity, density, and apparent density are important parameters when modeling cellular structures [57]. In a cellular structure, the volume fraction can be defined as the apparent density, which is the density of the cellular or porous material divided by the one corresponding to the solid material or matrix [60]. On the other hand, another key parameter that describes cellular structures mathematically is the fabric tensor (*FT*). The *FT* is a symmetric second-rank tensor. It characterizes the arrangement observed in the microstructural components of porous or multiphase materials [61].

Designing and fabricating with cellular structures has led to numerous strategies, often associated with parametric modeling techniques, either via optimal criterion (OC) tactics or intuitive heuristic approaches [53, 62–66]. Its focus is intrinsically tied to the constraints commonly encountered when implementing AM in civil structures, manufacturing processes, material choices, construction methodologies, and design paradigms [24]. Hereunder, one can find some of the cellular structures that have been used or have shown potential for the AEC industry.

2.1.1 Triply Periodic Minimal Surfaces (TPMS)

Minimal surface structures such as the triply periodic minimal surfaces (TPMS) have gained great attention in several fields, such as tissue engineering, aeronautics, and architecture, among others [38]. A minimal surface has an average curvature of zero at all points within its domain. If these surfaces are developed periodically within a 3D space, they are described as triply periodic minimal surfaces.

TPMS can be divided into ligament/solid-based and sheet/shell-based structures [41, 67]. They are infinite, non-self-intersecting periodic surface structures that extend independently in three principal directions and are associated with the crystallographic space group symmetry, which are smooth, without sharp edges or junctions compared to the lattice structures [68, 69]. The most common TPMS structures found in the literature are the primitive (P), gyroid (G), diamond (D), I-WP, and Neovius, characterized by a high surface area.

Their porous architecture is highly interconnected and can be found in several physicochemical and BS [70]. These features allow their mechanical traits to be adjustable and added to the fact that algebraic equations can mathematically model them, which have rendered them even more attractive. For instance, one can optimize specific material properties such as strength-to-weight and stiffness-to-weight ratios, or toughness by tuning the unit cell geometry to fit a particular purpose [59].

When working with TPMS, unit cell type, volume fraction, and unit cell size are considered geometric parameters. These parameters highly influence the manufacturing and, furthermore, the mechanical performance of TPMS lattice structures [71].

2.1.2 Lattice Structures

Lattice structures have also been explored in NbD approaches. A lattice is a porous structure composed of unit cells [37]. Lattices, hence, are open cell structures. Their mechanical performance in criteria such as elastic moduli and strength is highly related to the lattice parameters, e.g., the relativity density and cell size [58]. Such structures have been crafted using traditional techniques and, more recently, with AM. To ensure optimal performance, these structures demand high fidelity and minimal defects [41], which can be paired with the need for digital design of AM.

Large-scale cellular and lattice structures have been crafted primarily through 3D concrete printing and clay AM, with the former being the more prevalent material choice. For instance, Ahmed et al. [72] developed a study for two cases featuring double-curved surfaces with a lattice-shaped inner structure to improve ductility and strength. In the case of the works developed by Nguyen-Van et al. [73], the authors were inspired by TPMS to evaluate the design and performance of these structures for prefabricated construction. Their findings included that, according to their results, TPMS cellular shapes make structural deformation dominated by stretching rather than bending mechanisms.

Sangiorgio et al. [52] developed 18 clay bricks inspired by cellular structures, a notable first in the field. However, the study does not juxtapose their performance with traditional structures or benchmark them against key known parameters. Similarly, research by Abdallah and Estévez [74] ventured into a bioinspired topological design, drawing inspiration from the characteristics of clay and design forms that prioritize direct node connections. However, this study does not include the mechanical assessment of the produced bricks.

2.2 *Stochastic and Bioinspired Structures*

2.2.1 Nacre Structures

Nacre-based structures have been one of the most widely used in biodesign for architectural purposes in several fields since they exhibit exceptional mechanical performance, such as toughness and high strength. Nacre is characterized by a hierarchical multiscale architecture, in which the stacking arrangement, interlocking mechanisms, volume fraction of hard and soft parts at the micro-scale, as well as the mineral bridges, organic interlamellar matrixes, nanoasperities, and axial growth in the perpendicular direction are attributed as the factors leading to toughening mechanisms [75, 76]. These attributes have been explored with AM as a fabrication technique.

Jabir et al. [77] explore using nacre-based structures to show the material compliance gradients of mortar. The authors found that an increase in toughness, strength, and strain-to-break can be attained since the spatial grading of the mortar compliance can redistribute stress, reducing it at critical regions. Ko et al. [78] developed a parametric study to optimize the mechanical performance by testing nine 3D nacre-like patterns obtained by manipulating the Voronoi diagram. The parametrization was formulated through a dimensional model. A finite element analysis (FEA) was developed using LS-DYNA as software. This analysis led to a numerical model that linked the structural geometry, impact conditions, and the constitutive modeling of the material to obtain the mechanical response of the digitally fabricated specimens by performing impact tests.

More recently, in the study developed by Patadiya et al. [75], two nacreous architectures, a nacre columnar and a nacre sheet, and a pristine structure were designed and fabricated via fused deposition modeling (FDM). For this design, the authors used a parametric dimensional model previously developed by Ko et al. [78] to formulate the absorbed impact energy of the nacre-like structure and the fracture behavior. ASTM D638 and ASTM D790 were the standards for tensile and 3-point flexural bending tests, respectively. The authors found that controlling the structure's soft phase (intermediate component) could increase impact resistance, elastic and flexural modulus, and strength. Besides, they explored the AM capabilities of reproducing the bottom-up-like biomineralization process of nacre.

Further than these studies, nacre-like structures have been explored using other manufacturing techniques or field applications. However, most of these studies explore this structure's tensile-compression and impact performance. Nevertheless, this structure has great potential in AEC industries for envelope applications, whether they mimic the geometry or are used as a printing pattern by extrusion-based fabrication.

2.2.2 Bouligand Structures

The Bouligand structure is a hierarchical arrangement that exhibits uniaxial layers periodically assembled, forming a helicoidal pattern. It can be found in various biological organisms, such as the mantis shrimp (*Odontodactylus scyllarus*), lobsters, exoskeletons of shellfish, fish scales (*Arapaima gigas*), and crabs [79–83]. This structure has also been used as a source of inspiration since it exhibits in nature fracture resistance, energy dissipation, and a significant impact energy [82, 84]. This helicoidal pattern was tested as a printing pattern by Liu et al. [83], exploring four pitch angles, 0°, 15°, 30°, and 45°, showing that specimens printed with a helicoidal pattern exhibited better performance in impact duration, peak impact force, and energy absorption.

Although the shapes achieved at a geometric level can mimic those of nature, the effects of the material and the process can still influence the development of these pieces with nontraditional geometries. This translates into the possibility of collapses in the structures produced, mostly due to instability due to buckling or plastic collapse [85, 86], so the rheological performance, extrudability, and buildability tend to be a recurring topic of study in AM in the AEC industry [87–89].

2.2.3 Trabecular Structures

The trabecular structures can be found in trabecular bones, which are characterized by adapting their structure to the mechanical type of stress they withstand. Bones are highly hierarchical structures.

The trabecular architecture can be linked to their potential mechanical properties by means of the volume faction ρ_f and the fabric tensor (*FT*). When bones undergo anisotropic stresses, the tissue adapts to this and displays an oriented architecture, and thus, *FT*(*M*) is expressed through the orientation and shape of an ellipsoid, described by the radii M_1, M_2, and M_3, the eigenvalues of M. However, if this tissue is subjected to more isotropic stress, then the structure will display a less oriented architecture and *FT*(*M*) describes a shape closer to a sphere. Thus, it may be implied that *FT* characterizes porous materials' degree of anisotropy and orientation.

Cadoret et al. [90] proposed a mathematical model between bioinspired porous pattern parameters, the volume fraction, and the resulting fabric tensor. They aimed to develop a mass reduction method that calculates a set of parameters to obtain a targeted *FT* and volume fraction. For this, they proposed as factors the porosity shape ratios that covered the broadest range of possible architecture according to the consulted literature and the volume fraction. The expected responses of this model were FT shape ratios $Y_1 = \frac{M_1}{M_3}$, and $Y_2 = \frac{M_1}{M_2}$, expressed as a function on bioinspired porous pattern parameters (*c*/*a*, *c*/*b*, and ρ_f). The authors developed the mass reduction method to reduce the computing time, in which each porosity is computed as a function of the local stresses. Thus, to a given FCC unit cell, the porosities vary in shape, size, and orientation, mimicking the FT and ρ_f of a trabecular bone tissue. This method was tested using FEA and 3-point bending experiments, duplicating stiffness at equal mass if compared to a topology optimization method with similar targets.

Borunda and Anaya [85] explored the material aptitude of large-scale trabecular structures for AM use cases. While these architectural approaches can result in strong ultralight structures, identifying the right basic cell unit for the optimal architecture depends on extensive structural simulation; fulfilling the mechanical requirements of both the dynamics loads in the making and the static loads of the build is a tricky endeavor.

Thus, they proposed a morphological model that links the mechanical analysis with discrete data structure, and then, the discrete geometry units were mechanically characterized, originating a mechanically continuous membrane influenced by a stress field. Then, each lattice segment was categorized by using digital 3D printing. This methodology led to linked design, simulation, and fabrication processes. A space-filling polyhedral method allows this model to approximate the expected complex forms and capacity of AM, which can also be explored as a filling structure. Their results showed that tetrahedral, octahedral, and rhombic-dodecahedron tessellation performed the best for lightweight FDM technology.

Geometry creation through digital design, inspired by or modeled after nature, is a trending subject. Different sources are widely addressed, such as analytical models or pure mathematical techniques [85, 90], topological optimization [53, 62, 64, 91], parametric modeling [51, 92, 93], or bioinspired, stochastics methods [55, 94, 95] which are then combined with structural analysis, mostly FEA, to know in advance the behavior of the part to be manufactured. These strategies make it possible to propose complex internal structures, generally of the periodic minimum surface type, that can achieve better mechanical, thermal, and sustainability performances using limited materials.

Currently, parametric models rank among the most used techniques in bioinspired design studies. They offer a mechanism to produce precise models through parameter adjustments, presenting a swift and efficient method to craft intricate geometries. It also allows the user to include prior knowledge [9, 96, 97]. On the other hand, topology optimization allows the designer to obtain an optimal structure by distributing a given amount of material in the domain to maximize the structural performance [62, 64, 98]. A comprehensive review of diverse geometrical frameworks, their source of inspiration in nature, and corresponding mathematical models can be found in Table 1.

Table 1 Summary of geometries used in NbD with additive manufacturing.

Topology	Sources in nature	Illustration	Mathematical models	Applications	AM Techniques	References
TPMS	Sea urchin *Callophrys rubi* butterfly	TPMS structures from [59], license CC BY 4.0.	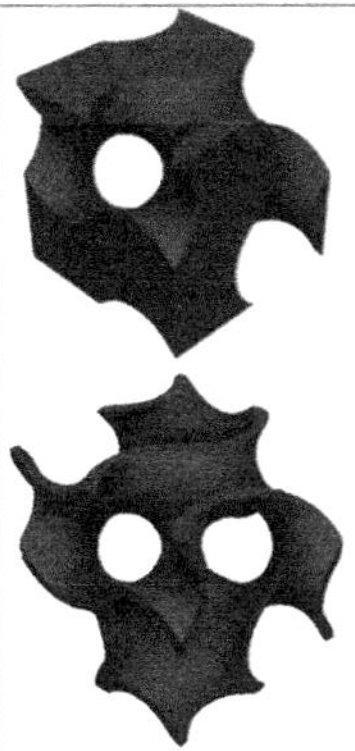 Topological description: Primitive (P): *coscos* (*X*) + *coscos* (*Y*) + *coscos* (*Z*) = *C* Gyroid (G): *sinsin* (*X*) *coscos* (*Y*) + *sinsin* (*Y*) *coscos* (*Z*) + *coscos* (*X*) *sinsin* (*Z*) = *C* Diamond (D): *sinsin* (*X*) *sinsin* (*Y*) *sinsin* (*Z*) + *sinsin* (*X*) *coscos* (*Y*) *coscos* (*Z*) + *cos* (*X*)*sinsin* (*Y*) *cos* (*Z*) + *coscos* (*X*) *coscos* (*Y*) *sinsin* (*Z*) = *C* Where $X = 2\alpha\pi x$, $Y = 2\beta\pi y$, and $Z = 2\gamma\pi z$, being x, y, and z the spatial coordinates, and α, β, and γ, the parameters that control the size of the unit cell in the three Cartesian directions, respectively [69, 79]. Gibson-Ashby model for young modulus: $E = \beta_1^{'}\left(1-\varphi\right)\rho_f + \beta_1\varphi^2\rho_f^2 + \beta_1^{''}\left(1-\varphi\right)^3\rho_f^2$ Where β_1, $\beta_1^{'}$, and, $\beta_1^{''}$ are constants and φ is a dimensionless parameter describing the ratio of solid material in the cell edges to the total solid material [58]	Fatigue resistance	Powder bed fusion	[59, 69, 79, 99]
				Impact resistance	SLA	
				Improvement in stiffness and strength-to-weight ratio	Material extrusion	
				Vibration isolation and thermal insulations	Material jetting	

Hierarchical nacre architecture	Mollusk shells	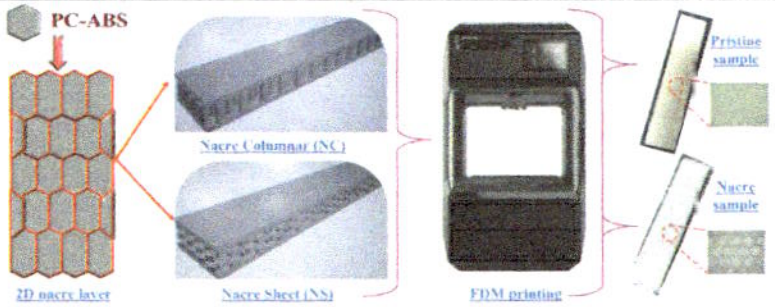 Nacreous-based structure from [75]. License CC BY 4.0.	Π-Buckingham modeling of absorbed impact energy "e": $e = f(A_M, A_B, v_i, D_i, m_i, \rho_c, E_c, \sigma_{yc}, I_c)$ Where A_M and A_B are the areas of mortar and brick, respectively, ρ_c is the composite density, E_c and σ_{yc} are the composite's elastic modulus and yield strength, and I_c is the Izod strength of the composite	Resistance to impact load Strength, elastic, and flexural modulus enhancement	Material extrusion (FDM)	[75, 78]
Hierarchical cellular material/lattice (network)	Insects' wings *Allomyrina dichotoma* beetle Blood vessels	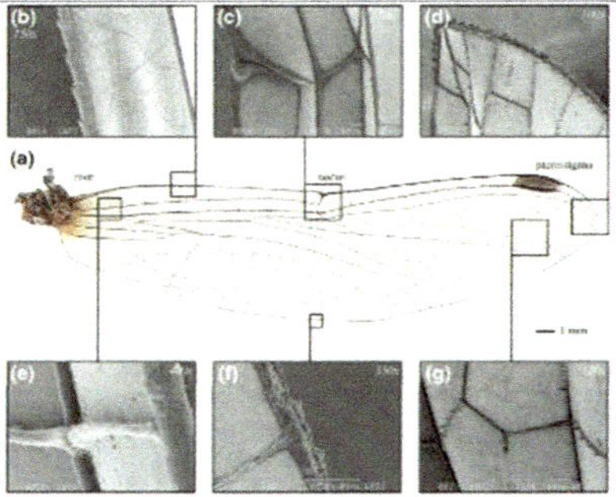 Dragonfly (*Sympetrum vulgatum*) wings' hierarchical structures [100]. License CC BY 4.0.	The Murray Law: For blood vessels, the cube of the radius of the parent vessel is equal to the cube of the radius of the parent vessel is equal to the sum of the cubes of the radii of daughter blood vessels [101]	Increase in stiffness Compression	Material extrusion (FDM, DIW) Material jetting Powder bed fusicn Sheet lamination Vat photopolymerization	[41, 100, 102–108]

(continued)

Table 1 (continued)

Topology	Sources in nature	Illustration	Mathematical models	Applications	AM Techniques	References
Irregular/ stochastic cellular patterns	Trabecular bone	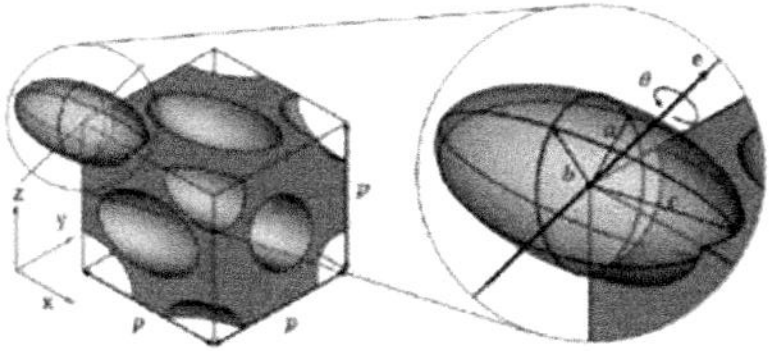 Cell parameters and FT on a trabecular tissue [90], license CC BY 4.0.	The Aboav-Weaire law: $\underline{m} = 5 + \frac{6}{n}$ Where n is the number of edges of the cell and $\underline{m}$ is the average number of edges of its n neighbors. The Lewis rule: $\frac{A(n)}{A(\underline{n})} = \frac{n - n_0}{\underline{n} - n_0}$ Where A(n) is the cell area with n sides, $A(\underline{n})$ is the cell area with an average number of sides $\underline{n}$, and n_0 is a constant. For trabecular bone tissue [90]: FT shape ratios $\left(Y_1 = \frac{M_1}{M_3}, \text{ and } Y_2 = \frac{M_1}{M_2} \right)$ expressed as a function on bioinspired porous pattern parameters. $\left\{ \frac{M_1}{M_3} = -0.0069 + 0.9433\frac{c}{a} + 1.019\rho_f - 0.937\frac{c}{a}\rho_f \; \frac{M_1}{M_2} \right\}$ $= 0.036 + 0.909\frac{c}{b} + 0.968\rho_f - 0.897\frac{c}{b}\rho_f$	Toughness under compressive and impact load types	Material extrusion SLM	[58, 109–111]

Bouligand structure	Mantis shrimp (*Odontodactylus scyllarus*) Lobsters, Exoskeletons of shellfish, Fish scales (*Arapaima gigas*) Crabs	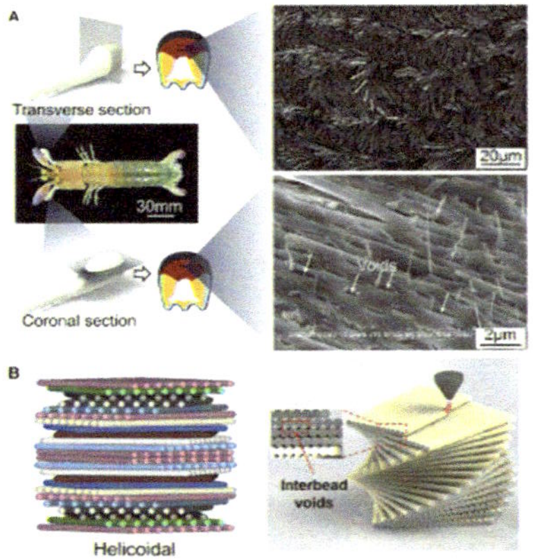 Bouligand structure from *Odontodactylus japonicas* [112], license CC BY 4.0.	The strain energy release rate: $G_0 = \frac{K_I^2}{E'} + \frac{K_{II}^2}{E'} + \frac{K_{III}^2}{\frac{E}{(1+v)}}$	Material extrusion	Better mechanical performance due to oriented-pattern approach	[82, 113, 114] [79–83].

3 Material-Based Approaches with Living Building Materials

NbD may also include the incorporation of biomaterials as construction materials. In some cases, these materials exhibit biological and architectural properties that may enhance the building materials' performance and can be found as an alternative to fossil-based materials and greenhouse emissions and waste generation pollution [115, 116]. Among them, one can find living organisms, such as mycelium and cyanobacteria, and nonliving materials, such as agricultural waste, biopolymers, geopolymers, and clay. Living building materials (LBM) introduce complexities not seen with conventional materials. During manufacturing, the incubation of living beings is included, leading to a more intricate interplay of physical, mechanical, and biological phenomena. Its application has been relatively minimal, yet there is significant potential for creating sustainable materials. Consequently, the authors have incorporated some case studies in this chapter for examination.

Fungi are living organisms recently used as construction material forming mycelium-based composites. Fungi are heterotrophic organisms that feed on organic compounds as parasites or saprophytes, with a chitin cell wall [117]. Mycelium-based composites are formed from growing living fungi onto organic substrates that become the source of nourishment, heat, and moisture. These substrates are usually plant-based [118] mostly from agricultural waste products. Their main components are cellulose, chitin, lignin, and hemicellulose. Therefore, it could be considered a solution to reduce the problems of managing these wastes [119].

To achieve these composites, fungal hyphae, a long-branching, three-dimensional, filamentous structure network, must be propagated into the fibrous substrate for several days under certain environmental conditions. Hyphae secretes enzymes that break the biopolymers into smaller units called monomers. These monomers are absorbed into a fungal tissue, the mycelium, which acts as a binding agent to the organic matter and forms a lightweight material [117, 120], leading to a material that can exhibit attractive mechanical properties such as heat insulation and stiffness [40].

In the last two decades, mycelium-based composites have caught attention for architectural applications and, more recently, as a building material for load-bearing applications. The most common fungi species used to form mycelium-based composites are the *Pleurotus* and *Ganoderma* genera.

Mycelium-based composites have been used as a construction material in some architectural projects, most using casting as their fabrication method. Among them is the MycoTree [121], based on *Pleurotus ostreatus* fungi and a bamboo steel structure. Wood chips, sawdust, and fibrous sugarcane and casava roots were used as substrates; the Shell Mycelium is a dismantlable and lightweight structure that used coir pith as substrate [122], and El Monolito Micelio pavilion is a grow-in-place monolithic structure [119], based on *Ganoderma lucidum* fungi. Similarly, this fungi species was used for the Growing pavilion, which consists of 88 mycelium panels and wood, bulrush, and cotton as substrates [123]. Recently, two Serpentine Wall Prototypes were developed by [124] La Parete Fungina and L'Orso Fungino.

However, its use in architectural or construction projects has been more limited when involving AM as a fabrication process. Although the casting process may be better suited for mass production and ease of fabrication, AM allows exploring more complex geometries and further expanding these structures' mechanical capabilities and customization. However, it is still challenging to use AM since growing living organisms depend on oxygen, controlled conditions, and spatial organization [125–127]. One way that is currently allowing to overcome some of the challenges of working with living organisms is using tools such as digital or CAD as a design tool since it allows to work at the mesoscale and the integration of micro-meso-macro scales, not only enhancing their biological features but also allowing a better understanding of microbiological processes [127, 128].

Using AM as the fabrication technique has recently caught great attention, and in most cases, the attention has been centered on formulating viscous living-based materials adapted to be extruded. An early study was conducted by Soh et al. [118], in which a mixture of chitosan with mycelium-enriched bamboo was used to develop mycelium composites adapted to be used in 3D printing. The authors used the workability, extrudability, and buildability properties frequently evaluated for fresh mixtures as performance criteria. The mixture that produced their best results was based on 500 μm mycelium-bamboo fibers mixed at chitosan, with a volume of fiber of 60% or 70% and with a 3 wt% chitosan solution at a pH ~6. These results were tested after 20 days of growth, yielding a compression modulus of 40 kPa, showing a decrease if compared to a mixture without chitosan (240 kPa). Despite this, the authors stated that the extrudability while working on complex shapes increased, demonstrating that it may have potential as a construction material using techniques such as DIW. Another early project on mycelium-based composites fabricated by AM was developed by [129], in which they developed a six-stage fabrication method from agricultural feedstock, describing the main requirements or effects during each stage until the final product is obtained. Since one stage consists of drying the structure, all fungi are killed in this stage.

The Live Pulp was a 3D-printed structure developed by [126], consisting of a substrate of fine woodchips, paper pulp that served as the nourishing source for the fungi, and kaolin clay mixed with water that offered bonding and stability during printing. Two fungal strains, *Byssomerulius corium* and *Gloeophyllum sepiarium*, were used since they are both wood decomposers. This study aimed to develop a fabrication process that can introduce fungal mycelium as a construction material. Their strategy was based on interconnecting the living system, the digital fabrication, and the computational design at multiple scales. They performed two colonization periods, one at the beginning for 1 week at an autoclave once the mycelial growth was sufficient and another one after the pulp was 3D printed; and the authors performed a second, more extended colonization, period, allowing the structure to transform into the desired biocomposite fully.

Later, Bhardwah et al. [130] further explored the relationship between mixture composition containing psyllium husk powder and the printing quality based on extrudability and shape stability. In this case, fungi material was procured from Ecovative Design, NY. Regarding the extrudability, two square samples were printed. Besides, the authors also tested the dimensional consistency of the filaments, considering that a good performance was achieved if no more than ±5% variation from the mean was observed. The mixtures containing a powder/water ratio of 1:40 and 2:40 exhibited good extrudability, while those above 3:40 did not. Besides, the mixture of 2:40 was the only one that met the shape stability criteria.

Furthermore, Modanloo et al. [117] developed a small-scale prototype of a tilted arch fabricated using digital fabrication and extruded biomass bound with biowelding. The authors used a material-driven design methodology considering the limitations and capabilities of the material used. Thus, their geometric design must have ensured compressive stress since the mycelium-based composites performed better in compression than tensile stresses. This is why a catenary shape was chosen. Besides, since AM allows more complexity compared to their traditional counterparts' forms, they could include forms with voids and divisions, increasing the mechanical capabilities with less material and weight and, simultaneously, allowing better surface-to-volume ratio if compared to formwork.

More recently, Elsacker et al. [131] used *Ganoderma lucidum* as a fungi source to develop a robotic additively manufactured process to obtain mycelium-based biocomposites. The substrates were composed of beechwood sawdust, psyllium husk, corn starch, xanthan gum, paper cellulose, guar gum, and locust bean gum. In this study, the process parameters were based on the robot parameters and the extrudability of the material, controlling the flow rate by the robot movement, the air pressure, the nozzle geometry, and the material velocity. The authors concluded that there is an interrelation between the deposition method, the nozzle, and the extruded material, in which the material is affected by the particle size of the fiber, the concentration of the constituents, the autoclave conditions, and the time, the two latter affecting the viscosity. The extrusion is mainly affected by the extrusion pressure and toolpath. The author also considered as critical geometric and process parameters, the nozzle diameter and geometry, and the printing speed.

In an updated version of Goidea et al.'s previous work [132], the authors focus on a research-by-design methodology led by transdisciplinarity, in which, through physical prototyping and testing in a demonstrator, they can explore the parameters that intervene during biofabrication to design a methodological design path, such as fungal species, substrate composition, printing parameters, inoculation methods, and growth time, among others. Even if this demonstrator, the so-called Protomycokion, can be explored at a macroscale, their analysis included from the micro- to the macroscale and their inherent interactions, since at the macroscale, the characteristics, and behaviors of the nano, milli, and mesoscale come into effect as carriers of architectural performance.

Mohseni et al. [116] explored using waste cardboard as substrate and *P. ostreatus* as the fungal species. Their work mainly focuses on the contamination challenges and their effects on mycelian growth. One challenge regarding contamination is that *Trichoderma* sp., also known as the green mold disease, is difficult to distinguish during the first colonization or incubation period since it first develops a thick white mycelium layer, similar to the fungi usually employed to develop the mycelium-based composite, and their presence affects mycelium-based composites' growth.

Their results showed that autoclaving effectively sterilized psyllium husk and water, but strict protocols must be followed to mitigate contamination.

Besides fungi, bacteria have been recently explored as a living building material; apart from their use as an addition to concrete mixtures to promote self-healing [133], cyanobacteria have been reportedly used as a construction material since they have shown to promote carbon fixation by photosynthesis and microbially induced carbonate precipitation (MICP) and soil stability. MICP consists of a natural calcification that occurs, for instance, when cyanobacteria grow exponentially and mineralize dissolved inorganic carbon (CO_3^{2-}) within the existence of Ca^{2+}, to calcium carbonate ($CaCO_3$) [134]. One study measuring the influence of bacteria in a living construction material composed of sand, a hydrogel binder, and bacteria was designed by [135], in which the authors stated that an improvement of the mechanical properties could be achieved if the manufacturing and processing of the material were well tailored.

Regarding the use of AM as a fabrication technique, Amarly et al. [127] used two genera of cyanobacteria (*Synechococcus* sp. strain PCC 7002 and *Synechocystis* sp. strain PCC 6803) to demonstrate the feasibility of using cyanobacteria as living building material by comparing nonliving materials consisting only on sand and agar mixtures and the living materials. Their results showed that the cyanobacteria adjusted to the media and were biologically active with it through the $CaCO_3$ precipitation. Besides, they found that sand particle size affects the distribution of cells within their sample. The authors updated their study on [128], in which they further evaluated the extent of linking digital fabrication and the material requirements, which they called the co-fabrication process. They use the AM capabilities of offering a great surface-to-volume ratio to promote light exposure since it encourages cell growth. Besides, the toolpath was also developed using a material-based approach since it was conceived to encourage bacterial motility, reduce sharp angles, and increase light exposure. Parametric design was included since having control of geometric parameters could be linked to the bacterial range of motility, obtaining suitable angles and curvatures in the design.

Although not directly studied as a structure in the study of Reinhardt et al. [136], the authors used *Synechococcus* sp. strain PCC 7002 as cyanobacteria and a hydrogel bioink based on alginate and methylcellulose (algMC) containing up to 50 wt% sea sand to explore the mechanical performance of cyanobacteria as environmentally friendly construction materials. Their rheological studies suggested that up to 50 wt% sand can be added to the algMC ink so as not to affect the shear-thinning behavior. Besides, the viscosity increased with the sand concentration at low shear rates of around 5 s^{-1}, although this behavior was not observed at higher shear rates of around 100 s^{-1}. A summary of the studies found in cyanobacteria and mycelium material-based AM for potential AEC applications can be found in Table 2.

Table 2 Selected case studies highlighting the use of living building materials in NbD combined with additive manufacturing

Year	Type of living material	Description	Figures	Reference
2020	Mycelium (*Byssomerulius corium* and *Gloeophyllum sepiarium)*	A fungi-based column consisting of the assemblage of separated components. The complex geometric design with vertical interstitial spaces promotes convective flow	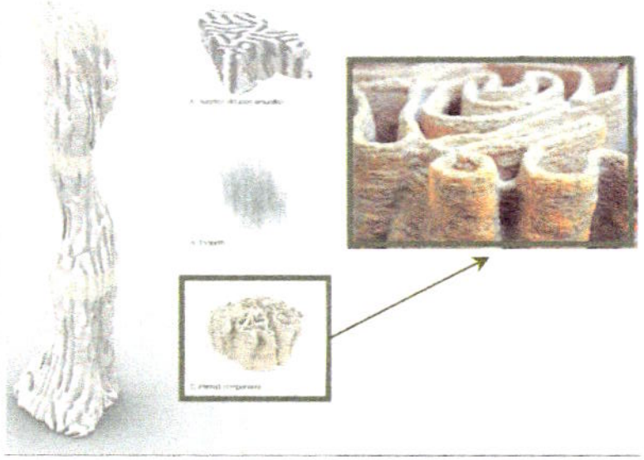	[126], license CC BY 4.0
2021	Biomass-fungi material	This project focused mainly on the effects of the mixture composition on the print quality and rheological behavior		[130], license CC BY 4.0

2021	Biomass-fungi material (*P. ostreatus*)	Material-driven design (MDD) methodology was used to produce a mycelium-based composite by following a computational design and digital fabrication		[130], license CC BY 4.0
2022	Mycelium (*Ganoderma lucidum*)	Robotic fabrication of mycelium-based material. The authors explored parameters such as components, toolpath geometry, nozzle size, printing speed, the impact of time and autoclave on viscosity, and mycelium growth	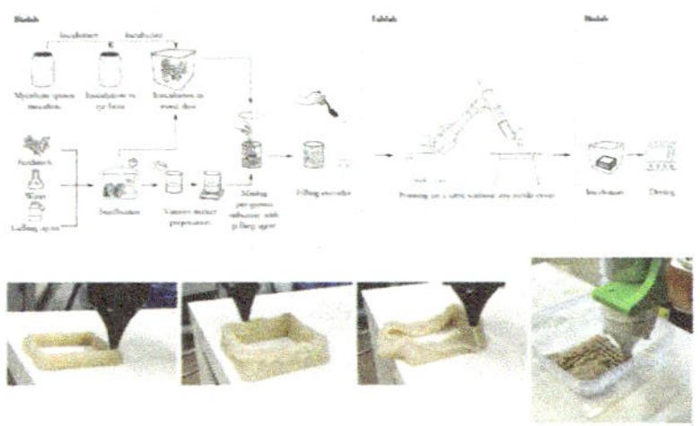	[131], license CC BY 4.0

(continue)

Table 2 (continue)

Year	Type of living material	Description	Figures	Reference
2023	Biomass-fungi material (*P. ostreatus*)	The challenges surrounding the sterile conditions during the biological processing of the fungal material are addressed in this study, in which a mycelium-based composite with waste cardboard as a substrate was explored	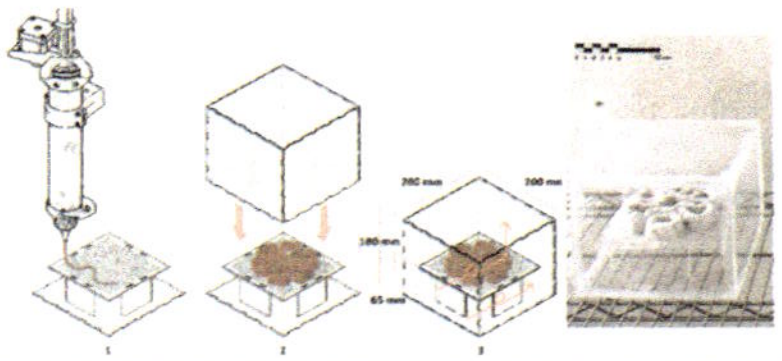	[116], license CC BY 4.0
2023	Cyanobacteria	Cyanobacteria was used as living building material to induce calcium carbonate precipitation. This study focuses on developing fabrication protocols when working with AM	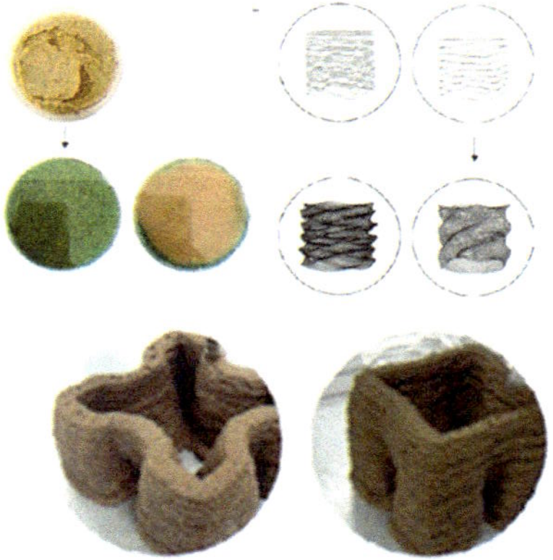	[127], license CC BY 4.0

4 Lessons, Challenges, and Opportunities

AM currently showcases its potential in efficiently fabricating structures inspired by NbD. The research explored in this chapter presents a broad panorama of applications and methodologies encompassing various manufacturing techniques. Drawing from this comprehensive review, the authors aim to distill insights from prior studies, highlighting both existing challenges and the immense potential for advancements in achieving high-performance structures suitable for the AEC sector.

However, it is essential to recognize that unlike traditional manufacturing methods, where design and shape expectations, along with the manufacturing sequence, can be derived manually, AM inherently demands a preceding digital design. Process, geometry, and material parameters highly influence the performance criteria of the final product, adding more constraints to the use of AM. According to Goss et al. [137], coupling NbD with AM has several constraints. First is the feature resolution and fidelity of the AM equipment. Cellular structures, for instance, are characterized by fine features and details and thin-wall structures. This could result in a change from the expected shape in the final product. Another constraint is the dimensional accuracy. For instance, in certain scenarios involving material extrusion, the material deposition can lead to nonuniform wall thicknesses along the path, deviating from the anticipated shape [138]. The final product's performance can also be influenced by orientation and scale dependencies. Products fabricated through AM, particularly those constructed layer-by-layer, exhibit pronounced anisotropy, but further than that, when working with cellular materials, the complex orientation may lead to significant variations in surface orientation. While the AEC industry typically deals with large-scale products, AM's behavior with smaller-scaled items does not necessarily mirror its performance at the macrolevel, leading to differences in the end product.

Living building materials and their applications, mainly including AM as a fabrication technique, have caught attention chiefly due to their implications in achieving sustainability while reducing greenhouse gas emissions, solid waste, and fossil-based materials use. However, biomaterials of this kind must meet several requirements to accomplish large-scale structure design and building with AM fully. One of the main challenges is to address the microstructural conditions and driving forces of living materials, such as fungi or cyanobacteria, ensuring the appropriate growth conditions while linking its performance to the product's macrostructural performance and the AM's inherent parameters. This leads to strategies such as digital fabrication, mainly digital additive manufacturing, that allow the development of complex morphological methodologies to meet all requirements by addressing the living system, the computational design strategy, and the digital and the physical fabrication. Digital fabrication plays a crucial role in providing a bridge between the computation design, the biological requirements, the physical complexities, and a fabrication process that can provide the conditions to ensure the expected mechanical performance while complying with the requirements.

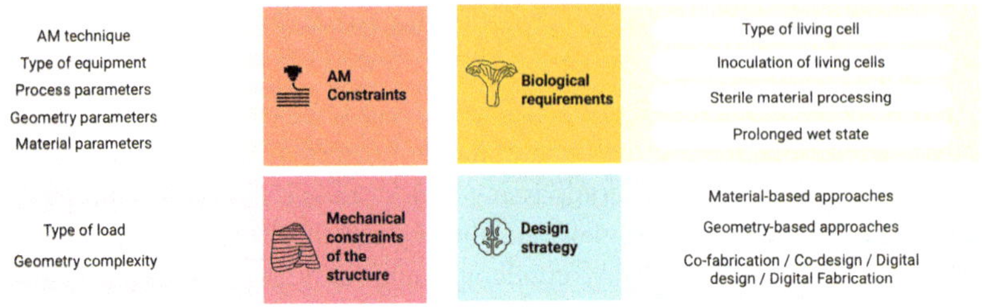

Fig. 2 Design guidelines and critical factors of AM with living building materials

Furthermore, designing and fabricating such structures with living material or adapting from bio-based shapes and geometries, such as lattices or cellular structures, may lead to multiscale problems. Computational tools also play a critical role in achieving appropriate transcalar relationships [132]. The algorithms must be able to reproduce large-scale behaviors, especially related to human requirements and mechanical performance, and a possibility of adapting to mass production while ensuring that biological processes and interactions at smaller scales is also addressed.

Biological requirements are one of the main challenges when working with living building materials, such as fungi. First, sterile environments are needed in several steps, including 3D printing; thus, sterile equipment such as an autoclave can be suitable to maintain minimal conditions. The use of an autoclave has proved to be beneficial in controlling contamination during the first phases [116, 126, 131], although strict protocols must be followed to mitigate fungal contamination. This condition requires specific equipment when implementing mycelium as a building material. Besides, moisture balance and a suitable growing temperature are critical factors to the adequate development of mycelium-based composites [40], which also add the need for specific setups during the primary and secondary colonization.

Figure 2 summarizes the main aspects that should be considered when designing and fabricating with living building materials, particularly with mycelium and cyanobacteria, using AM as a fabrication technique. Even though the projects explored in the literature and their experiences can give us some lessons and design guidelines, it is still on lab-scale experimental approaches, which render it necessary to maintain controlled development toward enhancing the process. All this still leaves excellent opportunities for exploration in this field, especially for the AEC industry.

The reviewed projects mostly used extrusion-based AM as fabrication techniques, such as direct ink writing and material extrusion and FDM for potential AEC industry applications. Process and geometry parameters such as printing speed, nozzle shape and diameter, extrusion pressure, and time between steps can yield varying performances based on material choices. Key performance indicators frequently assessed were the resulting structure's rheological properties, extrudability, buildability, and the mechanical performance with compressive and bending strength.

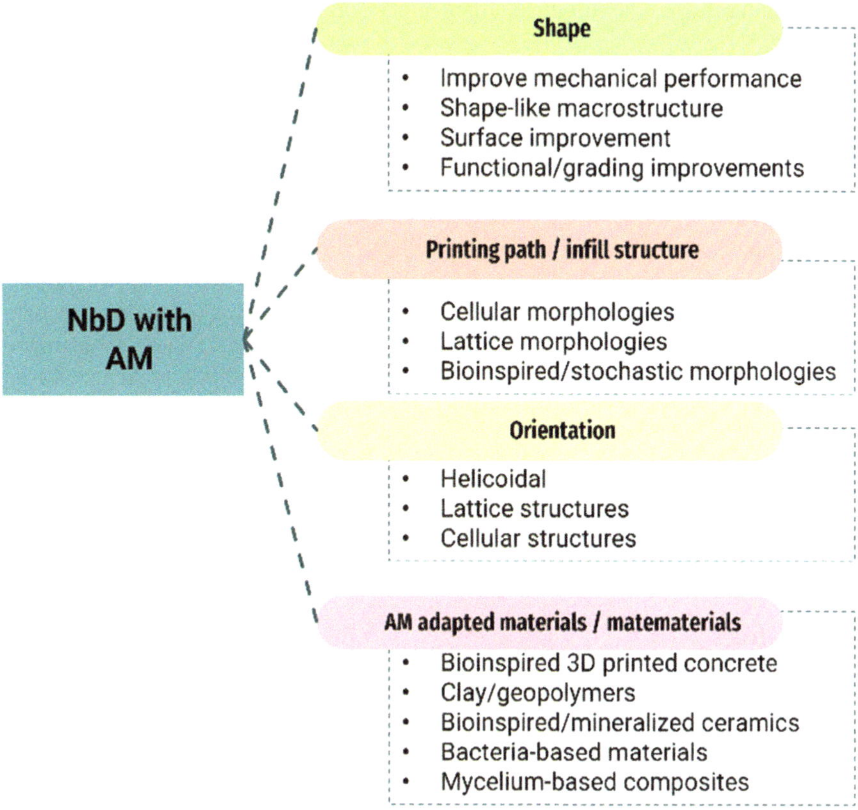

Fig. 3 Insights from pairing AM with NbD to produce structures suitable for the AEC industry

To fully harness the potential of AM and NbD within the AEC sector, there is a need to design and produce large-scale structures. These should be capable of integrating into building segments like walls and envelopes and ideally exhibit environmentally responsive or bioregenerative properties. Yet, the multiscale intricacy of structures derived from NbD approaches becomes increasingly challenging to investigate at larger scales. While AM faces difficulties in realizing completely responsive envelopes, its application for crafting walls and envelopes has been preliminarily investigated [4, 6, 83, 94].

A more thorough examination of robotic applications might offer a pathway to reach this objective. For instance, with LBM the development of large-scale products has been only developed in modular columns and arches [117, 136], and, in the case of geopolymers, clay and, similar materials, bricks, but aspects such as durability and mechanical performance have been barely explored [52]. Nevertheless, the possibilities offered by these kinds of materials align well with achieving sustainability and circularity while reducing fossil sources dependency. A summary of the guidelines for designers using NbD with AM can be found in Fig. 3.

The authors offer the subsequent final remarks as potential areas for enhancement:

1. AM has shown that it has the potential to develop complex structures akin to those found in nature. By mimicking these natural structures, there exists an opportunity to optimize mechanical properties at the macroscopic scale by fine-tuning the microstructure and its manufacturing technique.
2. There is a need for digital tools that facilitate the concurrent progression of design and manufacturing. While authors like Berman et al. [93] have ventured into this domain, it remains a promising area for enhancement, aiming for superior properties that are mechanical and more adaptive to specific applications.
3. Biodesign is predominantly employed to mimic shapes that bolster the mechanical integrity of the end product. Yet, there's ample scope to refine both the manufacturing procedure and the choice of materials to further this approach's efficacy.

While the challenges are significant, the swift advancements and heightened focus on AM have led to tremendous growth in this domain over the past 5 years. This progress paves the way for employing materials and NbD-inspired manufacturing strategies that are perfectly aligned with the needs of the AEC industry.

References

1. Yan, X., Bethers, B., Chen, H., et al. (2021). Recent advancements in biomimetic 3D printing materials with enhanced mechanical properties. *Frontiers in Materials, 8*.
2. Hershcovich, C., van Hout, R., Rinsky, V., et al. (2021). Thermal performance of sculptured tiles for building envelopes. *Building and Environment, 197*, 107809. https://doi.org/10.1016/j.buildenv.2021.107809
3. Mole, M. A., Rodrigues DÁraujo, S., van Aarde, R. J., et al. (2016). Coping with heat: Behavioural and physiological responses of savanna elephants in their natural habitat. *Conservation Physiology, 4*, cow044. https://doi.org/10.1093/conphys/cow044
4. Ahamed, M. K., Wang, H., & Hazell, P. J. (2022). From biology to biomimicry: Using nature to build better structures – A review. *Construction and Building Materials, 320*, 126195. https://doi.org/10.1016/j.conbuildmat.2021.126195
5. Badarnah Kadri, L. (2012). *Towards the LIVING envelope: Biomimetics for building envelope adaptation*. Wöhrmann Print Service B.V.
6. Ortega Del Rosario, M. D. L. Á., Beermann, K., & Chen Austin, M. (2023). Environmentally responsive materials for building envelopes: A review on manufacturing and biomimicry-based approaches. *Biomimetics, 8*, 52. https://doi.org/10.3390/biomimetics8010052
7. Savolainen, J., & Collan, M. (2020). How additive manufacturing technology changes business models? – Review of literature. *Additive Manufacturing, 32*, 101070. https://doi.org/10.1016/j.addma.2020.101070
8. Song, Y., Koeck, R., & Luo, S. (2021). Review and analysis of augmented reality (AR) literature for digital fabrication in architecture. *Automation in Construction, 128*, 103762. https://doi.org/10.1016/j.autcon.2021.103762
9. Dixit, S., & Stefańska, A. (2023). Bio-logic, a review on the biomimetic application in architectural and structural design. *Ain Shams Engineering Journal, 14*, 101822. https://doi.org/10.1016/j.asej.2022.101822

10. Nodehi, M., Ozbakkaloglu, T., & Gholampour, A. (2022). Effect of supplementary cementitious materials on properties of 3D printed conventional and alkali-activated concrete: A review. *Automation in Construction, 138*, 104215. https://doi.org/10.1016/j.autcon.2022.104215
11. Wohlers, T., & Gornet, T. (2014). *History of additive manufacturing*. Wohlers Associates.
12. Hulls, C. (1989). *Priority to US06638905*.
13. Al Rashid, A., Khan, S. A., Al-Ghamdi, S., & Koç, M. (2020). Additive manufacturing: Technology, applications, markets, and opportunities for the built environment. *Automation in Construction, 118*, 103268. https://doi.org/10.1016/j.autcon.2020.103268
14. ISO/ASTM 52900:2021(en). (2021). *Additive manufacturing – General principles – Fundamentals and vocabulary*.
15. Anton, A., Reiter, L., Wangler, T., et al. (2021). A 3D concrete printing prefabrication platform for bespoke columns. *Automation in Construction, 122*, 103467. https://doi.org/10.1016/j.autcon.2020.103467
16. Mohan, M. K., Rahul, A. V., De Schutter, G., & Van Tittelboom, K. (2021). Extrusion-based concrete 3D printing from a material perspective: A state-of-the-art review. *Cement and Concrete Composites, 115*, 103855. https://doi.org/10.1016/j.cemconcomp.2020.103855
17. Parusheva, S., & Aleksandrova, Y. (2021). *Technologies, tools, and resources - driving forces in construction sector digitalization*. In: 2021 Tenth international conference on intelligent computing and information systems (ICICIS), pp. 219–223.
18. Al-Qutaifi, S., Nazari, A., & Bagheri, A. (2018). Mechanical properties of layered geopolymer structures applicable in concrete 3D-printing. *Construction and Building Materials, 176*, 690–699. https://doi.org/10.1016/j.conbuildmat.2018.04.195
19. Agarwal, R., Chandrasekaran, S., Sridhar, M. (2016). *Imagining construction's digital future*.
20. Lasarte, N., Elguezabal, P., Sagarna, M., et al. (2021). Challenges for digitalisation in building renovation to enhance the efficiency of the process: A Spanish case study. *Sustainability, 13*, 12139. https://doi.org/10.3390/su132112139
21. Rahman, A. U., Alam, S. M., Dallasega, P., et al. (2020). Increasing control in construction processes: The role of digitalization. In O. A. Del Río, H. Leopold, & F. M. Santoro (Eds.), *Business process management workshops* (pp. 263–275). Springer.
22. Perera, S., Jin, X., Das, P., et al. (2023). A strategic framework for digital maturity of design and construction through a systematic review and application. *Journal of Industrial Information Integration, 31*, 100413. https://doi.org/10.1016/j.jii.2022.100413
23. Ashworth, A., & Perera, S. (2015). *Cost studies of buildings*. Routledge.
24. Menna, C., Mata-Falcón, J., Bos, F. P., et al. (2020). Opportunities and challenges for structural engineering of digitally fabricated concrete. *Cement and Concrete Research, 133*, 106079. https://doi.org/10.1016/j.cemconres.2020.106079
25. Stoyanova, M. (2020). Good practices and recommendations for success in construction digitalization. *TEM J, 9*, 42–47.
26. Mechtcherine, V., Nerella, V. N., Will, F., et al. (2019). Large-scale digital concrete construction – CONPrint3D concept for on-site, monolithic 3D-printing. *Automation in Construction, 107*, 102933. https://doi.org/10.1016/j.autcon.2019.102933
27. Peshkov, A. V. (2021). Construction 4.0: Immaterial assets types in the development of design estimates for effective digitalization of building projects. *IOP Conference Series: Earth and Environmental Science, 751*, 012107. https://doi.org/10.1088/1755-1315/751/1/012107
28. Lowke, D., Dini, E., Perrot, A., et al. (2018). Particle-bed 3D printing in concrete construction – Possibilities and challenges. *Cement and Concrete Research, 112*, 50–65. https://doi.org/10.1016/j.cemconres.2018.05.018
29. El-Sayegh, S., Romdhane, L., & Manjikian, S. (2020). A critical review of 3D printing in construction: Benefits, challenges, and risks. *Archives of Civil and Mechanical Engineering, 20*, 34. https://doi.org/10.1007/s43452-020-00038-w
30. Arango, A. M., & Acuña, L. E. (2018). *La Internacionalización del currículo y su relación con las condiciones de calidad en los programas académicos de educación superior para la obtención de registro calificado*. *2*, 15.

31. Sheina, S., Chubarova, K., Dementeev, D., & Kalitkin, A. (2023). Integration of BIM and GIS Technologies for Sustainable Development of the construction industry. In A. Guda (Ed.), *Networked control Systems for connected and automated vehicles* (pp. 1303–1311). Springer.
32. Wang, H., Du, W., & Li, S. (2022). Key issues for digital factory designing and planning: A survey. *Studies in Computational Intelligence, 1012*(SCI), 18–29. https://doi.org/10.1007/978-3-030-92317-4_2
33. Benedetti, A. C., Costantino, C., Gulli, R., & Predari, G. (2022). The process of digitalization of the urban environment for the development of sustainable and circular cities: A case study of Bologna, Italy. *Sustainability, 14*, 13740. https://doi.org/10.3390/su142113740
34. United Nations Environment Programme. (2021). *2021 Global status report for buildings and construction*. Towards a zero-emissions, efficient and resilient buildings and construction sector. Nairobi.
35. Craveiro, F., Duarte, J. P., Bartolo, H., & Bartolo, P. J. (2019). Additive manufacturing as an enabling technology for digital construction: A perspective on Construction 4.0. *Automation in Construction, 103*, 251–267. https://doi.org/10.1016/j.autcon.2019.03.011
36. Paolini, A., Kollmannsberger, S., & Rank, E. (2019). Additive manufacturing in construction: A review on processes, applications, and digital planning methods. *Additive Manufacturing, 30*, 100894. https://doi.org/10.1016/j.addma.2019.100894
37. Siddique, S. H., Hazell, P. J., Wang, H., et al. (2022). Lessons from nature: 3D printed bio-inspired porous structures for impact energy absorption – A review. *Additive Manufacturing, 58*, 103051. https://doi.org/10.1016/j.addma.2022.103051
38. Al-Ketan, O., Rowshan, R., & Alami, A. H. (2022). Biomimetic materials for engineering applications. In A.-G. Olabi (Ed.), *Encyclopedia of smart materials* (pp. 25–34). Elsevier.
39. Montana-Hoyos, C., Daneluzzo, M., Tchakerian, R., et al. (2022). Chapter one – biomimicry and biodesign for innovation in future space colonization. In V. Shyam, M. Eggermont, & A. F. Hepp (Eds.), *Biomimicry for aerospace* (pp. 3–39). Elsevier.
40. Kırdök, O., Akyol Altun, D., Dahy, H., et al. (2022). Chapter 17 – Design studies and applications of mycelium biocomposites in architecture. In M. Eggermont, V. Shyam, & A. F. Hepp (Eds.), *Biomimicry for materials, design and habitats* (pp. 489–527). Elsevier.
41. Almesmari, A., Alagha, A. N., Naji, M. M., et al. (2023). Recent advancements in design optimization of lattice-structured materials. *Advanced Engineering Materials, 25*, 2201780. https://doi.org/10.1002/adem.202201780
42. International Organization for Standardization. (2015). *ISO 18458:2015 Biomimetics – Terminology, concepts and methodology*.
43. Alsuwait, R. B., Souiyah, M., Momohjimoh, I., et al. (2023). Recent development in the processing, properties, and applications of epoxy-based natural fiber polymer biocomposites. *Polymers, 15*, 145. https://doi.org/10.3390/polym15010145
44. Judawisastra, H., & Refiadi, G. (2022). Permanganate treatment optimization on tensile properties and water absorption of Kenaf fiber-polypropylene biocomposites. *International Journal of Automotive and Mechanical Engineering, 19*, 9623–9633. https://doi.org/10.15282/ijame.19.1.2022.23.0742
45. Abidin, N. A. Z., Mahmud, J., Manssor, N. A. S., & Abd Rahim, N. N. C. (2022). Physical and mechanical properties of bamboo-silicone biocomposites (BaSiCs). *BioResources, 17*, 4432–4443. https://doi.org/10.15376/biores.17.3.4432-4443
46. Buehler, M. J. (2022). Generating 3D architectured nature-inspired materials and granular media using diffusion models based on language cues. *Oxford Open Materials Science, 2*, itac010. https://doi.org/10.1093/oxfmat/itac010
47. Vellayappan, M. V., Duarte, F., Sollogoub, C., et al. (2023). Fabrication of Architectured biomaterials by multilayer co-extrusion and additive manufacturing. *Advanced Functional Materials, 33*, 2301547. https://doi.org/10.1002/adfm.202301547
48. Mechtcherine, V., Bos, F. P., Perrot, A., et al. (2020). Extrusion-based additive manufacturing with cement-based materials – Production steps, processes, and their underly-

ing physics: A review. *Cement and Concrete Research, 132*. https://doi.org/10.1016/j.cemconres.2020.106037

49. Wolfs, R. R., Bos, F., & Salet, T. T. (2018). *Early age mechanical behaviour of 3D printed concrete: Numerical modelling and experimental testing*. https://doi.org/10.1016/J.CEMCONRES.2018.02.001.
50. Dielemans, G., Briels, D., Jaugstetter, F., et al. (2021). Additive manufacturing of thermally enhanced lightweight concrete wall elements with closed cellular structures. *Journal of Facade Design and Engineering, 9*, 59–72. https://doi.org/10.7480/jfde.2021.1.5418
51. Ooms, T., Vantyghem, G., Van Coile, R., & De Corte, W. (2021). A parametric modelling strategy for the numerical simulation of 3D concrete printing with complex geometries. *Additive Manufacturing, 38*, 101743. https://doi.org/10.1016/j.addma.2020.101743
52. Sangiorgio, V., Parisi, F., Fieni, F., & Parisi, N. (2022). The new Boundaries of 3D-printed clay bricks design: Printability of complex internal geometries. *Sustainability, 14*, 598. https://doi.org/10.3390/su14020598
53. Pastore, T., Menna, C., & Asprone, D. (2022). Bézier-based biased random-key genetic algorithm to address printability constraints in the topology optimization of concrete structures. *Structural and Multidisciplinary Optimization, 65*. https://doi.org/10.1007/s00158-021-03119-3
54. Wolfs, R. J. M., Salet, T. A. M., & Roussel, N. (2021). Filament geometry control in extrusion-based additive manufacturing of concrete: The good, the bad and the ugly. *Cement and Concrete Research, 150*, 106615. https://doi.org/10.1016/j.cemconres.2021.106615
55. Nguyen-Van, V. (2021). *Mechanical evaluations of bioinspired TPMS cellular cementitious structures manufactured by 3D printing formwork*.
56. Leary, M. (2019). *Design for additive manufacturing* (1st ed.). Elsevier.
57. Gladysz, G. M., & Chawla, K. K. (2015). Chapter 6 – Cellular materials. In G. M. Gladysz & K. K. Chawla (Eds.), *Voids in materials* (pp. 103–130). Elsevier.
58. Gibson, L. J., Ashby, M. F., & Harley, B. A. (2010). *Cellular materials in nature and medicine*. Cambridge University Press.
59. Pais, A. I., Belinha, J., & Alves, J. L. (2023). Advances in computational techniques for bio-inspired cellular materials in the field of biomechanics: Current trends and prospects. *Materials, 16*, 3946. https://doi.org/10.3390/ma16113946
60. Yang, L., Ferrucci, M., Mertens, R., et al. (2020). An investigation into the effect of gradients on the manufacturing fidelity of triply periodic minimal surface structures with graded density fabricated by selective laser melting. *Journal of Materials Processing Technology, 275*, 116367. https://doi.org/10.1016/j.jmatprotec.2019.116367
61. Cowin, S. C. (1985). The relationship between the elasticity tensor and the fabric tensor. *Mechanics of Materials, 4*, 137–147. https://doi.org/10.1016/0167-6636(85)90012-2
62. Bi, M., Tran, P., Xia, L., et al. (2022). Topology optimization for 3D concrete printing with various manufacturing constraints. *Additive Manufacturing, 57*. https://doi.org/10.1016/j.addma.2022.102982
63. Gan, N., & Wang, Q. (2022). Topology optimization design of porous infill structure with thermo-mechanical buckling criteria. *International Journal of Mechanics and Materials in Design, 18*, 267–288. https://doi.org/10.1007/s10999-021-09575-5
64. Kontovourkis, O., Tryfonos, G., & Georgiou, C. (2020). Robotic additive manufacturing (RAM) with clay using topology optimization principles for toolpath planning: The example of a building element. *Architectural Science Review, 63*, 105–118. https://doi.org/10.1080/00038628.2019.1620170
65. Tafakkori, K., Tavakkoli-Moghaddam, R., & Siadat, A. (2022). Sustainable negotiation-based nesting and scheduling in additive manufacturing systems: A case study and multi-objective meta-heuristic algorithms. *Engineering Applications of Artificial Intelligence, 112*, 104836. https://doi.org/10.1016/j.engappai.2022.104836

66. Valjak, F., & Lindwall, A. (2021). Review of design heuristics and design principles in design for additive manufacturing. *Proceedings of the Design Society, 1*, 2571–2580. https://doi.org/10.1017/pds.2021.518
67. Al-Ketan, O., Rowshan, R., & Abu Al-Rub, R. K. (2018). Topology-mechanical property relationship of 3D printed strut, skeletal, and sheet based periodic metallic cellular materials. *Additive Manufacturing, 19*, 167–183. https://doi.org/10.1016/j.addma.2017.12.006
68. Han, L., & Che, S. (2018). An overview of materials with triply periodic minimal surfaces and related geometry: From biological structures to self-assembled systems. *Advanced Materials, 30*, 1705708. https://doi.org/10.1002/adma.201705708
69. Feng, J., Fu, J., Yao, X., & He, Y. (2022). Triply periodic minimal surface (TPMS) porous structures: From multi-scale design, precise additive manufacturing to multidisciplinary applications. *International Journal of Extreme Manufacturing, 4*, 022001. https://doi.org/10.1088/2631-7990/ac5be6
70. Michielsen, K., & De Raedt, H. (2001). Integral-geometry morphological image analysis. *Physics Reports, 347*, 461–538. https://doi.org/10.1016/S0370-1573(00)00106-X
71. Yang, X., Yang, Q., Shi, Y., et al. (2022). Effect of volume fraction and unit cell size on manufacturability and compressive behaviors of Ni-Ti triply periodic minimal surface lattices. *Additive Manufacturing, 54*, 102737. https://doi.org/10.1016/j.addma.2022.102737
72. Ahmed, Z., Biffi, A., Hass, L., et al. (2020). 3D concrete printing – Free form geometries with improved ductility and strength. In F. P. Bos, S. S. Lucas, R. J. M. Wolfs, & T. A. M. Salet (Eds.), *Second RILEM international conference on concrete and digital fabrication* (pp. 741–756). Springer.
73. Nguyen-Van, V., Tran, P., Peng, C., et al. (2020). Bioinspired cellular cementitious structures for prefabricated construction: Hybrid design & performance evaluations. *Automation in Construction, 119*, 103324. https://doi.org/10.1016/j.autcon.2020.103324
74. Abdallah, Y. K., & Estévez, A. T. (2021). 3D-printed biodigital clay bricks. *Biomimetics, 6*, 59. https://doi.org/10.3390/biomimetics6040059
75. Patadiya, J., Wang, X., Joshi, G., et al. (2023). 3D-printed biomimetic hierarchical nacre architecture: Fracture behavior and analysis. *ACS Omega, 8*, 18449–18461. https://doi.org/10.1021/acsomega.2c08076
76. Mishra, N., & Kandasubramanian, B. (2018). Biomimetic Design of Artificial Materials Inspired by iridescent nacre structure and its growth mechanism. *Polymer-Plastics Technology and Engineering, 57*, 1592–1606. https://doi.org/10.1080/03602559.2017.1326139
77. Ubaid, J., Wardle, B. L., & Kumar, S. (2020). Bioinspired compliance grading motif of mortar in nacreous materials. *ACS Applied Materials & Interfaces, 12*, 33256–33266. https://doi.org/10.1021/acsami.0c08181
78. Ko, K., Lee, S., Hwang, Y. K., et al. (2022). Investigation on the impact resistance of 3D printed nacre-like composites. *Thin-Walled Structures, 177*, 109392. https://doi.org/10.1016/j.tws.2022.109392
79. Zhao, Z., Liu, Y., & Wang, P. (2023). Computational design of bio-inspired mechanical metamaterials based on lipidic cubic phases. *JOM, 75*, 2126–2136. https://doi.org/10.1007/s11837-023-05866-8
80. Yang, R., Zaheri, A., Gao, W., et al. (2017). Exoskeletons: AFM identification of beetle exocuticle: Bouligand structure and nanofiber anisotropic elastic properties (Adv. Funct. Mater. 6/2017). *Advanced Functional Materials, 27*. https://doi.org/10.1002/adfm.201770031
81. Yang, W., Sherman, V. R., Gludovatz, B., et al. (2014). Protective role of Arapaima gigas fish scales: Structure and mechanical behavior. *Acta Biomaterialia, 10*, 3599–3614. https://doi.org/10.1016/j.actbio.2014.04.009
82. Suksangpanya, N., Yaraghi, N. A., Kisailus, D., & Zavattieri, P. (2017). Twisting cracks in Bouligand structures. *Journal of the Mechanical Behavior of Biomedical Materials, 76*, 38–57. https://doi.org/10.1016/j.jmbbm.2017.06.010

83. Liu, J., Li, S., Fox, K., & Tran, P. (2022). 3D concrete printing of bioinspired Bouligand structure: A study on impact resistance. *Additive Manufacturing, 50*, 102544. https://doi.org/10.1016/j.addma.2021.102544
84. Moini, M., Olek, J., Youngblood, J. P., et al. (2018). Additive manufacturing and performance of architectured cement-based materials. *Advanced Materials, 30*, 1802123. https://doi.org/10.1002/adma.201802123
85. Nguyen-Van, V., Nguyen-Xuan, H., Panda, B., & Tran, P. (2022). 3D concrete printing modelling of thin-walled structures. *Structure, 39*, 496–511. https://doi.org/10.1016/j.istruc.2022.03.049
86. Suiker, A. S. J. (2022). Effect of accelerated curing and layer deformations on structural failure during extrusion-based 3D printing. *Cement and Concrete Research, 151*, 106586. https://doi.org/10.1016/j.cemconres.2021.106586
87. Ilcan, H., Sahin, O., Kul, A., et al. (2022). Rheological properties and compressive strength of construction and demolition waste-based geopolymer mortars for 3D-printing. *Construction and Building Materials, 328*, 127114. https://doi.org/10.1016/j.conbuildmat.2022.127114
88. Lee, H., Jay Kim, J. H., Moon, J. H., et al. (2020). Experimental analysis on rheological properties for control of concrete extrudability. *Advances in Concrete Construction, 9*, 93–102. https://doi.org/10.12989/acc.2020.9.1.093
89. Wu, Y., Liu, C., Liu, H., et al. (2021). Study on the rheology and buildability of 3D printed concrete with recycled coarse aggregates. *Journal of Building Engineering, 42*, 103030. https://doi.org/10.1016/j.jobe.2021.103030
90. Cadoret, N., Chaves-Jacob, J., & Linares, J.-M. (2023). Structural additive manufacturing parts bio-inspired from trabecular bone form-function relationship. *Materials and Design, 231*, 112029. https://doi.org/10.1016/j.matdes.2023.112029
91. Kladovasilakis, N., Tsongas, K., Kostavelis, I., et al. (2022). Effective mechanical properties of additive manufactured strut-lattice structures: Experimental and finite element study. *Advanced Engineering Materials, 24*. https://doi.org/10.1002/adem.202100879
92. Doodi, R., & Balamurali, G. (2023). Experimental and analytical investigation of bio-inspired lattice structures under compressive loading. *Engineering Research Express, 5*, 035035. https://doi.org/10.1088/2631-8695/aceed1
93. Berman, O., Weizman, M., Oren, A., et al. (2023). Design and application of a novel 3D printing method for bio-inspired artificial reefs. *Ecological Engineering, 188*, 106892. https://doi.org/10.1016/j.ecoleng.2023.106892
94. Nguyen-Van, V., Choudhry, N. K., Panda, B., et al. (2022). Performance of concrete beam reinforced with 3D printed bioinspired primitive scaffold subjected to three-point bending. *Automation in Construction, 134*. https://doi.org/10.1016/j.autcon.2021.104060
95. Ming, X., Huang, J. C., & Li, Z. (2022). Materials-oriented integrated design and construction of structures in civil engineering – A review. *Frontiers of Structural and Civil Engineering, 16*, 24–44. https://doi.org/10.1007/s11709-021-0794-9
96. Ruusuvuori, P. (2022). Chapter 2 – Parametric modeling in biomedical image synthesis. In N. Burgos & D. Svoboda (Eds.), *Biomedical image synthesis and simulation* (pp. 7–21). Academic Press.
97. Zhou, T., Xiong, W., Obata, Y., et al. (2022). Chapter 2 – Digital product design and engineering analysis techniques. In C. D. Patel & C.-H. Chen (Eds.), *Digital manufacturing* (pp. 57–96). Elsevier.
98. Hoang, V.-N., Tran, P., Vu, V.-T., & Nguyen-Xuan, H. (2020). Design of lattice structures with direct multiscale topology optimization. *Composite Structures, 252*, 112718. https://doi.org/10.1016/j.compstruct.2020.112718
99. Al-Ketan, O., & Abu Al-Rub, R. K. (2019). Multifunctional mechanical metamaterials based on triply periodic minimal surface lattices. *Advanced Engineering Materials, 21*, 1900524. https://doi.org/10.1002/adem.201900524
100. Jongerius, S. R., & Lentink, D. (2010). Structural analysis of a dragonfly wing. *Experimental Mechanics, 50*, 1323–1334. https://doi.org/10.1007/s11340-010-9411-x

101. Murray, C. D. (1926). The physiological principle of minimum work. I. The vascular system and the cost of blood volume. *Proceedings of the National Academy of Sciences of the United States of America, 12*, 2017–2014.
102. Hoffmann, J., Donoughe, S., Li, K., et al. (2018). A simple developmental model recapitulates complex insect wing venation patterns. *Proceedings of the National Academy of Sciences, 115*, 9905–9910. https://doi.org/10.1073/pnas.1721248115
103. Musenich, L., Stagni, A., & Libonati, F. (2023). Hierarchical bioinspired architected materials and structures. *Extreme Mechanics Letters, 58*, 101945. https://doi.org/10.1016/j.eml.2022.101945
104. Bowen, J. J., Mooraj, S., Goodman, J. A., et al. (2022). Hierarchically porous ceramics via direct writing of preceramic polymer-triblock copolymer inks. *Materials Today, 58*, 71–79. https://doi.org/10.1016/j.mattod.2022.07.002
105. Mora, S., Pugno, N. M., & Misseroni, D. (2022). 3D printed architected lattice structures by material jetting. *Materials Today, 59*, 107–132. https://doi.org/10.1016/j.mattod.2022.05.008
106. Lin, Y., Shi, W., Li, J., et al. (2023). Evaluation of mechanical properties of Ti–6Al–4 V BCC lattice structure with different density gradient variations prepared by L-PBF. *Materials Science and Engineering A, 872*, 144986. https://doi.org/10.1016/j.msea.2023.144986
107. Yang, L., Han, C., Wu, H., et al. (2020). Insights into unit cell size effect on mechanical responses and energy absorption capability of titanium graded porous structures manufactured by laser powder bed fusion. *Journal of the Mechanical Behavior of Biomedical Materials, 109*, 103843. https://doi.org/10.1016/j.jmbbm.2020.103843
108. Du, Y., Gu, D., Xi, L., et al. (2020). Laser additive manufacturing of bio-inspired lattice structure: Forming quality, microstructure and energy absorption behavior. *Materials Science and Engineering A, 773*, 138857. https://doi.org/10.1016/j.msea.2019.138857
109. Borunda, L., & Anaya, J. (2023). Hierarchical structures computational design and digital 3D printing. *Journal of the International Association for Shell and Spatial Structures, 64*, 5–18. https://doi.org/10.20898/j.iass.2022.015
110. Gibson, L. J. (2003). Cellular solids. *MRS Bulletin, 28*, 270–274. https://doi.org/10.1557/mrs2003.79
111. Danielli, F., Berti, F., Nespoli, A., et al. (2023). Towards the development of a custom talus prosthesis produced by SLM: Design rules and verification. *Journal of Mechanical Science and Technology, 37*, 1125–1130. https://doi.org/10.1007/s12206-022-2109-z
112. Yin, S., Chen, H., Yang, R., et al. (2020). Tough nature-inspired helicoidal composites with printing-induced voids. *Cell Reports Physical Science, 1*, 100109. https://doi.org/10.1016/j.xcrp.2020.100109
113. Guarín-Zapata, N., Gomez, J., Yaraghi, N., et al. (2015). Shear wave filtering in naturally-occurring Bouligand structures. *Acta Biomaterialia, 23*, 11–20. https://doi.org/10.1016/j.actbio.2015.04.039
114. He, M. Y., Evans, A. G., & Hutchinson, J. W. (1994). Crack deflection at an interface between dissimilar elastic materials: Role of residual stresses. *International Journal of Solids and Structures, 31*, 3443–3455. https://doi.org/10.1016/0020-7683(94)90025-6
115. Almpani-Lekka, D., Pfeiffer, S., Schmidts, C., & Seo, S. (2021). A review on architecture with fungal biomaterials: The desired and the feasible. *Fungal Biology and Biotechnology, 8*, 17. https://doi.org/10.1186/s40694-021-00124-5
116. Mohseni, A., Vieira, F. R., Pecchia, J. A., & Gürsoy, B. (2023). Three-dimensional printing of living mycelium-based composites: Material compositions, workflows, and ways to mitigate contamination. *Biomimetics, 8*, 257. https://doi.org/10.3390/biomimetics8020257
117. Modanloo, B., Ghazvinian, A., Matini, M., & Andaroodi, E. (2021). Tilted arch; implementation of additive manufacturing and bio-welding of mycelium-based composites. *Biomimetics, 6*, 68. https://doi.org/10.3390/biomimetics6040068
118. Soh, E., Chew, Z. Y., Saeidi, N., et al. (2020). Development of an extrudable paste to build mycelium-bound composites. *Materials and Design, 195*, 109058. https://doi.org/10.1016/j.matdes.2020.109058

119. Dessi-Olive, J. (2019). *Monolithic mycelium: growing vault structures*. In 18th International conference on non-conventional materials and technologies "construction materials & technologies for sustainability".
120. Appels, F. V. W., Camere, S., Montalti, M., et al. (2019). Fabrication factors influencing mechanical, moisture- and water-related properties of mycelium-based composites. *Materials and Design, 161*, 64–71. https://doi.org/10.1016/j.matdes.2018.11.027
121. Heisel, F., Schlesier, K., Lee, J., et al (2017). *Design of a load-bearing mycelium structure through informed structural engineering*.
122. Syed, S. (2017). *This pavillion lives and dies through its sustainable agenda | ArchDaily*. In Arch Dly. https://www.archdaily.com/878519/this-pavillion-lives-and-dies-through-its-sustainable-agenda. Accessed 17 Sep 2023.
123. Buro Meta. (2021). Home – The growing pavilion. In *Home – Grow*. Pavilion. https://thegrowingpavilion.com/. Accessed 17 Sep 2023.
124. Dessi-Olive, J. (2022). Strategies for growing large-scale mycelium structures. *Biomimetics, 7*, 129. https://doi.org/10.3390/biomimetics7030129
125. Kalantari, S., & Saleh Tabari, M. H. (2017). *Growmorph: Bacteria growth algorithm and design*. In Proceedings of the 22nd international conference of the association for computer-aided architectural design research in Asia (CAADRIA) 2017. The Association for Computer-Aided Architectural Design Research in Asia (CAADRIA).
126. Goidea, A., Floudas, D., & Andréen, D. (2020). Pulp faction: 3D printed material assemblies through microbial biotransformation. In *Fabricate 2020*. UCL Press.
127. Amarly, P., Kirzner, S., Khasi, Y., & Barath, S. (2023). Proceedings of the 28th international conference of the association for computer-aided architectural design research in Asia (CAADRIA), Hong Kong, pp. 149–158.
128. Armaly, P., Iliassafov, L., Kirzner, S., et al. (2023). Biologically informed design – Towards additive biofabrication with cyanobacteria. In M. Turrin, C. Andriotis, & A. Rafiee (Eds.), *Computer-aided architectural design. INTERCONNECTIONS: Co-computing beyond Boundaries* (pp. 425–436). Springer.
129. Bhardwaj, A., Vasselli, J., Lucht, M., et al. (2020). 3D printing of biomass-fungi composite material: A preliminary study. *Manufacturing Letters, 24*, 96–99. https://doi.org/10.1016/j.mfglet.2020.04.005
130. Bhardwaj, A., Rahman, A. M., Wei, X., et al. (2021). 3D printing of biomass–fungi composite material: Effects of mixture composition on print quality. *Journal Of Manufacturing And Materials Processing, 5*, 112. https://doi.org/10.3390/jmmp5040112
131. Elsacker, E., Peeters, E., & De Laet, L. (2022). Large-scale robotic extrusion-based additive manufacturing with living mycelium materials. *Sustainable Futures, 4*, 100085. https://doi.org/10.1016/j.sftr.2022.100085
132. Goidea, A., Floudas, D., & Andréen, D. (2022). Transcalar design: An approach to biodesign in the built environment. *Infrastructures, 7*, 50. https://doi.org/10.3390/infrastructures7040050
133. Nodehi, M., Ozbakkaloglu, T., & Gholampour, A. (2022). A systematic review of bacteria-based self-healing concrete: Biomineralization, mechanical, and durability properties. *Journal of Building Engineering, 49*, 104038. https://doi.org/10.1016/j.jobe.2022.104038
134. Wang, J., Ersan, Y. C., Boon, N., & De Belie, N. (2016). Application of microorganisms in concrete: A promising sustainable strategy to improve concrete durability. *Applied Microbiology and Biotechnology, 100*, 2993–3007. https://doi.org/10.1007/s00253-016-7370-6
135. Qiu, J., Artier, J., Cook, S., et al. (2021). Engineering living building materials for enhanced bacterial viability and mechanical properties. *iScience, 24*. https://doi.org/10.1016/j.isci.2021.102083
136. Reinhardt, O., Ihmann, S., Ahlhelm, M., & Gelinsky, M. (2023). 3D bioprinting of mineralizing cyanobacteria as novel approach for the fabrication of living building materials. *Frontiers in Bioengineering and Biotechnology, 11*, 1145177.

137. Goss, D., Penick, C. A., Grishin, A., & Bhate, D. (2022). Chapter six – Bio-inspired design and additive manufacturing of cellular materials. In V. Shyam, M. Eggermont, & A. F. Hepp (Eds.), *Biomimicry for aerospace* (pp. 141–185). Elsevier.
138. Long, O. D. A., Del Rosario M. D. L. Á. O., & Brischetto, S. (2022). *Mechanical properties enhancement for additive manufactured short fiber composites with salt remelting post-processing Mejora de las propiedades mecánicas para compuestos de fibras cortas fabricados por manufactura aditiva con post-procesamiento de recocido en sal*. In 2022 8th International Engineering, Sciences and Technology Conference (IESTEC), pp. 723–727.

Maria De Los Angeles Ortega Del Rosario, is an electromechanical engineer with a PhD in Physics and Engineering Sciences from the University of Bordeaux, France. She also holds a Master of Science in Mechanical Engineering, from Arts et Métiers, France. She is a full-time researcher and lecturer at UTP, Faculty of Mechanical Engineering, focusing on Materials Sciences. Her research interests include additive manufacturing, bioinspired design, and sustainable approaches.

Carmen Castaño, is a mechanical and industrial engineer with a doctorate in Production System Engineering from RWTH Aachen University in Germany. She is a full-time lecturer and researcher at the Faculty of Industrial Engineering at UTP. At home in model-based system engineering, she specializes in IT-based quality and risk management with an application focus in sustainable lifecycle management as well as medical technology.

Miguel Chen Austin, completed his Ph.D. at the University of Bordeaux in France, specializing in Energy in Buildings. He is an Electromechanical Engineer from Universidad Tecnológica de Panamá (UTP) with a Master's in Science and Technology, specializing in Energetics and Mechanics, from ENSAM, ParisTech. He's a full-time researcher and lecturer at UTP Faculty of Mechanical Engineering (FIM). He is the technical director of the Research Group called "Energy and Comfort in Bioclimatic Buildings."

Index

A. J. V. Arruda, F. L. Palombini (eds.), *Biomimetics, Biodesign and Bionics*, Environmental Footprints and Eco-design of Products and Processes,
https://doi.org/10.1007/978-3-031-51311-4

The manufacturer's authorised representative in the EU is Springer Nature Customer Service Centre GmbH, Europaplatz 3, 69115 Heidelberg, Germany. If you have any concerns regarding our products, please contact ProductSafety@springernature.com

Printed and bound by CPI Group (UK) Ltd, Croydon, CR0 4YY
30/09/2026
02225571-0007